PHARMA-ECOLOGY

PHARMA-ECOLOGY

The Occurrence and Fate of Pharmaceuticals and Personal Care Products in the Environment

PATRICK K. JJEMBA
American Water, Deltran, New Jersey

WILEY

A JOHN WILEY & SONS, INC., PUBLICATION

Library of Congress Cataloging-in-Publication Data:

Jjemba, Patrick K.
  Pharma-ecology : the occurrence and fate of pharmaceuticals and personal care products in the
environment / Patrick K. Jjemba.
      p. ; cm.
  Includes bibliographical references and index.
  ISBN 978-0-470-04630-2 (cloth)
1. Drugs--Toxicology.   2. Hygiene products--Toxicology.   3. Drugs--Environmental aspects.
4. Hygiene products--Environmental aspects.   5.   Environmental toxicology.   I.   Title.
      [DNLM: 1.   Environmental Pollutants--adverse effects.   2.   Cosmetics--analysis.
3. Environmental Monitoring.   4.   Environmental Pollutants--analysis.
5. Pharmaceutical Preparations--analysis.   WA 671 J61p 2008]
    RA1238.J54 2008
    615′.1--dc22

                                                                        2008002716

CONTENTS

PREFACE

Pharma-ecology aims at studying and minimizing the impact of pharmaceutical and personal care products (PPCPs) on the environment. In this context, personal care products broadly include a number of compounds that we use in our daily lives ranging from soaps, detergents, perfumes, aftershaves, and other maintenance (e.g., cleaning agents, disinfectants, sprays, deodorants, etc.) products. The initial interest in these compounds was expressed as early as the 1960s (Stumm-Zolinger and Fair, 1965) but this issue did not gain until the review by Richardson and Bowron (1985) was published. Since then, an exponential number of studies reported the presence of these compounds in the environment, with most reports focusing on the presence of these compounds in aquatic systems.

PPCPs are currently designed and indeed used to target our individual ailments, a usage that may inadvertently disregard their effects on the ecosystem. An ecosystem is a self-sustaining structure that is comprised of various members of a community exerting a similar set of functions. Communities are in turn comprised of populations that exploit the same resources. Populations are composed of individuals of the same species. Under this intentional usage, PPCPs are increasingly being realized as micro-pollutants. In the recent past, the debate about PPCPs in the environment has expanded to ask serious questions about the impact of these compounds in those settings. This book was conceived when I was a Research Associate in the Civil and Environmental Engineering Department at the University of Cincinnati after I realized a need to forge linkages between health and environmental scientists about the occurrence and fate of PPCPs in the environment. Medical professionals are trained to primarily minimize or eliminate our pain and suffering from disease. However, this recognition has remained elusive to the healthcare providing professions, possibly as a result of apparent differences in the "language" as well as relatively dichotomous sources of literature. This book is cognizant of the fact that PPCPs are an important and, indeed indispensable, part of our individual well-being but also attempts to bridge some of the gaps facilitated by such individualized usage to highlight its ecotoxicological implications.

The book is organized with the recognition that its subject material, and thus readership, will inevitably come from different training backgrounds. It starts off with the range of pharmaceutical compounds categorized by mode of action and common usage, displaying the volumes (or number of prescriptions) that are regularly dispersed. In the second chapter, the complexities of detecting these compounds in the environment are discussed, giving the reader an appreciation of what is behind the values that have been reported, including the fact that these can vary greatly depending on the analytical method that is chosen. Chapter 3 draws some linkage between the medical community and environmental science community starting from terminology to emphasize the fact that some basic concepts of pharmacokinetics and pharmacodynamics which are all too familiar with the former group can be applied (with some modifications) to environmental sciences so as to look at the persistence and degradation of PPCPs in the environment. Merely detecting these compounds in the environment, which has been greatly advanced by continuous improvements in analytical techniques and has received a lot of attention in recent books that discuss this subject, may not be sufficient from a biological perspective. Thus, Chapter 4 introduces ecological aspects and the range of approaches that have been used to assess the risks from PPCPs in the environment. Unlike other books that deal with the issue of PPCPs in the environment, the risks from these compounds are presented by looking at a range of organisms starting from the simplest forms of life (i.e., microorganisms) to more complex ones. Chapter 5 focuses on the engineering and treatment techniques that could minimize the impact of PPCPs in the environment. The last chapter discusses some general needs for the future so as to enhance our understanding of the biological implications of this subject and strategies to minimize the impact of these compounds on the environment.

It is important to note that not all questions and suspicions about the impacts of PPCPs in the environment will be exhaustively answered in this book. However, this book makes a headway in initiating the process of getting to some of those answers. It is intended for a broad audience of undergraduate and graduate students in toxicology, ecology, microbiology (mostly environmental), chemistry (including medical chemists), agriculture, and healthcare delivery (i.e., public health, nursing, pharmacy, veterinarians, and physicians). Some policy makers and members of the general public will also find some parts of the book informative. Considering the range of these seemingly fragmented disciplines, individual readers may be dissatisfied with the level of coverage of one aspect or another, particularly aspects that directly relate to their respective discipline. However, it is my sincere hope that such dissatisfaction can ultimately be used to inform the other stakeholders of differing fields, a trend that will truly serve the original purpose of this book which is to collectively and critically examine the issue of PPCP micropollutants and how best to minimize its impact on the environment.

I would not have been able to complete this book without the continued direct and indirect support from my previous employer (i.e., the University of Cincinnati) as well as my current employer. However, the analyses and opinions expressed in this book are purely my own and do not necessarily represent those respective entities. I also need to acknowledge the support and encouragement that I have received from my wife, Enid, daughter Patricia, and sons Daniel and Eric while writing this book. My promise to them was that this one should take a much shorter duration compared to *Environmental Microbiology: Principles and Applications*—a promise that I have kept. This book is dedicated to my parents Daniel (deceased) and Racheal Kayondo as well as my uncle Bethel Mulondo for their love and sacrifices to ensure that we get a decent education.

PATRICK K. JJEMBA PH.D.

American Water, Delran, NJ
Email: pjjemba@msn.com

1

USAGE OF PHARMACEUTICAL AND PERSONAL CARE PRODUCTS FOR DIFFERENT THERAPEUTIC END POINTS IN RELATION TO ENVIRONMENT

The human impulse for a cure runs quite deep, and our first instinct whenever we feel sick or heading toward sickness is to medicate. As the Baby Boomers age, there is an increased demand for state-of-the-art medical care. The pharma–patient transaction has transformed itself from the previous practice of selling pharmaceutical products to selling a life-style. Amiss from that transformation, however, is the lack of recognition of the need to appreciate the intertwined relationship between the health of ecology and the ecology of health. Both of these concepts collectively refer to the health of humans as determined, at least in part, by the condition of their surroundings in an ecological fashion. These considerations have led to the emergence of what is referred to as ecosystem health, a science that aims at integrating our desire to assess and monitor ecosystems and health-related problems in a more holistic fashion, environmental degradation, and ecology (Rapport et al., 2001; Jjemba and Robertson, 2005). Ecology is the study of the distribution, activities, and interactions of organisms with their habitat. Thus, ecosystem health necessitates the identification and characterization of the natural and anthropogenic sources of environmental contaminants that can compromise our health, a need to predict their movement, and persistence both in time and space, as well as determining how pathogens and target as well as nontarget organism respond to the presence of such compounds. Pharmaceutical and personal care products (PPCPs) are increasingly being recognized as emerging contaminants in the environment. PPCPs are a diverse group of chemicals that include prescription and nonprescription medications, veterinary drugs, nutritional supplements, diagnostic agents, as well as a variety of consumer

Pharma-Ecology: The Occurrence and Fate of Pharmaceuticals and Personal Care Products in the Environment. By P.K. Jjemba
Copyright © 2008 John Wiley & Sons, Inc.

products such as fragrances, sunscreens, and costmetics. This book is intended to examine the usage of these chemicals, their occurrence in the environment, their ecotoxicity, and efforts to remove them from various matrices in the environment.

As a former U.S. Food and Drug Administration (FDA) chief, David Kessler, once indicated at a direct-to-consumer (DTC) national conference that the more the pharmaceutical industry wears the public health hat, the more drugs that industry will ultimately sell. The pharmaceutical industry has traditionally included medical chemists, pharmacists, physicians, nurses, marketing experts, and other public health professionals. The industry has traditionally excluded other disciplines such as engineers and ecologists. Microbiologists and other biologists have had a limited role in understanding physiological processes as they relate to disorders, pathogens, and their control, particularly through the use of antibiotics. Over time, the per capita consumption of pharmaceutical compounds and the range of choices has steadily increased, especially in developed countries, as more natural and synthetic compounds are discovered. For example, total drug sales in Canada have doubled from $6.6 billion in 1996 to $13.8 billion in 2004 (Campbell, 2007). That increase in pharmaceutical use has also coincided with the detection of these compounds in the environment. First brought to the attention of the scientific community by the work of Richardson and Bowron (1985), focus on the fate of these compounds did not really catch on until the late 1990s when Halling-Sørensen et al. (1998) and Jørgensen and Halling-Sørensen (2000) published extensive reviews about the issue of drugs in the environment.

The consumption of pharmaceutical products is currently mostly driven through advertising, with more and more individuals becoming aware of conditions that were once less noticeable as significant or even of concern. Such consumption is typically not accompanied by basic but seemingly fundamental questions about:

1. How a particular drug is able to achieve what it does to make one feel relieved (i.e., mode of action).
2. How much of the active ingredient that is consumed is actually used to make one feel better or even get cured.
3. If not all of the drug is used by our ailing bodies, what happens to the excess.

A similar complaisance prevailed during the early days of the Green Revolution when unlimited quantities of agrochemicals (i.e., pesticides, herbicides, fungicides, and fertilizers) were applied, generating tremendous increases in plant yield. Whereas those yield increases were appreciated in stemming world hunger, it ultimately become clear that continued use of agrochemicals without proper precautions can be detrimental to the ecosystem and to our well-being. Those realizations were prompted by celebrated publications such as Rachael Carson's *Silent Spring* (1962). It is important to realize that PPCPs are not very different from agrochemicals and, in a number of instances, they are actually used in quantities that are equal (if not exceeding) those of agrochemicals (Hirsch et al., 1999). However, while there are some similarities between PPCPs and other organic pollutants, there are also some dramatic differences. For one, PPCPs tend to be more polar, and in most instances they have acidic or basic functional groups. This attribute poses challenges when it comes to efforts to completely remove PPCPs from the environment once they are introduced and also contributes to the difficulties we face in trying to detect their presence in the environment. Besides being biologically active, PPCPs also have other unique attributes:

1. They are typically composed of large chemically complex molecular structures.
2. They have parent-neutral compounds that are associated with salts to form polymorphic solid states.
3. Generally, they have multiple ionizable sites that are spread throughout the molecule.

These attributes enable PPCPs to serve their therapeutic purposes but are also important when one is considering their fate and impact on the environment as the compounds in the environment can be the parent, metabolites, or glucoronide moieties. Thus, the lessons learnt from other organic pollutants cannot be transplanted wholesale to address issues of PPCPs in the environment. PPCPs are characterized or classified according to chemical structure, their effects (i.e., mode of action), or their use (i.e., therapeutic purpose). That stated, however, it is important to note that even within those classifications, PPCPs are quite diverse and are therefore not expected to have a homogenous set of characteristics once they get into the environment. This is in contrast to other conventional pollutants such as polycyclic aromatic hydrocarbons (PAHs), polychlorinated biphenyls (PCBs), dioxins, BTEX, herbicides, and pesticides, which are, within each group (or class) not very variable even with a variation in the number of carbons or type of substations at a position within the molecule of a basic backbone. This diversity in PPCPs is very apparent even in classes of compounds that target the same organ and/or are for the same therapeutic use. They are deliberately designed to be biologically very active, which in plain terms means that they have exceptional ability to affect biochemical and physiological functions of biological systems and ecosystems. All of these observations lead us into ecological issues and, by virtue, the need to develop a clear understanding of how various organisms in the environment interact with compounds.

It is widely agreed that the properties of the molecule are important determinants of its biological activity. The specific mode of action, which is widely researched during drug development, may provide relevant information about likely effects on nontarget organisms in the environment. The primary focus of medical science is, first and foremost, to concentrate on relieving pain and suffering. However, some of the practices currently in place to achieve this noble cause seem to set up a chain reaction that relieves pain for an individual but exposes the ecosystem to even more aggressive or subtle maladies even across generations (i.e., multigenerational exposure). Although not a new concept, making the leap from an individual patient to an ecosystem may seem mind-boggling for a medical practitioner who is trained to address the issues of individuals as they file through the clinic. However, it is important to remember that a group of individuals (e.g., consuming a particular antibiotic) of the same species comprise a population. Beyond that, a group of populations in the same locale may be genetically related (e.g., humans and other primates) or unrelated (livestock and earthworms; fish and algae) but can perform a similar function. Populations assemble into a community exploiting the same resources, usually competing for those resources. In that sense, members of a community exert a similar set of functions, ultimately comprising a self-sustaining but complex ecosystem. From this brief individual–population–community–ecosystem outlay, it is apparent that linking our understanding of community, culture, and health with ecology requires us to build bridges across disciplines, disciplines that are still mostly quite fragmented and driven by specialization. Building such bridges will enable members of the respective disciplines to appreciate the complexity of issues pertaining to the presence and fate of PPCPs in the environment and to start elucidating seriously whether they are detrimental in those settings and what to do about such detriments as an informed society. This book attempts to put these issues in the limelight to expand the already increasing interest in this subject.

The use of pharmaceuticals has also become an integral part of livestock production. In industrialized countries, livestock production similar to other sectors of agribusiness, involves the maintenance of large flocks or herds in very close quarters, otherwise referred to as confined animal feeding operations (CAFOs; Fig. 1.1).

Figure 1.1 Examples of confined animal feeding operations (CAFOs). Such operations typically rely on subtherapeutic doses of antibiotics and other forms of pharmaceutical compounds to ensure healthy and fast growing herds or flocks.

Within the United State, CAFOs are defined as having ≥ 1000 animal units (U.S. EPA, 2000), and there are currently more than 6600 CAFO units within the United States. Such conditions can be a prime avenue for the rapid spread of diseases, and such animal husbandry scenarios have led to an increased use of pharmaceuticals to maintain viable livestock. Thus, it is common practice to regularly administer a whole range of pharmaceuticals including antibiotics, antacids, anesthetics, antihelminths, anti-inflammatory steroids, antiparasitic compounds, emetics, estrous synchronizers, growth promoters, sedatives, tranquilizers, insecticides (against ticks and flies), and nutritional supplements to the livestock. Most commonly used are antibiotics for specific therapeutic and subtherapeutic reasons as listed in Table 1.1. A number of these products may be administered to the herd or flock for relatively long durations, whereas some are used occasionally. Currently, data about the

TABLE 1.1 Pharmaceutical and Growth Promoters Routinely Used in Livestock Industry

Livestock	Product	Purpose
Poultry	Coccocidiostats such as Monensin, Lasalocids, Salinomycin, Narasin	Antiprotozoan and antibiotics to guard against coccidiosis. These ionophores are also used in cattle and swine as growth promoters. They generally have a different mode of action compared to other antibiotics.
	Arsenical, e.g., Roxarsone, Arsenilic acid	Improve growth performance and bird pigmentation.
	Antibotics such as tylosin, bacitracin, virginiamycin	Control bacterial infection and improve feed consumption leading to large/heavier birds.
Swine	Antibiotics such as apramycin, tylosin, bacitracin, carbadox, olaquindox, tiamulin, avoparcin	Controls enteritis, dysentery, and collibacacillosis? Also generally improves growth possibly due to better feed consumption. Avoparcin has also been used in cattle and poultry but it has been banned in the EU and Australia.
Cattle	Hormones such as oestradiol, testosterone, progesterone. Brand names include Zeranol, Melengestrol acetate, Trenbolone acetate	Used in the beef industry to increase the rate of weight gain and feed use efficiency. Some of these are applied in the feedstuff, as suppositories or as implantable pellets (subcutenously).
	Hormones such as bovine somatotrophin (BST)	Improving milk production in dairy.
	Long-term antibiotics such as tylosin	Provided to control liver abscesses.

(Continued)

TABLE 1.1 *Continued*

Livestock	Product	Purpose
	Short-term antibiotics such as tetracycline, sulfamethazine, and oxytetracycline	Used periodically to control/prevent bacterial infections.
	Antihelminths such as Iverectins, Fenbendazole	To control parasites.
	Lactams such as amoxicillin, cyclosporin, erythromycin, novobiocin, penicillin, etc.	Treatment of mastitis, a major infection that can cripple the dairy industry.
	Nonsteroid anti-inflammatory drugs (NSAIDs) such as diclofenac, meloxicam (Metacam), ketoprofen, etc.	A variety of veterinary ailments such as pain, including the pain exerted by mastitis. Also used in other domestic animals including pigs and dogs.
Aquaculture	Antibiotics such as sulfadiamethoxine, ormathoprim, and oxytetracycline	Applied to the water or as part of the feedstuff for fish, shrimp, shellfish. Some are applied by injecting individual animals.
	Tricaine methanesulfonate	Used in fish as a chorionic gonadotropin to enhance spawning. Applied intramuscularly, i.e., as an injectable.
Apiary (Bee-keeping)	Oxytetracycline (Terramycin), tylosin, and lincomycin	Used to control Foulbrood larva disease caused by *Paenibacillus larvae* in honey bees.

quantities of antibiotics used in livestock production in various countries are not systematically collected in a standardized fashion. Thus, Jensen (2001) estimated that 150,000 kg of antibiotics were used in Denmark in 1997 of which >100,000 kg were primarily used as growth promoters, whereas the Union of Concerned Scientists estimated that 16 million kilograms of antibiotics are used in U.S. livestock annually, with 70% used for nontherapeutic purposes (UCS, 2001). By contrast, the Animal Health Institute estimates that only 9.3 million kilograms of antibiotics are used in the United States of which only 1.3 million kilograms are for nontherapeutic purposes (AHI, 2002).

Reports indicate that tylosin, tetracycline, and bacitracin are three of the most used antibiotics in livestock production within the United State (Sarmah et al., 2006). The macrolide tylosin is a broad-spectrum antibiotic with excellent antibacterial activity against most gram-positive (including *Mycobacterium* sp.) and some gram-negative bacteria, vibrios, coccidian, and spirochete. In vitro, it acts by inhibiting the synthesis of proteins as it binds on the ribosomes (McGuire et al., 1961; Weisblum, 1995). It consists of mainly tylosin A, which comprises approximately 80–90% together with three other constituents, that is, tylosin B (desmycosin), tylosin C (microcin), and tylosin D (relomycin) on a 16-membered lactone ring attached to an amino sugar (mycaminose) and two neutral sugars called mycarose and mycinose (McGuire et al., 1961). It is very stable at neutral pH but becomes very unstable under acidic or alkaline conditions. This attribute may have very significant effects on its stability once it gets into the environment. It targets the 50S ribosomal subunit, inhibiting the transcription and eventually leading to death of the cell (Retsema and Fu, 2001). More than 634 million poultry are exposed to macrolides such as tylosin and tilmicosin in the United States annually (Hurd et al., 2004).

Sulfonamides are widely used in humans and livestock against gram-positive and some gram-negative pathogens. In livestock, they are in some instances used at prophylactic levels to prevent disease outbreaks. As a matter of fact, sulfonamides are some of the most widely used antibiotics in the livestock industry. Their attributes and mode of action will be more extensively discussed in Section 1.2.4.4. They are excreted as parent compound or acetic acid conjugates, which eventually reconvert to the parent compound. Bacitracin is a polypeptide antibiotic commonly added to livestock (i.e., chicken, turkey, cattle, swine) feedstock. It is very soluble in water and has a high molecular weight. Similarly, the β-lactam monenomycin A is also used widely as a growth promoter in livestock feed.

In general, these drugs are administered to the livestock through water and foodstuff, although some may be injected, applied in dips, or during spraying events. They can be administered to individual animals or to the entire herd. In the United States, some of the antibiotics are approved for use in livestock for the treatment and prevention of diseases, whereas others are approved for use as growth promoters. For example, virginiamycin is approved for use in cattle, turkeys, and swine but mostly in chickens, primarily as a growth promoter and also to prevent or control diseases. It has been licenced for use in the U.S. livestock industry since 1975. The wide use of this specific compound has raised concern in some circles as it is very similar to other streptogramins such as Synercid (see Section 1.2.4.3), which are dependable antibiotics used against enterococci infections (Werner et al., 1998; Claycamp and Hooberman, 2004). Such transfer of resistance is possible as animal-derived resistant enterococci may colonize humans directly when humans interact with animals (e.g., farm workers), consume tainted animal products, or consume other farm produce that have had contact with animal products such as improperly treated animal manure. Enterococci are otherwise part of the normal human enteric microflora, although they occur in low abundance (i.e., <1% of the enteric bacterial population). They are also widely distributed in other animals and a common contaminant on unprepared foods. They are recognized as an important causative agent of nosocomial infections (Murray, 1997; Witte, 2001).

The exact mechanisms of how pharmaceuticals, especially antibiotics, exert growth promotion attributes in livestock are not clearly known, but it is suspected that the antibiotics ease minor infections that do not make the animals sick, ultimately increasing feed utilization (Ferber, 2003). In practice, most of the antibiotics are used for both purposes. In the European Union, the nontherapeutic use of most antibiotics for agriculture has been banned, leaving only four compounds, that is, avilamycin, monesin, flavophsopholipol, and salinomycin, in use for nontherapeutic purposes (Kümmerer, 2004a). These four have been spared so far as they are not deemed close to those used in human medicine. Estimates of the quantities of pharmaceuticals in livestock within developing countries are not very precise, and the few studies that have been done show that antibiotics are used mainly for therapeutic purposes rather than growth promotion (Jin, 1997; Mitema et al., 2001), although there is still much disagreement about that conclusion. Nonsteroidal anti-inflammatory drugs (NSAIDs) have also been used in livestock, particularly in southern Asia (Oaks et al., 2004; Cuthbert et al., 2006; Swan et al, 2006). Whether for livestock or human needs, the usage of pharmaceutical compounds in most developing countries is even harder to track precisely as some pharmaceuticals that are typically available through prescription only (e.g., most antibiotics) can be easily obtained over the counter without a prescription in developing countries (WHO, 2001).

Some antibiotics are also used in horticulture to control contamination of micropropagation, plant tissue culture, and in controlling bacterial diseases of fruit trees (Levy, 1992; Falkiner, 1998). Commonly used in horticulture are cephalosporins, neomycin, novobiocin, polymyxin, and sulfaguanidine. More than 20 tons of streptomycin and tetracycline are used by the horticulture industry in the United States per annum. Substantial amounts of antibiotics are also used in aquaculture, whereby they are either directly added to the water (therapy) or as part of the feed, resulting in high concentrations in the water and adjoining sediments. An examination of the levels of use of various PPCPs for various purposes is outline next.

1.1   PERSONAL CARE PRODUCTS

Personal care products broadly include a number of compounds that we use in our daily lives ranging from soaps, detergents, perfumes, aftershaves, and other maintenance (e.g., cleaning agents, disinfectants, sprays, deodorants, etc.) products. The use of some of these products has gone on possibly since the beginning of humankind. By all means, their use steadily increased from around the time when Anton van Leeuwenhoek discovered bacteria, which he, at that time, referred to as "wee animalcules." The work of Louis Pasteur almost two centuries later showed that the air contains microorganisms that cause putrefaction. Soon thereafter, Robert Koch, through what is known today as Koch's postulate clearly demonstrated that microorganisms can cause diseases. This series of discoveries led to the systematic development of cleanliness (hygiene) to minimize human contact with microorganism in order to stay healthy. As a matter of fact, the word *hygiene* originated from the Greek word *Hygieia*, the goddess of health. With improved hygiene, life expectancy has doubled in developed countries over the past two centuries and infectious diseases have been greatly reduced. However, these have been replaced by an increase in deaths due to cancers, strokes, heart disease, and diabetes. Some of these, together with a number of subtle changes within populations, may at least in part be attributed to the effects of PPCPs and other pollutants in the environment.

1.1.1 Fragrances and Musks

The use of fragrance has been around for many centuries. Fragrances are used quite widely in perfumes, cosmetics, deodorants, sunscreens, washing and cleaning agents, and a whole range of other personal care products. Fragrances provide confidence, joy, and a sense of well-being. Several lines of personal care product fragrances are directly applied on the skin or used in fabric cleaners. Cadby et al. (2002) published the typical percent levels of fragrances in various personal care products (Table 1.2), and these range from as low as 0.3% in face creams to as high as 20% in perfume extracts. It is noticeable that most of these musks are, through normal usage, dispensed down the drain either directly (as sullage) or subsequently after routine use.

While some fragrances are derived from plants, others were, in the past, derived from animals. Most popular from animals were musk from deer and other related species. However, with increased awareness about animal rights and fears of wiping out some of these animal species, that source of fragrances has diminished. Industry learned very fast how to synthesize these compounds, leading to the advent of what is now referred to as synthetic musks. Synthetic musks belong to two main categories: nitro musks and polycyclic musks (Fig. 1.2). Nitro musks are further classified as musk xylene, musk ketone, musk tibetene, and musk moskene. The last two classes of nitro musks are not very common, and in fact their production has been phased out in some countries (IFRA, 2006). Nitro musks are used almost exclusively in cosmetics. Polycyclic musks are substituted tetralins and indanes. They are used in both cosmetics and detergents. Most widely used among the polycyclic musks are 7-acetyl-1,1,3,4,4,6-hexamethyl-1,2,3,4-tetrahydronaphthalene (AHTN) and 1,3,4,6,7,8-hexahydro-4,6,6,7,8,8-hexamethylcyclopenta-γ-2-benzopyran (HHCB). These comprise about 95% of polycyclic musks sold in Europe

TABLE 1.2 Upper Limit Fragrance Concentrations in Various Personal Care Products

Types of Product	Fragrance Level (%)
Perfume extract	20
Toilet waters[a]	8
Fragrance cream	4
Bath products	2
Toilet soap	1.5
Shower gels	1.2
Deodrants/antiperspirants	1
Hair spray	0.5
Shampoo	0.5
Body lotion	0.4
Face cream	0.3

[a]Includes perfumes and aftershave lotions.
Source: Cadby et al. (2002).

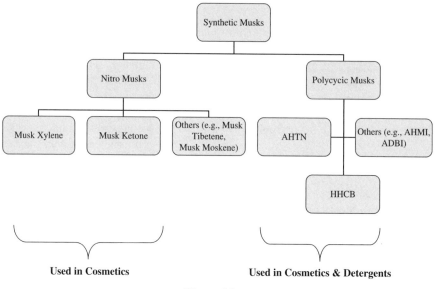

Figure 1.2

and approximately 90% of the sales in the United States (HERA, 2004). Current production of AHTN in the United States stands at more than 4500 tons/year (Kannan et al., 2005). The characteristics of the some of these compounds are summarized in Table 1.3.

Polycyclic musk fragrances (PMFs) are a major class of compounds. More than 5000 tons of polycyclic musks are synthesized worldwide annually (Kannan et al., 2005). HHCB and AHTN are two of the most frequently used PMFs worldwide. They are semi-volatile compounds with a log K_{ow} of 5.4–6.3 (Ricking et al., 2003). They are poorly soluble in water and are quite hydrophobic and very lipophilic by design, with a high propensity to sorb to organic materials (including the skin) and organic matter. Polycyclic musks are widely used in washing and cleaning agents and in a variety of personal care products including perfumes and deodorants. The breakdown by usage in the European Union (EU) where the data are more available is shown in Figure 1.3. Per capita use in the EU is estimated at 4.4 mg AHTN/day and 11.1 mg HHCB/day (Balk and Ford, 1999). They have a characteristic indane or tetraline skeleton that is highly substituted by methyl groups. The content of AHTN in fragrances has been limited by the EU Scientific Committee on Consumer Products (SCCP) to a maximum of 12%, whereas no maximum has been recommended for HHCB. No restrictions in the usage of musks are in place in any other part of the world.

Synthetic musks are also desirable in consumer products because of their fixative properties. Much as they are classified into these two groups, the members within each group of synthetic musks are quite diverse and do not have a common chemical structure. Thus, even within each group, musks cannot be treated as a single entity in terms of characteristics except for the musky scent common to all musks. Cosmetics application and absorption from textiles is a major source of HHCB and AHTN, materials that end up in the environment (Cadby et al., 2002).

TABLE 1.4 Characteritics of a Few Commonly Used Organic UV Filters

Abbreviation and CAS Number	Common Scientific Name	Structure[a]	Solubility	log K_{ow}[b]	Remarks
BMDBM (CAS 70356-09-1)	Butyl methoxydibenzoylmethane (avebenzone)		Insoluble	4.51[b]	Has a molecular weight (MW) of 310.4 and is effective against UVA (380–315 nm) that are associated with skin damage and UVB (315–280 nm) that cause sunburn. It has a roselike ordor.
BP-3 (CAS 131-57-7)	Benzophenone-3 (phenylketone)		Insoluble (<0.1 g/100 mL at 20°C)	3.8[b]	Traces have been detected in the bloodstream and urine after topical application.
DHB (CAS 131-56-6)	Dihydroxybenzophenone		Insoluble	?	Has MW of 214.2. Reduces damage by blocking UVA and UVB rays.
DHMB	Dihydroxy methyoxybenzophenone		Insoluble	3.82[c]	MW = 274.3
EHMC (CAS 5466-77-3)	Ethylhexyl methoxycinnamate		<0.1 g/100 mL at 27°C	6[b]	MW = 290.4

(Continued)

TABLE 1.4 *Continued*

Abbreviation and CAS Number	Common Scientific Name	Structure[a]	Solubility	$\log K_{ow}$ [b]	Remarks
4-MBC (CAS 36861-47-9)	4-Methylbenzylidene camphor	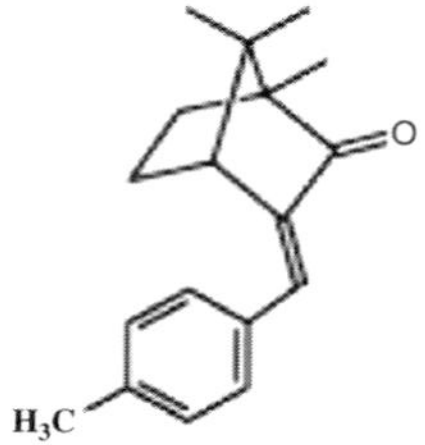		5.9[b]	
OC (CAS 6197-30-4)	Octocrylene	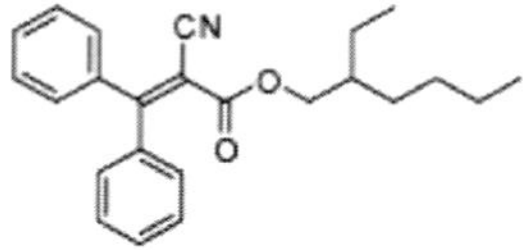		6.88[b]	An ethylhexyl ester
PBSA (CAS 27503-81-7)	Phenyl-benzimidazole sulfonic acid	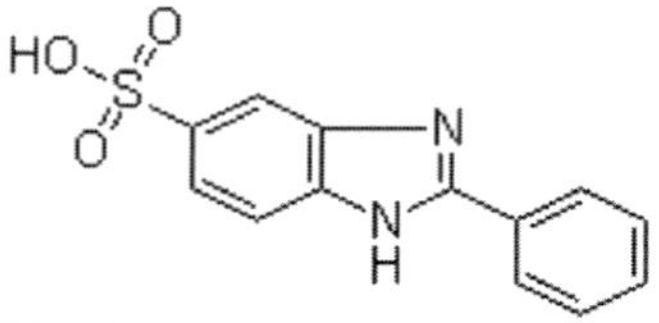	Soluble	−0.16[b]	MW = 274.3

[a]Most structures and solubility information obtained from http://www.chemblink.com.
[b]Log K_{ow} values were calculated from the Syracuse Research Corporation demo program available at http://www.syrres.com/esc/est_kowdemo.htm or obtained from Jeon et al. (2006).

14

widely used detergent in soaps, shampoos, shower gels, bubble bath, foot pastes, and the like, and is, by nature of its use, thus discharged into aquatic environments in large amounts. SDS is a primarily amphoteric compound consisting of a hydrophilic (alkyl chain, alkylphenyl ethers, alkylbenzene, etc.) and a hydrophilic (i.e., carboxyl, sulfate, sulfonate, phosphate, etc.) part. The hydrophilic and hydrophilic moieties readily interact with both polar and nonpolar structures in macromolecules such as proteins, cellulose, and a variety of other molecules. This ability to interact enables them to decrease the energy of interaction and energy of solvation between various biological systems, absorbing on oil–water and air–water interfaces. These characteristics also enable detergents to enhance the solubility of sparingly soluble compounds in water. SDS also has tensioactive properties that can lead to an increase in the surface tension of phosphatidylcholine monolayers, reducing the protection against oxidative stress, causing toxicity (Cserháti et al., 2002). Surfactants have also been linked to the toxicity of several organisms in the environment.

Surfactants are also important ingredients for pharmaceuticals, influencing the adsorption and absorption processes of drugs as well as the partitioning of the drugs between hydrophilic and hydrophobic components in organisms. For example, SDS dramatically improves the absorption of the antihelminthic drug albendazole in the intestines (Del Estal et al., 1993). Just like pharmaceuticals, surfactants also have a marked biological activity.

1.1.3 Disinfectants

Disinfectants include sodium hypochlorite (bleach), triclocarban (TCC) and triclosan (TCS). Triclocarban and triclosan are major components, together or separately in the soaps, cosmetics, and laundry detergents industry at concentrations of 0.1–3% (w/w). They are also used as over-the-counter topical antimicrobial drugs in humans (Halden and Paull, 2004). Disinfectants are used in hospitals and other medical facilities to clean the skin, equipment, clothing, and surfaces, ultimately getting washed into the sewage treatment plant (STP) (Kümmerer et al., 1997a). Both triclocarban and triclosan have limited solubility in water and a considerably high lipophilicity that renders them to bioaccumulate and bioconcentrate in tissues.

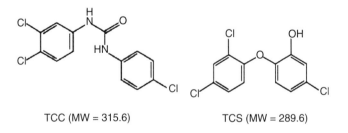

TCC (MW = 315.6)   TCS (MW = 289.6)

Triclosan also has a broad-range bacteriostatic activity against gram-positive and gram-negative bacteria, fungi, molds, and yeasts. It is metabolized to its glucuronide or sulfate conjugates and is sparingly soluble in water. TCC solubility at 25°C is 0.65–1.55 mL/L, whereas log K_{ow} at that same temperature is 4.9. By comparison, TCS is slightly more soluble (1.97–4.6 mL/L) and a log K_{ow} of 4.8.

Other disinfectants include quarternary ammonium compounds (QACs) such as benzalkonium chloride (BACl) and didecyldimethylammonium chloride (DDMAC). Disinfectants have been shown to denitrifying bacteria at concentrations as low as 1–2 ng L^{-1} (Kümmerer, 2004b). QACs are specifically not used for disinfecting clothing, but their use is central in other purposes. Triclocarban and triclosan have an aromatic structure that is rich in chlorides, which contribute to their persistence in the environment. QACs are also quite persistent in the environment (Al-Ahmad et al., 2000). Other disinfectants include chloramines and 1,3-dichlorosucyanuric acid as well as PVP-iodide-based disinfectants.

1.2   PHARMACEUTICAL COMPOUNDS

Accurate statistics about the production and consumption of the individual pharmaceutical compounds are not readily available because of privacy and industry competition issues. However, some crude estimates can be assembled based on the number of prescriptions that are handed out. For example, in the United States, which uses more than half of the world's medications, the most dispensed 200 drugs registered 2.13, 2.82, and 2.32 billion prescriptions in 2003, 2004, and 2005, respectively. Based on these statistics, antihypertensive and cardiovascular drugs are the most prescribed, contributing 26–27% of the prescriptions for the top 200 most prescribed drugs (Fig. 1.4). Sedatives, hypnotics, and antipsychotic drugs rank second (19–23% of the top 200 most prescribed drugs) followed by analgesics and anti-inflammatory agents (14–15%) and antimicrobial agents (10–11%). As a parallel comparison, antihypertensive and cardiovascular drugs as well as antipsychotic drugs were the three most frequently purchased drugs in Canada in 2004, collectively accounting for 54% of the expenditure on prescription medicine in that country (Morgan et al., 2005). In the United States 110–140 million gastrointestinal medication prescriptions were dispensed between 2003 and 2005 (5–6% of the top 200 prescriptions). During that time period, 2.5–4.3% of the prescriptions were medications that are used mostly for respiratory infections (2.5–3.6%), oral contraceptive and reproductive therapy (2.5–4.3%), thyroid hormones (2.9–3.5%), diuretics and electrolytes (3.9–4.1%), or antidiabetics (3.3–4.2%) (Fig. 1.4). Other key prescriptions belonged to biophosphonates and anti-bone-loss (1–1.4%), steroids (1–1.5%), hematology (1%), and nutritional (0.2–1.2%) categories. Antineoplasts, dipaminergics and immunomodulators, anesthetics, and triptans were least prescribed among the leading 200 prescriptions during those 3 years. These data will be more closely examined below for each category of medications.

1.2.1   Antihypertensive and Cardiovascular Medications

The heart pumps blood to all parts of the body following well-timed rthymic contractions. However, if this contractility is decreased for any reason, the blood

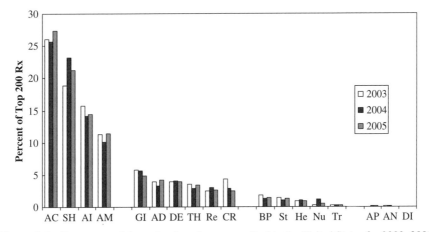

Figure 1.4 Percentage of drugs that have been prescribed in the United States for 2003–2005 as a fraction of the top 200 most prescribed drugs. Note that the total of the top 200 most prescribed was 2.1 billion, 2.8 billion, and 2.3 billion for 2003, 2004 and 2005, respectively. AC = Antihypertension/cardiovascular medication, SH = sedatives/antipsychotics, AI = analgesics/anti-inflammatory, AM = antimicrobial, GI = gastrointestinal, AD = antidiabetic, DE = diuretics/electrolytes, TH = thyroid drugs, Re = respiratory, CR = contraceptives/ reproductive therapy, BP = biophosphonates and other anti-bone loss, St = steroids, He = hematology, Nu = nutritional, Tr = triptan, AP = antineoplast, AN = anesthetic, and DI = dopaminergics and immunomodulators.

output (i.e., the total amount of blood that the heart is able to pump in a given time frame) is also equally decreased. Such a decrease can ultimately lead to a decrease in the perfusion of renal fluids into the kidney or an increase in sympathetic activity. Both of these activities can lead to an increased concentration of the circulating angiotensin I, and the increased sympathetic activity can also either stimulate cardiac beta receptors or vascular alpha receptors. All of these actions can collectively or individually lead to an increase blood volume in the arteries and increase the resistance as well as increased heart rate. A range of medications, some of which are listed in Table 1.5, are often prescribed to alleviate some of these circulatory system shortcomings. These are primarily designed to dilate the arteries (vasodilators), block the stimulation of alpha receptors (alpha blockers) or beta receptors (beta blockers), block the transformation of agiotensin I to a much more potent angiotensin II [angiotensin converting enzyme (ACE)], or improve the retention of salts and water in the body (i.e., diuretics).

High lipid concentrations in the plasma are manifested by cholesterol, a waxy fatlike substance found in animal fats and oils. Gradual buildup of these substances, particularly those that have a low density, that is, low-density lipoproteins (LDL), in the arteries can cause atherosclerotic plaque (blockage of the arteries) for which, besides exercise and adapting better diets, antilipemic drugs are increasingly prescribed. If not attended to, such blockage increases pressure within the arteries due to more restricted passage for the blood, thus leading to cardiovascular maladies that are similar to those caused by decreased renal perfusion and increased sympathetic activity described above.

TABLE 1.5   Most Frequently Used Antihypertensive and Cardiovascular Drugs in the United States (2003–2005)

Antihypertensives and Cardiovacular Drugs	Specific Usage	% of the Top 200 Prescriptions in		
		2003	2004	2005
Lipitor	Antilipemic	3.1	2.5	2.7
Norvasc (Amlodipine, Lotrel)	Calcium channel blocker	2.1	1.7	2.0
Metoprolol tartrate	Beta blocker	2.1	2.6	2.4
Atenolol (Chlorthalidone)	Beta blocker	1.9	1.6	1.9
Hydrochlorothiazide	Duiretic	1.6	1.5	1.8
Lisinopril (Zetril)	ACE inhibitor	1.5	2.0	2.6
Zocor (simavastatin)	Antilipemic	1.3	1.0	1.3
Diovan HCT (Valsartan)	Angiotensin II receptor inhibitor combination	1.0	0.8	1.0
Viagra (sildenafil)	Vasodilator	0.7	1.0	0.5
Pravachol (pravastatin)	Antilipemic	0.7	1.0	0.4
Plavix (clopidogrel)	Anticoagulant	0.7	1.3	0.8
Digitek (digoxin, lanoxin)	Cardiac glycoside	0.7	0.6	0.6
Cozaar (hyzaar, losartan)	Angiotensin II receptor blocker combination	0.7	0.6	0.6
Quinapril (Accupril)	ACE inhibitor	0.6	0.4	0.3
Altace (ramipril)	ACE inhibitor	0.6	0.5	0.5
Verapamil HCl	Calcium channel blocker	0.5	0.4	0.4
Diltiazem HCl (diltiazem CD, Cartia)	Calcium channel blocker	0.5	0.5	0.5
Lotensin (Benazepril)	ACE inhibitor	0.4	0.2	0.3
Clonidine HCl	Centrally acting antihypertensive	0.3	0.3	0.4
Tricor (Fenofibrate)	Antilipemic	0.3	0.3	0.4
Isosorbide monnitrate (Monoket, Ismo)	Vasodilator	0.3	0.3	0.4
Avapro (Irbesartan)	Angiotensin II receptor blocker	0.3	0.2	0.2
Proporanolol HCl (Inderal LA)	Beta blocker	0.3	0.2	0.2
Nifedipine ER (Nifediac CC)	Calcium channel blocker	0.3	[a]	0.1
Coreg (Carvedilol)	Beta blocker	0.2	0.2	0.4
Enalapril maleate (Vasotec)	ACE inhibitor	0.2	0.5	0.5
Gemfibrozil (Lopid)	Antilipemic	0.2	0.2	0.2
Monopril (fosinopril sodium)	ACE inhibitor	0.2	0.1	[a]
Zetia (ezetimibe, ezetrol)	Antilipemic	0.2	0.3	0.5
Niaspan (niacin)	Antilipemic	0.2	0.2	0.2

(*Continued*)

TABLE 1.5 *Continued*

Antihypertensives and Cardiovacular Drugs	Specific Usage	% of the Top 200 Prescriptions in		
		2003	2004	2005
Timolol maleate (Blocadren)	Beta blocker	0.1	[a]	[a]
Atacand HCT	Angiotensin II blocker combination	0.1	[a]	[a]
Lescol XL (fluvastatin)	Antilipemic	0.1	[a]	[a]
Plendil (Felodipine)	Calcium channel blocker	0.1	[a]	[a]
Nitroquick (nitroglycerin)	Vasodilator	0.1	[a]	0.1
Avalide	Angiotensin II receptor blocker	0.1	[a]	0.2
Lovastatin (Mevacor)	Antilipemic	[a]	0.3	0.4
Crestor (rosuvastatin calcium)	Antilipemic	[a]	0.2	0.3
Benicar HCT (olmesartan medoxomil)	Angiotensin II receptor blocker	[a]	0.1	0.4
Bisoprolol/HCTZ (Zebeta)	Beta blocker	[a]	0.2	0.2
Cialis	Vasodilator	[a]	[a]	0.2
Percent as total Rx of top 200 drugs		24.8 (527.4)[b]	23.7 (691)	25.9 (601.4)

[a]Not among the top 200 most prescribed drugs in the United States that year.
[b]Numbers in parentheses are the total number of prescriptions in millions.

Antihypertensive and cardiovascular drugs are the most prescribed drugs for human therapy in the United States, and among those drugs that comprise the top 200 prescriptions there were 527.4, 691, and 601.4 million prescriptions in 2003, 2004, and 2005, respectively (Table 1.5). By contrast, most of the use of pharmaceuticals in the livestock industry is antimicrobials (Zielezny et al., 2006). Some of these antihypertensive and cardiovascular drugs are referred to as beta blockers, whereas others are diuretics, antilipemic, ACE inhibitors, vasodilators, or anticoagulants. Beta blockers act by competitively inhibiting beta-andrenergic receptors, a system that is involved in various physiological functions such as the need for oxygen by the heart and also its rate of beating, the dilation mechanisms of blood vessels, and the dilation of bronchioles. Adrenergic receptors that are similar to those targeted by beta blockers have been found in fish (Nickerson et al., 2001), frogs (Devic et al., 1997), and birds (Yardeny et al., 1986). These findings have implications to the potential effects of beta blockers to nontarget organisms in the environment, as will be discussed in Chapter 4. The adrenergic system also interacts with the metabolism of lipids and carbohydrates in response to stress conditions such as starvation. In a nutshell, beta blockers are used to prevent heart attacks or their reoccurrence. By contrast, diuretics block the reabsorption of sodium and chloride ions. Furosemide and hydrochlorothiazide are two of the most widely used diuretics in managing hypertension and cardiovascular ailments. They are extensively used in the treatment of cardiovascular diseases. Antilipid drugs are also categorized into two types: fibrates and statins. They are both used to primarily lower the concentration of cholesterol and triglycerides. Commonly used statins include Zocor (simvastatin), Lipitor, Pravachol, fluvastatin, lovastatin, and Crestor. These statins are basically derivatives of fibric acid (fibrates) and 3-hydroxyl-3-methylglutaryl coenzyme A (HMG-CoA). Lipitor (atorvastatin) is the most prescribed in the United States and Canada. Clofribrate, etiofibrate, and etofibriclofibrate are some of the other lipid-regulating drugs that are widely prescribed. Their active metabolite, clofibric acid, is actually a stereo isomer to the herbicide mecaprop (MCPP; 2-[4-chloro-2-methylphenoxy]-propionic acid) and has been detected in various Swiss lakes by a group that was strictly looking at herbicide contamination in water (Buser et al., 1998a). They impose their effect by decreasing the plasmatic content of cholesterol and triglycerides. Statins were discovered in extracts of filamentous fungi. They have a characteristic polyketide portion and a hydrohexahydronaphthalene system to which various side chains connect at the C8 (R1) and C6 (R2). Some of these drugs are referred to as antilipemic in reference to the condition that they treat, that is, hyperlipidemia. Hyperlipidemia is a term that generally refers to a high concentration of lipids in the plasma. This condition is mostly acquired from poor dietary habits, although it can also rarely be acquired through heredity. The most widely used antilipemic drugs listed in Table 1.5 act by inhibiting the synthesis of cholesterol in the liver. Most of them are susceptible to first-pass biotransformation by the liver with only small amounts ($<20\%$) eliminated in the urine. They inhibit the synthesis of cholesterol by preventing the HMG-CoA reductase enzyme, which in turn affects the conversion of HMG-CoA to mevalonate. Cholesterol modulates the fluidity of membranes in eukaryotes and is an important precursor of steroid hormones such as progesterone, testosterone, estradiol, and cortisol. It comes from the food we eat but can also be synthesized de novo in eukaroytes from acetate when acetate is converted to acetyl-CoA and the abbreviated pathway is displayed in the pathway underneath.

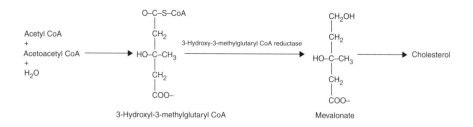

The most distinctive product from this pathway in vertebrates is cholesterol. The HMG-CoA reductase is the rate-limiting enzyme in this pathway. The formation of mevalonate is the step that commits HMG-CoA to form cholesterol and is catalyzed irreversibly by 3-hydroxyl-3-methylglutaryl reductase. Studying the activity of this enzyme has therefore provided very good insights about the effects of various statins on cholesterol formation. These interactions occur in the liver and can lead to a buildup of cholesterol that can ultimately clog the arteries. More recently marketed statins, notably atorvastatin and fluvastatin are entirely synthetic and structurally different from the conventional natural and semisynthetic, but they still retain the ability to inhibit HMG-CoA reductase, the enzyme that is very central in the mevalonate pathway. Statins are effective in humans, lowering LDLs, but they may also target similar sites in other organisms. For example, unlike the case in vertebrates, the mevalonate pathway generates juvenile hormone, dolichol, and ubiquinone in insects. Ubiquinone is involved in the electron transport system, whereas dolichol is a donor of oligosaccharide residues in the glycosylation of protein. The juvenile hormones have important roles in the development and reproduction in insects (Zapata et al., 2002, 2003) and lobsters (Li et al., 2003). For this activity, mevalonate pathway inhibitors are increasingly recognized for their pesticidal activity. The ecological impacts of this aspect as it relates to our routine use of statins and their occurrence in the environment will be discussed in Chapter 4.

The fibrates are believed to act by activating the lipoprotein lipase enzyme, reducing very low density lipoproteins (VLDL), which in turn increases the high density of lipoproteins (HDL) cholesterol. The net result from this chain of events is a decrease in the levels of triglycerides in the plasma and a reduced risk of cardiovascular attacks (Staels et al., 1998). Thus, fibrates stimulate the uptake of fatty acids, converting them into acetyl-CoA derivatives. The derivatives are then converted through β-oxidation leading to a decrease in the production of very low density lipoproteins. Experiments in rats have shown that chronic exposure to fibrates can lead to hepatic damage, possibly due to inhibition of mitochondrial oxidative phosphorylation (Qu et al., 2001; Keller et al., 1992) and/or massive proliferation of peroxisomes (Hess et al., 1965). Increases in peroxisomes lead to an increase in fatty acid peroxisomal β-oxidation due to the activation of the nuclear peroxisome proliferator activated receptors (PPARs) (Nunes et al., 2006). This activation is followed by an enhanced production of hydrogen peroxide and a complementary production of catalase, the enzyme that degrades the hydrogen peroxide. Some species such as rodents and red algae (*Kappaphyaus alvarezii*) are particularly sensitive to high levels of hydrogen peroxide (H_2O_2) and can experience oxidative stress on exposure to fibrates due to the peroxide that such exposure generates (Mannaerts et al., 1979; Qu et al., 2001; Barros et al., 2003).

Anticoagulants work by preventing the aggregation of platelet that can otherwise cause blockage of the arteries. Most extensively used in this category of medications in the United States is Plavix, which has steadily commanded approximately 0.6% of the top 200 most prescribed drugs in that country between 2003 and 2005. It is rapidly absorbed, with a bioavailability of about 50%. Half of it is eliminated through the renal system, whereas the other half is excreted through the fecal route (Wallis and Sciacca, 2003).

1.2.2 Anxiolytic Sedatives, Hypnotics, and Antipsychotics

Anxiolytic sedatives, hypnotics, and antipsychotics are lumped together in this section because all of them act by some form of intervention with the brain and nervous system. They are addressed through a field commonly referred to as pyscho-pharmacotherapy. Three monoaminergic neurotransmitters, commonly referred to as monoamines, namely dopamine, serotonin, and noradrenaline, have been implicated in the psychotic disorders that these drugs combat. Thus, conditions such as depression have been associated with deficiencies in one or more of these mono-amines (Hensiek and Trimble, 2002). The medications currently available for these conditions include serotonin reuptake inhibitors (SSRIs), tricyclic antidepressants (TCAs), monoamine oxidase inhibitors (MAOs), and benzodiazepines. The consumption of anxiolytic sedatives, hypnotics, and antipsychotic drugs is one of the fastest growing in developed countries. As noted from Table 1.6, the range of

TABLE 1.6   Most Prescribed Pharmaceuticals for Psychotic Disorders in United States (2003–2005)

Anxiolytic Compound	Purpose	% of Top 200 Prescriptions in		
		2003	2004	2005
Alprazolam (Xanax)	Sedative	1.4	1.2	1.5
Zoloft (sertraline)	Antipsychotic	1.4	1.1	1.2
Paroxetine (Paxil)	Antipsychotic	1.1	0.8	0.9
Ambien (zolpidem)	Sedative	1.0	1.7	1.0
Fluoxentine HCl (Prozac)	Antipsychotic	0.8	0.8	0.9
Effexor XR (venlafaxine)	Antidepressant	0.8	0.7	0.7
Neurontin (gabapentin)	Antipsychotic	0.8	0.5	0.7
Lorezepam (Ativan)	Antipsychotic	0.8	0.7	0.8
Clonazepam (Klonopin, Rivotril)	Anxiolytic/ sedative	0.8	0.6	0.7
Celexa (citalopram)	Antidepressant	0.7	0.4	0.4
Wellbutrin (Bupropion SR, Zyban)	Antidepressant	0.7	0.7	0.6
Amitriptyline HCl (Elavil)	Antidepressant	0.6	0.5	0.6
Lexapro (escitalopram)	Antidepressant	0.6	0.8	1.1
Cyclobenzaprine HCl (Flexeril)	Hypnotic/ sedative	0.5	0.5	0.6
Trazodone HCl (Desyrel, Trialodine)	Antidepressant	0.4	0.5	0.6
Risperdal (risperidone)	Antipsychotic	0.4	0.3	0.3
Flomax (tamsulosin)	Sedative	0.4	0.3	0.3
Promethazine HCl	Antipsychotic/ sedative	0.4	0.5	0.6
Zyprexa (olanzapine)	Antipsychotic	0.4	0.2	0.2
Dilantin (phenytoin sodium)	Anticonvulsant	0.3	0.1	0.1
Concerta (Ritalin)	Antipsychotic	0.3	0.3	0.4
Adderall XR (amphetamine)	Antipsychotic	0.3	0.3	0.4
Seroquel (quetiapine)	Antipsychotic	0.3	0.3	0.4
Depakote (Epival)	Antipsychotic	0.3	0.2	0.2

(*Continued*)

TABLE 1.6 *Continued*

Anxiolytic Compound	Purpose	% of Top 200 Prescriptions in		
		2003	2004	2005
Carisoprodol (Soma)	Opiod/sedative	0.3	0.4	0.4
Diazepam (Valium)	Sedative	0.3	0.4	0.5
Detrol LA (tolterodine)	Sedative	0.2	0.2	0.2
Topamax (topiramate)	Antipsychotic	0.2	0.2	0.3
Skelaxin (Metaxalone)	Sedative	0.2	0.2	0.2
Temazepam (Restoril)	Sedative	0.2	0.3	0.3
Aricept (donepezil)	Sedative	0.2	0.2	0.2
Ditropan XL (oxybutynin)	Sedative	0.2	[a]	[a]
Strattera (atomoxetine HCl)	Antipsychotic	0.2	0.2	0.2
Tussionex	Antipsychotic	0.1	[a]	0.1
Terazosin HCl (Hytrin)	Sedative	0.1	0.2	0.2
Percocet (oxycodene)	Sedative	0.1	[a]	[a]
Mirtazapine (Remeron)	Antidepressant	0.1	0.2	0.2
Hydroxyzine	Sedative	[a]	0.2	0.2
Doxazosin	Sedative	[a]	0.2	0.2
Cymbalta (duloxentine HCl)	Antidepressant	[a]	[a]	0.2
Butalbital	Sedative	[a]	[a]	0.2
Lamictal	Antipsychotic	[a]	0.1	0.2
Benzonatate	Antutissive (anticough)	[a]	[a]	0.2
Amphetamine mixed salts	Antipsychotic	[a]	[a]	0.2
Nortriptyline (Aventyl)	Antidepressant	[a]	0.1	0.1
Buspirone	Anxiolytic/ sedative	[a]	0.2	0.2
Tizanidine	Sedative	[a]	[a]	0.1
Total Rx anxiolytic sedatives, hypnotics and antipsychotics		18.1	17.0	19.9

[a]Not among the top 200 most prescribed in the United States at that time.

choices of drugs in this category is quite extensive, with a good part of the market share captured by alprazolam (Xanax), Zoloft (sertraline), paroxetine (Paxil), and Ambien (zolpidem). More recently, Lexapro (escitalopram) also became quite popular. Also widely used in developed countries, but not the United States, is carbamazepine (Ambrósio et al., 2002). Carbamazepine was initially used almost exclusively for epilepsy; its use has also been extended to schizophrenia, acute mania, and depressive episodes of bipolar disorder due to its mood-stabilizing properties (Kudoh et al., 1998). It is also used in the treatment of alcoholism (Sternebring et al., 1992) as well as drug addiction (Bertschy et al., 1997). It acts in mammals in a receptor-mediated manner by inhibiting neurotransmitters (Ambrósio et al., 2002). It is extensively metabolized in the hepatic portal vein by cytochrome P450 forming a variety of metabolites such as 10,11-dihydro-10,11-epoxycarbamazepine (through oxidation), which are further metabolized to glucuronide conjugates.

These metabolites are also therapeutically potent with some neurological effects in some instances (Metcalfe et al., 2004).

Antidepressants have contributed 17%–20% of the top 200 most prescribed drugs in the United States during 2003–2005 (Table 1.6). They are generally used for an extended period of time (4–6 months) after the resolution of acute symptoms at therapeutic dosages so as to avoid relapses (Donoghue and Hylan, 2001). Their sales in children in some developed countries have recently surpassed those of antibiotics (Steinbrook, 2002; Moyniham et al., 2002). Antidepressants act on the central nervous system (CNS), decreasing the neuronal activity by blocking voltage-dependent sodium channels of neurons that excite those neurons or by increasing the inhibition of the γ-aminobutyric acid (GABA), an inhibitory neurotransmitter of the CNS, when they bind onto specific γ-subunit sites. The GABA system is also present in fish and can certainly be targeted by these excitatory compounds. The evidence and ramification of this are discussed in Chapter 4.

The chemical structures and chemical characteristics of these leading compounds are shown below:

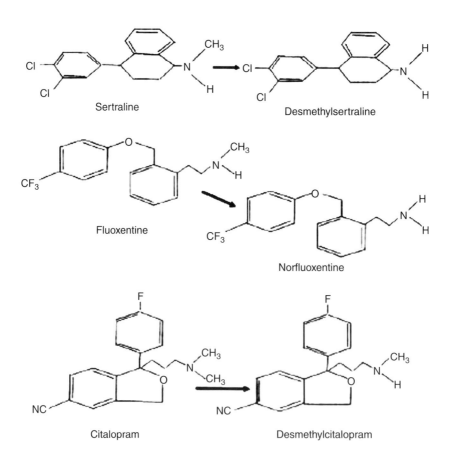

Sertraline

Desmethylsertraline

Fluoxentine

Norfluoxentine

Citalopram

Desmethylcitalopram

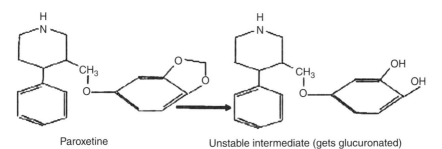

Paroxetine   Unstable intermediate (gets glucuronated)

The increased popularity of these compounds can, at least in part, be attributed to the aggressive marketing campaign that has been pursued by the industry in their favor (Lacasse and Leo, 2005). Their development has centered around manipulating serotonin levels in the body. Serotonin (5-hydroxytroptamine, 5-HT) has been implicated in the relaxation of body posture, reducing the excitation of muscles, and subsequent locomotion (Ranganathan et al., 2001). As a vasodilator and neuro-transmitter, serotonin has also been implicated in a variety of functions including sleep, sexuality, learning, appetite, and mood control. Thus, low serotonin levels have been associated with disorders such as anxiety, bipolar, sleep apnea, migraine, autism, and depression, with losses in its levels reportedly leading to neurons that are morphologically immature, possibly due to their limited ability to take up calcium-binding protein (Azmitia, 1999).

This brief explanation is not intended to explore all the complexities, enigmas, and controversies surrounding serotonin but rather to introduce the fact that various anxio-lytic sedatives, hypnotics, and antipsychotics have been developed to modulate the serotonin system by affecting serotonin levels and/or their receptors. Thus, the reuptake of serotonin can be inhibited by antidepressants and SSRIs. Interestingly, serotonin is present in a whole range of other organisms including plants, vertebrates, and invertebrates (Azmitia, 1999). This observation may have important implications on these nontarget organisms once the sedatives and antidepressants that we inno-cently use get into the environment. This perspective will be discussed further in Chapter 4. Unlike TCAs, SSRIs selectively or exclusively block monoamine uptake sites. SSRIs are examples of these and are currently more preferred in clinical practice because of their high affinity to serotonin reuptake sites and low affinity for neurotransmitter receptors. The better safety associated with SSRIs (as compared to TCAs) made the treatment of psychotic conditions easier to handle even on an out-patient basis but also expanded the usage of antipsychotic medication for minor cases of depression and other psychiatric disorders that are associated with a dysfunc-tional state of the serotonin system including premenstrual dysphoric disorders, anxiety, panic attacks, and obsessive compulsive disorder.

Most SSRIs inhibit cytochrome P450 (CYP) isoenzymes, a family of about 30 enzymes that catalyze the oxidase metabolism of multiple drugs in humans. They are eliminated mostly through hepatic metabolism by oxidase demethylation or oxidative deamination. SSRIs generally act by inhibiting the pumps that send serotonin across synapses, thus locally increasing the levels of serotonin in those synapse spaces. However, it is important to point out from the onset that the responses to serotonin are extremely complex, and they are mediated by a diverse range of serotonin receptors that are located in various parts of the body. For example, monamine oxidase inhibitors (MAOIs) are believed to increase serotonin (and other monoamine neurotransmitters) in the brain by preventing the breakdown of serotonin. By contrast, some antidepressants act by inhibiting the reuptake of serotonin, making it persist in the synapses for longer. As one of the popular antidepressants, fluoxetine (Prozac) acts by blocking serotonin reuptake, a neurotransmitter that is involved in sexual behavior, food intake, and a variety of other behaviors. Studies in nematodes (*Caenorhabiditis elegans*) have shown that it blocks the uptake of serotonin from the sympathetic nerve cleft (Ranganathan et al., 2001).

Fluoxantine (Prozac) is a lipophilic drug composed of a racemic mixture of an S- and R-enantiomer with the former having more potency than the latter. On oral administration, most of the drug is adsorbed, extensively accumulating in the tissue. It is also extensively metabolized to an active metabolite norfluoxentine. Fluoxetine is easily demethylated on metabolism by cytochrome P450 isoenzymes to norfluoxentine with 10% excreted as parent compound or fluoxentine *N*-glucuronide in urine (Hiemke and Härther, 2000). It is primarily metabolized in the liver by demethylation to norfluoxentine (Ring et al., 2001; Renshaw et al., 1992). At high concentrations (i.e., 0.25–1 mg serotonin/L), fluoxetine paralyzed these organisms, causing their nose muscles to contract. The cleft in nematodes has a gene (*mod-5*) that is similar to the one encoding human serotonin reuptake transport (Ranganathan et al., 2001).

Introduced in the late 1980s SSRIs were an alternative to TCAs, which have more severe side effects (see below). To that end, SSRI compounds are regarded as the gold standard for the treatment of depression. They are believed to have more benign adverse effects that are transient with continued therapy or easily managed, for example, by altering the time of administration (morning or evening), taking the medication with food, and so forth. The current fear in the medical community is that they are probably overprescribed primarily by physicians with little or no training in psychiatry.

Unlike TCAs, SSRIs have a more selective effect on the reuptake of serotonin. They are also better tolerated, safer in overdose, and have modest side effects. They are designed to derive their therapeutic benefits by acting on specific receptors. Examples include Zoloft, Paxil, Prozac, and Ritalin. Zoloft is second in potency as a serotonin reuptake inhibitor only to Paxil. It is absorbed from the gastrointestinal tract (GIT) completely, albeit slowly. It is eliminated at a much lower rate in females than males, suggesting sex-dependent differences in metabolisms and/or distribution. It is mainly eliminated through hepatic metabolism in the urine as the parent compound (Murdoch and McTavish, 1992) but approximately 50% is excreted in feces (Warrington et al., 1992). These substantial amounts of parent compound in the feces suggest extensive transport of the metabolites and their conjugates into the bile duct. It is mainly transported by *N*-demethylation to *N*-desmethylasertraline, the metabolite having a longer half-life in the body (i.e., 60–100 h compared to 22–37 h for the parent compound). Sertraline is extensively metabolized in the liver, with less than 0.2% ending up in the urine as the parent compound (Kobayashi et al., 1999).

Paxil, another popular SSRI that has had about 1% of the top 200 most prescribed drugs, is one of the most potent SSRIs clinically available. However, it has a lower selectivity for SSR sites compared to its marketed rival Zoloft (Hiemke and Härtter, 2000). It is a lipophilic drug that undergoes extensive metabolism in the liver, forming more lipophilic excretable compounds that are further conjugated with sulfuric acid or glucuronic acid. It is effectively absorbed from the GIT but metabolized readily during first pass through the liver. Thus, approximately 36% of the compound is excreted in feces but with less than 1% of that as the parent compound (Hiemke and Härtter, 2000). It is currently unclear as to whether these metabolites are active. Normal dosage for clinical efficacy is about 20 mg/day, and its bioavailability increases with increasing frequency of consumption. It primarily acts by inhibiting the CYP2D6, with inhibition from a single dose lasting 3–7 days.

Celexa (citalopram) is a lipophilic drug that has the highest selectivity for inhibiting serotonin reuptake compared to all the other leading SSRIs (Hensiek and Trimble, 2002). In the body, it is initially metabolized to N-desmethylcitalopram, which is also as active as the parent compound. Approximately 80% of the drug is bioavailable and 50% of the dose is excreted in urine and the rest through the fecal route. The N-demethylated metabolite has a longer half-life than the parent compound. Sexual dysfunction is associated with some SSRIs. For example, paroxentine has been associated with higher incidences of antimuscurinic side effects due to its affinity for cholinergic receptors (Donoghue and Hylan, 2001). By contrast, another SSRI, Celexa, has been associated with a higher degree of sedation resulting from its degree of binding to H1 receptors (Donoghue and Hylan, 2001).

Tricyclic antidepressants were designed to interact with the three monoamines. They are the first generation of drugs that produce antidepressant actions by inhibiting monoamine uptake. However, as a side effect, they interact nonselectively with other neurotransmitter receptors such as cholinergic and histaminergic, producing a wide range of clinical side effects that range from drowsiness, impaired coordination, memory impairment, constipation, difficulty in urinating, cardiotoxicity, and outright occurrence of seizures. TCAs are thought to exert their effects by inhibiting the reuptake of serotonin, norepinepherine, or both (Kwok et al., 2003). The risks of seizures with TCAs is dose related, increasing with increasing plasma concentrations (Stimmel, 1996). They are reputed to have more adverse side effects such as CNS toxicity, cardiac arrest, and delir. From a clinical perspective, Donoghue and Hylan (2001) have argued that the lower efficacy of TCAs compared to SSRIs is due to the fact that they are prescribed at dosages that are lower than those initially recommended because of evidence from randomized controlled clinical trials. Such low dosages have been perpetuated because of the adverse side effects generally associated with TCAs, as well as the logical notion that the higher the dose, the greater the incidence and severity of adverse effects. Those effects do not decrease in severity with continued therapy. The metabolism of TCAs also leads to multiple metabolites whose pharmacological properties are very different from those of the parent compounds.

Monoamine oxidases (MAOs) are involved in the metabolism of catecholamine and serotonin neurotransmitters such as dopamine, norepinephrine, and epinephrine.

The inhibition of MAO enzymes increases the concentrations of these neurotransmitters at storage sites throughout the CNS. Side effects of MAOIs include hypotensive effects from the inhibition of central vasomotor centers and cholinergic effects.

Benzodiazepines such as alprazolam (Xanax), clonazepam, and diazepam act on the CNS, producing effects ranging from mild sedation to hypnosis and even comas. As a matter of fact, the benzodiazepine alprazolam has been the most prescribed drug among all anxiolytic sedatives, hypnotics, and antipsychotics in the United States during 2003–2005 (Table 1.6). They enhance the effects of GABA, with such inhibition in turn leading to an increased opening of chloride channels in the neuronal membrane. The increased flow of chloride ions allows for hyperpolarization of the membrane, subsequently inhibiting firing of the neurons. In rats, the benzodiazepine diazepam has been shown to have a high propensity to alter cellular redox systems (Musavi and Kakker, 2003).

1.2.3   Analgesics and Anti-inflammatory Drugs

Analgesics and anti-inflammatory drugs are primarily used for relieving pain. A number of them are mostly available over the counter (OTC) but those that are prescribed stood at 34.2, 40.5, and 35.8 million prescriptions in the United States during 2003, 2004, and 2005, respectively. These volumes of prescriptions correspond to 16.1, 14.4 and 15.4% of the top 200 most prescribed drugs in that country. The frequency of prescription for these drugs is summarized in Figure 1.5. Most prescribed is hydrocodone (3.3–4.4% of the total top 200 prescriptions during 2003–2005). The other most prescribed drugs in this category, that is, Zyrtec, propoxyphene, ibuprofen (Motrin), Celebrex, Allegra, Singulair, Flonase, oxycontin, and acetaminophen (Tylenol, Paracetamol, or Panadol) have been prescribed with a frequency of 0.5–1% during 2003–2005. Vioxx (rofecoxib) and Bextra (valdecoxib) had also acquired the same frequency of prescribing by the time they were banned or withdrawn in late 2004 due to severe side effects. Other frequently prescribed drugs for pain and anti-inflammation include naproxen, Nasonex, Endocet, alluprinol, Patanol, Mobic, and tramadol. Diclofenac, Clarinex, nabumetone (Relafen), Etodolac, aspirin (acetylsalcylic acid), and loratadine (Claritin) are also commonly prescribed to a lesser extent, but it should be noted that some of these, notably aspirin, naproxen, and Claritin are available over the counter and are, therefore, almost household companions.

Consumption of analgesics and anti-inflammatory drugs in other developed countries is also quite high as is reflected in Table 1.7. Most widely used almost across the board in those countries is acetylsalicyclic acid and paracetamol. Beyond these two, it is apparent that the pattern of use of the other pain medications is quite diverse, with some pain relievers more popular in some countries than in others. Although the statistics are not readily available, pain relief medications are also the most consumed per capita in developing countries.

Analgesics and anti-inflammatories generally act based on the fact that inflammation increases the synthesis of cyclooxgenase (i.e., COX)-dependent prostaglandins, which in turn sensitizes receptor terminals, producing localized hypersensitivity to pain.

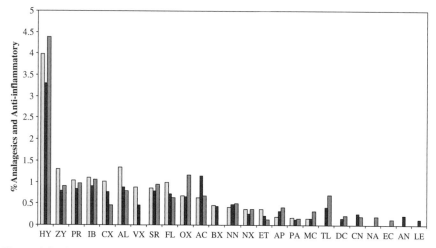

Figure 1.5 Common analgesic and anti-inflammatory drugs prescribed in the United States in 2003 (■), 2004 (■), and 2005 (■) as a percentage of the top 200 most prescribed drugs in that country. The total number of prescriptions for the top 200 drugs were 2.1, 2.8. and 2.3 billion for 2003, 2004, and 2005, respectively. HY = hydrocodone w/APAP, ZY = Zyrtec, PR = propoxyphene napsylate w/APAP (Darvocet), IB = ibuprofen (Motrin), CX = Celebrex (calecoxib), AL = Allegra, VX = Vioxx (rofecoxib), SR = Singulair, FL = Flonase, OX = oxycontin, AC = acetaminophen (Paracetamol or Panadol), BX = Bextra (valdecoxib), NN = naproxen, NX = Nasonex, ET = Endocet, AP = alluprinol, PA = Patanol, MC = Mobic, TL = tramadol, DC = diclofenac, CN = Clarinex, NA = Nabutone (relafen), EC = Etodolac, AN = aspirin (acetylsalcylic acid), and LE = loratadine (Claritin).

Thus, cyclooxygluconases are a set of key enzymes that catalyze the biosynthesis of prostaglandins, leucotrienes, and other compounds involved in the genesis of pain and inflammations (Vane, 1971; Vane and Botting, 1998). Most, but not all, of the analgesics and anti-inflammatory drugs listed in Figure 1.5 and Table 1.7 are nonsteroid anti-inflammatory drugs (NSAIDs), which act by inhibiting the enzyme cyclooxygenase (COX). Most of them also relieve pain (analgesic) and reduce fevers (antipyretic). The COX enzyme leads to the formation of prostaglandins (PG), which cause inflammation, swelling, fever, and pain. Prostaglandins are synthesized from arachidonic acid. The pain is specifically caused by the release of arachidonic acids after a cell is damaged. At least two COX genes have been identified. COX-1 is constitutively produced in most tissues, whereas the expression of COX-2 is regulated by a broad spectrum of mediators that are involved in inflammation. By definition, COX-2 inhibitors do not cause any significant clinical suppression of the formation of platelet thromboxane during blood clotting or synthesis of gastric prostaglandins when they are used at typical therapeutic doses.

Common NSAIDs include aspirin, ibuprofen, indomethacin, naproxen, and ketoprofen. Even though anti-inflammatories generally target cyclooxygenase, there are apparent differences in the details of how they relieve pain. For example, aspirin acts by primarily inhibiting the COX-dependent synthesis of eicosanoids, which are end products of metabolism of essential fatty acids including prostaglandin

TABLE 1.7 Annual Consumption of Various Pain-Relief Medications in Europe and Australia

			Consumption (Tons per year)[a]			
Country and Year	Acetylsalicylic Acid	Salicylic Acid	Paracetamol	Naproxen	Ibuprofen	Diclofenac
Germany (1999)	902.27 (1)	89.7 (12)	654.42 (2)		259.85 (5)	81.79 (16)
Germany (2000)	862.6 (1)	76.98 (17)	641.86 (2)		300.09 (5)	82.2 (14)
Germany (2001)	836.26 (1)	71.67 (17)	621.65 (2)		344.89 (5)	85.8 (14)
Austria (1997)	78.45 (1)	9.57 (11)	35.08 (2)	4.63 (16)	6.7 (13)	6.14 (15)
Denmark (1997)	0.21 (7)		0.24 (6)		0.03 (19)	
Australia (1998)	20.4 (9)		295.9 (1)	22.8 (7)	14.2 (13)	
England (2000)			390.9 (1)	35.07 (12)	162.2 (3)	26.12 (16)
Italy (2001)					1.9 (15)	
Switzerland (2004)	43.8 (3)	5.3 (6)	95.2 (1)	1.7 (12)	25 (4)	4.5 (7)

[a]Data obtained from Fent et al. (2006) as compiled from various sources listed in that reference. The numbers in parentheses represent the ranking of the medication in question among the top 20 most popular drugs in that country during that year.

in preserving the dynamic balance between thrombosis, homeostasis, and the fluidity of blood. It irreversibly inactivates both COX-1 and COX-2 by acetylating the active site, thus interfering with the binding of arachidonic acid at the COX active site. Aspirin is the most widely used drug worldwide (Vane and Botting, 2003). It is almost completely metabolized to salicylic acid with less than 1% excreted as parent compound compared to 85% paracetamol excreted as conjugates (Schowanek and Webb, 2005). By contrast, some other NSAIDs such as ibuprofen and mefenamate compete with arachinodic acid for the COX active site. Yet others such as indomethacin and rofecoxib slowly cause a reversible inhibition of COX-1 and COX-2, forming salt bridges between the drug and the enzyme (Hinz and Brune, 2002).

Unlike COX-1, COX-2 is inducible and is responsible for inflammatory reactions unlike COX-1. These two enzymes differ in mechanism due to differences in the size of their binding sites. This difference provides selectivity. COX-2 drugs selectively inhibit COX-2 enzymes and not COX-1, thus alleviating pain without causing gastric problems that occur if COX-1 enzymes are inhibited. However, COX-1 is also responsible for inhibiting blood clots. This selective COX-2 inhibitor elevates the incidence of blood clots, increasing the incidence of heart attacks (Marcus et al., 2002). To that effect, some of the most potent COX-2 inhibitors, notably Vioxx and Bextra, have been withdrawn or banned. Some of the side effects associated with COX inhibitors is related to their mechanism of action, that is, inhibiting the production of prostaglandins. Prostaglandins play a role in maintaining the equilibrium between vasodilatation and vasoconstriction of the blood vessels in the kidneys. Thus, sustained absence of prostaglandins in the kidney vessels can trigger vasodilation, which can ultimately lead to renal failure in mammals (Murray and Brater, 1993) and have more recently been shown to have similar effects in some birds (Meteyer et al., 2005). Prostaglandins are also involved in the production of acids in the stomach. By inhibiting this key enzyme in the synthesis of prostaglandins, NSAIDs also prevent the production of physiologically important prostaglandins that otherwise protect the stomach mucosa (lining) from damage by stomach acids. Thus, NSAID therapy has been frequently associated with an increased incidence of gastrointestinal events. In search of alternative approaches to dealing with pain and inflammation, another COX gene, that is, COX-2, was discovered more recently, and further investigation has revealed that it is induced by pro-inflammatory stimuli, becoming up-regulated by inflammatory mediators. Based on this, more selective NSAIDs have been developed. These, unlike the nonselective NSAIDs, selectively inhibit COX-2 to generate prostaglandins that mediate inflammation and pain without inducing COX-1, involved in generating prostanoids that are responsible for maintaining the integrity of gastrointestinal mucosa and platelet aggregation.

Some of the popularly used anti-inflammatory drugs, such as paracetamol, are not an NSAID but have analgesic/antipyretic properties. Paracetamol specifically has relatively weak anti-inflammatory activity. Once excreted, it gets reactivated in the environment through some microbially mediated transformation (Henschel et al., 1997). The exact mechanisms by which paracetamol relieves pain are not very clear. It has a chemical structure that resembles several estronegic compounds, including bisphenol, diethylstilbestrol, 17β-estradiol, and *p*-nonylphenol all of which have a characteristic *p*-phenol moiety. It can act adversely by enhancing the formation of *N*-acetyl-*p*-benzoquinone imine, which is present in hepatotoxic metabolites and has also been linked to some proliferation of breast cancer cells reflected by stimulated DNA (deoxyribonucleic acid) synthesis quantified with the rate of incorporation of [3H]-thiamidine], in culture with MCF7, T47D, and ZR-75-1 cell lines, all of which had an estrogen receptor (ER$^+$) gene. The acetaminophen used was at therapeutic concentrations of 0.03–0.1 mM, which are typically encountered in the serum (Harnagea-Theophilus and Miller, 1998). In that same study, no significant effects were observed in two other breast cell lines, that is, MDA-MB-231 and HS578T, which lacked estrogen receptors (ER$^-$), clearly indicating that this compound has some estrogenic activity. Human breast cancer cells in culture are a well-established in vitro test system for assessing compounds that display estrogenic and antiestrogenic properties. It is not clear at this point whether acetaminophen has the same effects in vivo.

Prostanglandins, the main target of anti-inflammatory and antipyretic drugs, are formed in a wide array of vertebrates and invertebrates. In birds, prostaglandins are involved in the eggshell synthesis. A gene that is homologous to COX-2 has been found in fish (Zou et al., 1999). Such findings and ways in which some side effects for several popular NSAIDs and steroidal anti-inflammatory drugs such as acetaminophen may impact other nontarget organisms in the environment will be discussed in Chapter 4.

1.2.4 Antimicrobial Compounds

Antimicrobial compounds generally include antibiotics, antifungal agents, antiprotozoan agents as well as antiviral agents. Antibiotics stop microorganisms, primarily bacteria and fungi, from growing (i.e., bacteriostatic) and/or kill these organisms outright (i.e., bactericidal). Some antibiotics have both bacteriostatic and bactericidal effects. Most antibacterial agents are ineffective against fungi, and, similarly, most antifungal agents are not effective against bacteria. Most classes of antibiotics are from natural products, although the individual antibiotics currently on the market are semisynthetic as they have been greatly modified or engineered to incorporate desirable traits. Desirable traits that have been introduced through engineering include enhanced oral bioavailability, a wider range of activity, increased stability, and increased efficacy against target microorganisms.

Wise (2002) estimates that 100,000–200,000 tons of antibiotics are used annually worldwide. However, usage estimates greatly vary (Raloff, 2002). Production of antibiotics in the United States is estimated at more than 22.7 million kilograms of antibiotics, with more than 40% used as animal feed supplements (Kumar et al., 2004). Furthermore, use patterns greatly vary from one country to another (Huchon et al., 1996; Mölstad et al., 2002; Tesař et al., 2004), clearly indicating differences in patient expectations, attitudes toward taking drugs, physician practices, and differences in health care systems as well as differences in drug marketing activities. The World Health Organization (WHO) developed an anatomical

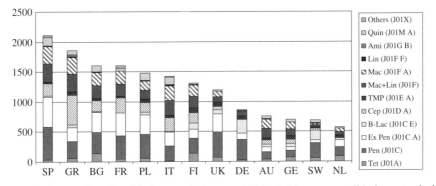

Figure 1.6 Number of antibiotic prescriptions per 1000 inhabitants per antibiotic anatomical therapeutic chemical (ATC) classification in 13 European countries in 1997. In parentheses are the ATCs used by the WHO. Tet = tetracyclines, Pen = penicillin, Ex-Pen = extended-spectrum penicillins, B-Lac = β-lactamase-sensitive penicillins, Cep = cephalosporins, TMP = trimethoprim (alone or in combination), Mac + Lin = macrolides and lincosamides, Mac = macrolides, Lin = lincosamides, Ami = aminoglycosides, and Qui = quinolone. The 13 countries are SP = Spain, GR = Greece, BG = Belgium, FR = France, PL = Portugal, IT = Italy, FI = Finland, UK = United Kingdom, DE = Denmark, AU = Austria, GE = Germany, SW = Switzerland, and NL = Netherlands. (Based on data from Mölstad et al., 2002.)

therapeutic chemical (ATC) classification system for pharmaceuticals, of which the classifications for antibiotics are increasingly becoming adopted. Based on that classification, antibiotic usage patterns in 13 European countries in the late 1990s are shown in Figure 1.6. Antibiotic prescription per 1000 inhabitants in those countries is highest in Spain and lowest in the Netherlands. In general, per capita prescription of antibiotics is higher in southern Europe. Exceptions to this generalization are Belgium and Finland, which have the third and seventh highest per capita antibiotic prescriptions, respectively. In the United States approximately 70% of the antibiotics dispensed are given to healthy livestock (mostly poultry, pigs, and cattle) to prevent infections and/or promote growth (UCS, 2001). By contrast, the use of antibiotics for this same purpose has been primarily outlawed in the EU countries (FEDESA, 1999). Similarly, antibiotics such as vancomycin are of last resort in some EU countries but are readily used in the United States. By comparison, in developing countries most antibiotics are fairly readily available over the counter and therefore widely used whenever deemed necessary and affordable.

A more conventional classification of antibiotics is based on targets or mechanisms of action, categorizing antibiotics into five main groups:

1. Antibiotics that inhibit bacterial cell wall biosynthesis
2. Antibiotics that selectively block the 30S or 50S ribosomes
3. Antibiotics that block the replication of DNA

4. Antibiotics that disrupt the integrity of cell membranes
5. Antibiotics that antagonize metabolic processes

Typical classes and examples within these categories as they apply to what is currently most prescribed on the U.S. market are summarized in Table 1.8. The targets in groups 1 and 4 are unique in bacteria and absent in humans and other animals, whereas groups 2, 3, and 5 have human counterparts that are structurally different between prokaryotes and eukaryotes. These differences in targets make the use of antibiotics selective for bacteria with little or no effect on eukaryotic cells from a therapeutic perspective. However, that does not mean that antimicrobial compounds are completely inert to eukaryotes. The mechanisms that block bacterial protein synthesis, block DNA replication, and those that disrupt membrane integrity affect membrane pores.

1.2.4.1 Cell Wall Synthesis Inhibiting Antibiotics Included in this category are β-lactams, cephalosporins, glycopeptides, and everninomycins. Over the recent past, the β-lactam amoxicillin has consistently been most prescribed in the United States among all antibiotics, followed by the macrolide zithromax and the first-generation cepharosporin, cephalexin (Fig. 1.7). The most recent prescriptions of the individual antibiotics across Europe are not readily available, but based on Figure 1.6, β-lactams, macrolides, and cephalosporins are also among the most prescribed. Thus, in essence, antibiotics that either inhibit bacterial cell wall synthesis or block the synthesis of proteins are the ones that are most widely used.

In this category of antibiotics, β-lactams have been most widely used, among which penicillin was the pioneer to the extent that in some circles penicillin has almost become synonymous with lactams. Penicillins were accidentally discovered by Alexander Fleming when he noticed that some natural substance produced by a contaminating fungus was inhibiting growth of the bacterial culture in some areas on a solid laboratory medium. The fungal contaminant was later determined to be *Penicillium* sp. and the compound produced was later named penicillin. This discovery remained in the doldrums for several years until massive production of the compound was made possible and boosted its use to control bacterial infections. That capability changed medicine and the pharmaceutical industry forever, and soon thereafter other compounds that inhibit bacterial growth were eventually discovered.

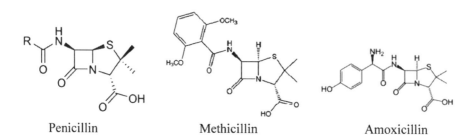

Penicillin   Methicillin   Amoxicillin

TABLE 1.8 Examples of Antibiotics That Operate Using One of the Four Common Mechanisms

General Mechanism	Class	Characteristics and Specific Mechanisms	Examples[a]	Remarks
Cell wall synthesis inhibition basically by impacting the enzymes involved in the assembly of the peptidoglycan layer and its cross-linkages.	β-Lactams	β-Lactams have a characteristic four-membered β-lactam ring fused to a five-membered ring characterized by the first antibiotic in this class, i.e., penicillin. Others include oxacillin, methicillin (and its succesors nafcillin and cloxacillin), and ampicillin. Bacitracin, which is frequently used in the livestock industry, is also in this class.	**Amoxicillin**

Penicillin V | Generally well absorbed and is quite bioavailable (95%). It is a moderate spectrum antibiotic effective against a range of gram-positive and a limited number of gram-negative bacteria.

Is resistant to acidic conditions and can therefore be given orally. Penicillins are generally the least toxic of antibiotics to humans. |
| | Cephalosporins | Have a nucleus that is similar to that of β-lactams and similarly disrupt the synthesis of the peptidoglycan layer. This layer is important in maintaining cell wall integrity in gram-positive bacteria. | **Omnicef (Cefdinir)**

Cefzil (Cefprozil)

Cephalexin (Keflex) | Bioavailability is dose dependent is only about 16–21%.

About 90% is excreted as the parent compound.

It is well absorbed by the body and 90% of the drug is excreted unchanged. |
| | Glycopeptides | Inhibit the correct synthesis of cell walls in gram-positive bacteria by specifically preventing the incorporation of N-acetylmuramic acid (NAM) and N-acetylglucosamine (NAG). Both NAM and NAG are important peptide subunits that are present in the peptidoglycan layer of gram-positive bacteria. | Vancomycin | Similar to avoparcin, which is commonly used in livestock as a growth promoter. Has been relied on for infections caused by staphylococci and other gram-positive bacteria that are resistant to β-lactams and macrolides. However, increasing cases of vancomycin-resistant enterococci (VRE), vancomycin-intermediate $S.$ $aureus$ (VISA) and vancomycin-resistant $S.$ $aureus$ (VRSA) are a major concern in medicine. |
| Antibiotics that block bacterial protein biosynthesis. | Macrolides | Target the 50S ribosome subunit. Tylosin, which is widely used in the livestock | **Clarithromycin (biaxin)** | Semisynthetic and has better intracellular penetration compared to earlier |

(*Continued*)

35

TABLE 1.8 *Continued*

36

General Mechanism	Class	Characteristics and Specific Mechanisms	Examples[a]	Remarks
		industry (Table 1.1), also belongs in this class.		macrolides such as erythromycin. Bioavailability is 55 and 40% is excreted unchanged (Queener and Gutierrez, 2003).
			Bactroban (mupirocin)	Naturally derived from fermentation using *Pseudomonas fluorescens* and active against MRSA and some other gram-positive bacteria. Acts by blocking the activity of isoleucyl-tRNA synthetase in bacteria. This enzyme is necessary for bacteria to synthesize proteins.
			Azithromycin (zinthromax)	Bioavailability of 38% and has similar antimicrobial spectrum as erythromycin but is more effective than erythromycin on gram-negative bacteria. Mostly excreted in feces although some low levels (approximately 6%) of the parent compound are excreted through the urine.
			Erythromycin	Introduced in 1952 as the first macrolide antibiotic (Roberts et al., 1999). Inactivates bacterial ribosomal proteins leading to a loss of vital processes of synthesizing proteins, DNA, RNA, and cell wall. Also leads to the loss of aerobic energy metabolism. It has been around since the 1950s without serious emergence of resistance because of its broad-based range of modes of action.
			Macrobid (nitrofurantion)	

Class	Description	Drug	Notes
Tetracyclines	Target the 30S subunit. Also in this class are oxytetracycline and chlorotetracycline which are frequently used in the livestock industry	Doxycline hyclate (vibramycin)	Highly soluble in lipids and thus still penetrates bacterial cells that lack the pumping mechanism necessary for uptake of most tetracycline-resistant strains (same for minocycline too).
		Minocycline (Minocin)	Extensively biotransformed in the liver.
		TobraDex	A combination of the aminoglycoside tobramycin and the corticosteroid dexamethasone.
Aminoglycosides	Target the 30S subunit, primarily causing a frame-shift in the codon, leading to the failure of the synthesis of the correct proteins or the incorrecting folding of proteins. The susceptible bacteria are destroyed as they cannot make the correct proteins when and as needed.	Kanamycin	Much resistance to kanamycin has developed over time and these antibiotics are not commonly used compared to what was the case in the past.
		Streptomycin Gentamicin	Other aminoglycosides include tunilamycin, moenomycin, and fosfomycin.
Lincosamides	Similar to macrolides, lincosamides also target the 50S ribosome subunit. More specifically, they inhibit the enzyme peptidyl transferase, which in turn inhibits the activity of ribosomes, preventing the binding of amino acyl-tRNA to the A site on the 50S subunit. That activity blocks the synthesis of proteins.	Clindamycin (Dalacin)	Semisynthetic antibiotics for treatment against anaerobic bacteria and these are responsible for respiratory infections and septicemia. Also for S. aureus infections and topically for acne. Less than 20% of the parent compound is excreted (Queener and Gutierrez, 2003) but its metabolites such as N-dimethyl clindamycin and clindamycin sulfoxide are also active.
		Lincomycin	Structurally similar to macrolides and also binds to 50S ribosomes like macrolides.

(Continued)

TABLE 1.8 *Continued*

General Mechanism	Class	Characteristics and Specific Mechanisms	Examples[a]	Remarks
			Chloramphenicol	Originally derived from *Streptomyces venezuelae*, it was the first antibiotic to be manufactured synthetically on a large scale. It can directly attack mitochondria in humans, leading to bone marrow suppression, a condition that is manifested by a reduction in hemoglobin levels. This series of events can lead to aplastic anemia even several weeks after actively taking the antibiotic.
	Streptogramins	Target the 50S ribosome subunit. Also included in this class is virginiamycin which has been used in animal husbandry for more than 2 decades (see Table 1.1).	Synercid	Synercid is really a combination of quinupristin and delfopristin approved by the U.S. FDA in 1999 (Claycamp and Hooberman, 2004).
			Pristinamycin	Derived from the bacterial species *Streptomyces pristine spiralis*. Like Synercid and virginiamycin, it also has two components that act synergistically. One of the components is a macrolide that has similar spectrum of action that is similar to erythromycin, whereas the other one is depsipeptide. It is effective against MRSA but has poor solubility.
	Oxazolidinones	Target the 50S ribosome subunit. These antibiotics are relatively new on the market and their long-term efficacy is still unknown.	Linezolid (Zyvox)	Used as a last generation against gram-positive bacteria including MRSA, MRSE, PRSP, and VRE. Approved by the FDA in the late 1990s, it is 100% bioavailable and 30–35% of the parent compound is excreted in the urine (Moellering, 2003).

Mechanism	Class description	Class	Drug	Description
Antibiotics that inhibit nucleic acid synthesis.	Interfere with protein and metabolic processes.	Furantoin	Eperezolid	Has demonstrated good oral bioavailability and inhibits many clinically significant bacterial species.
			Nitrofurantoin (Macrobid)	30–40% is excreted in urine (Queener and Gutierrez, 2003).
	Good selectivity for topoisomerases in bacteria.	Quinolones		The topoisomerases enzymes are essential in prokaryotic and eukaryotic cell visibility. Most quinolones have a characteristic core planar heterocyclic nucleus.
			Levofloxacin (Levaquin)	Third-generation quinolone; 60–80% excreted unchanged.
			Ciprofloxacin	Second-generation quinolone.
			Avelox (moxifloxacin)	Third-generation quinolone; >96% excreted unchanged (Stass and Kubitza, 1999).
			Nalidixic acid	Pioneer in this class which was discovered as a by-product of the antimalarial drug, chloroquine.
	Suppress the synthesis of RNA, inhibiting the synthesis of proteins.	Rifamycins	Rifampin	Effective against gram-positive bacteria, including *Mycobacterium* sp. in combination treatment regimens.
			Rifabutin	Also effective against various gram-positive bacteria and is even more potent that rifampin.
Antagonize metabolic processes.	Also used widely in the livestock industry. Others include nisin, daptomycin, and gramicidin A.	Sulfonamides	Sulfamethaxazole[c]	Mostly used in combination with trimethoprin (i.e., SMZ–TMP).
			Cotrim (Bactrim; Septra)	Combination preparation of sulfamethaxazole and trimethoprim.
		Trimethoprims	Trimethoprim[c]	Mostly used in combination with sulfamethoxazole (i.e., SMZ–TMP).

(Continued)

TABLE 1.8 *Continued*

General Mechanism	Class	Characteristics and Specific Mechanisms	Examples[a]	Remarks
Disruption of membrane integrity.	Azole antifungal	Inhibit synthesis of ergosterol, an important component of fungal membranes.	**Elidel**	Interact with antigens increasing the concentration of calcium in cells which in turn induces transcription of the interleukin that stimulates growth and differentiation of T-cells. Interleukin is a hormone of the immune system that is important in the body's response to microbial infections.
			Diflucan (Fluconazole)	Synthetic triazole antifungal agent that inhibits cytochrome P450 sterol C-14 α-demethylation. The resulting high concentrations of C-14 α-methyl sterols are fungistatic. Approximately 90% is bioavailable and most of the drug is excreted unchanged (Queener and Gutierrez, 2003).
	Polyenes	React with ergosterol thus destroying fungal membrane integrity.	**Ketoconazole (Nizoral)** **Nystatin (Mycostatin)** **Clotrimazole (Mycelex)**	Used for many systemic fungal infections. Used for many systemic fungal infections.

Class	Drug	Mechanism	Notes
Others. Antiprotozoan and antihelminthic	**Metronidazole (Flagyl)**	Attack and disrupt DNA.	Active against anaerobic organisms and under such conditions it is reduced to a short-lived metabolite that damages DNA. Oral doses are highly bioavailable but 40–80% of the parent compound can be excreted.
	Quinine		Mostly for chloroquine-resistant malarial infections. Extensively transformed before excretion.
	Chloroquine	Binds tightly to double-stranded DNA, thus altering its physical properties.	Mostly for malarial infections and is one of the most used drug worldwide. Resistance has been an issue.
Acylic nucleoside analogs	**Valtrex [Zovirax; Valacyclovir (includes Acyclovir)]**	Completely inhibits viral DNA polymerase terminating the growing viral chain and inactivating viral DNA polymerase.	Antiviral mostly used against herpes simplex virus (HSV1 and 2) and varicella-zoster virus (VZV).

[a]The antibiotics in boldface were among the top 200 most prescribed drugs in the United States (see Fig. 1.7).

[b]MRSA = methicillin-resistant *Streptococcus aureus*, MRSE = methicillin-resistant *Staphylococcus epidermitis*, PRSP = penicillin-resistant *Streptococcus pneumoniae*, and VRE = vancomycin-resistant enterococci.

[c]These two have been mostly as combination drug, i.e., SMZ-TMP (see Fig. 1.7).

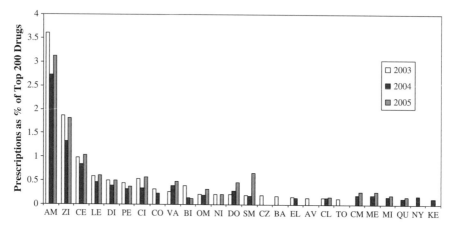

Figure 1.7   Top 200 most prescribed antibiotics in the United States during 2003–2005. AM = amoxicillin (Augmentin), ZI = zithromax (azithromycin), CE = cephalexin, LE = Levaquin (levofloxacin), DI = Diflucan (fluconazole), PE = penicillin, CI = ciprofloxacin, CO = Cotrim, VA = Valtrex (including Acylclovir), BI = Biaxin (clarithomycin), OM = Omnicef (Cefdinir), NI = nitrofurantoin (Macrobid), DO = doxyclycline hyclate, SM = sulfamethoxazole/trimethoprim, i.e., SMZ-TMP, CZ = Cefzil, BA = Bactroban (Mupirocin), EL = Elidel, AV = Avelox (monofloxacin), CL = Clotrimazole (betamethasone), TO = TobraDex, CM = clindamycin, ME = metronidazole (Flagyl), MI = Minocycline, QU = quinine, NY = Nystatin, and KE = ketoconazole.

The potency of the early penicillin was greatly compromised by the development of resistance for which reports surfaced as early as the late 1950s. To combat that deficiency research generated the first semisynthetic penicillin, methicillin, which was approved by the U.S. FDA in 1958, followed by cephalexin soon thereafter. Currently, there are about 40 different penicillins in use worldwide, with penicillin as the nucleus (see the structures for methicillin and amoxicillin above). The use of these earlier semisynthetics, particularly methicillin, has been eclipsed by the emergence of methicillin-resistant *Staphylococcus aureus* (MRSA). To date, amoxicillin is one of the most prescribed antibiotics and is certainly the most prescribed β-lactam (Fig. 1.7). It is an extended-spectrum penicillin that is able to penetrate gram-negative bacterial cell walls effectively. In Europe, the per capita prescription of β-lactams is highest in Denmark, followed by Switzerland (Fig. 1.6). β-lactam usage per capita is lowest in Belgium, France, and Italy.

Cephalosporins were first isolated from a fungus *Cephalosporium* sp. in the late 1940s. They have a nucleus that is analogous to that of penicillins and other β-lactams. As a matter of fact, they have the same mode of action but disrupt the synthesis of the peptidoglycan layer of gram-positive bacterial cells. They are some of the most widely prescribed antibiotics, and, because of their broad-spectrum nature, cephalosporins are frequently dispensed to people who are allergic to penicillins.

Most notably prescribed in the United States in this class are cephalexin, omnicef, and cefzil (Fig. 1.7). Modifications in the side chain to the four-ring and the six-ring moieties give them varied antibacterial and pharmacokinetic properties. Cephalexin and other first-generation cephalosporins are primarily effective against gram-positive bacteria. A third-generation cephalosporin, omnicef, and a second-generation cephalosporin, cefzil (Cefprozil), have also been among the top 200 most prescribed drugs (Fig. 1.7):

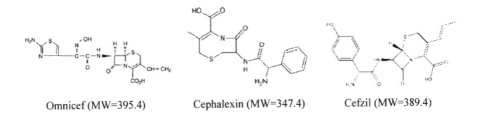

Omnicef (MW=395.4)   Cephalexin (MW=347.4)   Cefzil (MW=389.4)

Cephalosporins are generally well absorbed from the gastrointestinal tract and have a longer half-life compared to penicillins. They are mostly excreted through the kidney and are, therefore, highly concentrated in the urine, and, in some instances, more than 90% of the drug is excreted unchanged. This route of excretion, however, makes them very suitable candidates for controlling urinary tract infections. Other cephalosporins include cephalothin, cefamandole, cefoxitin, and Kefipime.

The first glycopeptide, vancomycin, was first isolated from a soil-based *Streptomyces orientalis* and originally indicated for the treatment of penicillin-resistant *S. aureus*. Vancomycin emerged as a dominant alternative to infections that are caused by tetracycline-resistant staphylococci and other gram-positive pathogens in individuals that are not tolerant to β-lactams and macrolides. It has also been used against enterococci that are resistant to penicillins. First approved by the U.S. FDA in the late 1950s, the drug was not frequently used as initial tests had shown it to have ototoxic and nephrotoxic effects. It was primarily reserved for more complicated cases that could not be resolved with penicillins, and this relegation characterized it as an antibiotic of last resort for controlling infections in humans. The emergence of methicillin-resistant staphylococci in the 1970s revived interest in using vancomycin as an alternative. Thus, its use within the United States and Europe increased more than a 100-fold between the early 1980s and 1995 (Levine, 2006) as to mostly counter the increased incidences of methicillin-resistant *Streptococcus aureus* and penicillin-resistant *Streptococcus pneumoniae* (PRSP).

Vancomycin acts by inhibiting the correct synthesis of cell walls in gram-positive bacteria by specifically inhibiting the incorporation of *N*-acetylmuramic acid (NAM) and *N*-acetylglucosamine (NAG), two important peptide subunits that are present in the peptidoglycan layer of these types of bacteria.

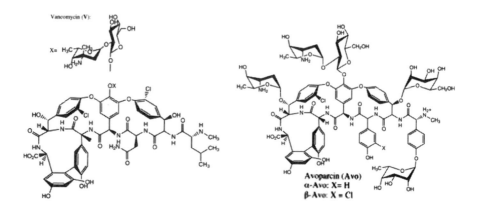

The drug is mostly applied intravenously as it is not absorbed orally. Since the late 1980s, resistance to vancomycin emerged, and, more recently, there have been numerous cases of vancomycin-resistant enterococci (VRE) (Levine, 2006). The resistance to vancomycin has since become an ever-increasing problem to the extent that vancomycin-intermediate *S. aureus* (VISA) and vancomycin-resistant *S. aureus* (VRSA) have also been reported (Howe et al., 2004). All of these resistances are of particular concern as very few antibiotics are effective against these bacteria. Antibiotic resistance will be discussed extensively in Chapter 4, but for purposes of the present discussion, some of these resistances are suspected to be associated with the use of avorparcin, another glycopeptide that is frequently used in livestock management as a growth promoter (Table 1.1). Other glycopeptides include actaplatin, daptmycin, teicoplanin, and restocetin, but none of these featured in the 200 most prescribed drugs in the United States within the recent past. In general, all of these glycopeptides are quite hydrophobic and have an increased affinity for the cytoplasmic membrane, and this enhances the ability of the glyocoproteins to interact with their membrane-associated target. The specific target is a carrier-bound peptide–disaccharide peptidoglycan precursor (Williams and Waltho, 1988).

More recently, everninomycins have also been introduced as antibiotics that inhibit the synthesis of bacterial cell walls. They show excellent activity against fluoroquinolone-resistant bacteria and all gram-positive vancomycin-resistant bacteria (Lin et al., 2000).

1.2.4.2 Inhibitors of Protein Synthesis

1.2.4.2 Inhibitors of Protein Synthesis Several antimicrobial agents act by inhibiting the synthesis of proteins by target ribosomes. These include macrolides, streptogramins, chlorophenicol, tetracyclines, lincosamides, aminoglycosides, oxazolidinones, fusidic acid, and pleuromutilins. As a matter of fact, some, notably macrolides, lincosamides, and streptogramins (MLS), functionally very related—with all three specifically targeting the 50S ribosomal subunits—result in the inhibition of proteins. These three are specifically referred to as the MLS superfamily, and due to their relatedness can enable the development of the cross resistance as they share overlapping binding sites. The blockage of protein synthesis totally disrupt, or at least slow down, the growth of bacteria. They also take advantage of the architectural differences in the 23S ribonucleic acid (RNA) of bacterial ribosomes compared to that of eukaryotes, and this provides selective effect on the former and not the latter for some of them. Activity is improved by blocking the sites in the peptidyl transferase center of the ribosome, which presents the first peptide bond-forming step in protein synthesis. They act by specifically inhibiting peptide chain elongation after binding tightly (but reversibly) to the large, that is, 50S, ribosome subunit. The binding occurs at the ribosomal peptidyl transferase center.

At least four macrolides, that is, zithromax, Biaxin, nitrofurantion, and bactroban, are among the top 200 most prescribed drugs in the United States (Fig. 1.7).

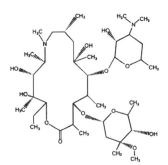

Zithromax (MW = 748.9)

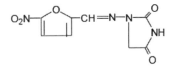

Clarithromycin (Biaxin; MW = 747.96)

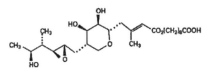

Bactroban (Mupirocin; MW = 500.63)   Nitrofurantion (Macrobid; MW = 238.16)

They contain 12-, 14-, or 16-membered macrocyclic lactone rings attached to amino acids and/or a neutral sugar moiety attached via glycoside bonds. Beyond this, their individual chemical structures and sizes as reflected by molecular weight are dramatically different. Thus, despite exerting similar mechanisms on bacteria, they are very chemically different from each other.

Macrolides such as zithromax have been characterized as penicillin substitutes as they are effective against gram-positive bacteria, including those that are responsible for respiratory tract, throat, skin, and ear infections. It can also be effectively used in individuals who are allergic to penicillins. The bioavailability of zithromax is about 37%, and 50% of the drug is excreted unchanged, although it is retained in the body over a long period of time, that is, a half-life of 14 h to 4 days (Queener and Gutierrez, 2003). In Europe, macrolides are, on a per capita basis, also extensively prescribed both individually and in combination formulations with lincosamides (Fig. 1.6).

Macrolides are produced by a soil organism *Streptomyces* spp. Macrolides are quite lipophilic and enter bacterial cells by passive diffusion, making them more effective against gram-positive bacteria as these have a more elaborate murein layer. By contrast, the gram-negative bacteria are less permeable to macrolides because of a protective outer membrane, although they can still be highly potent inhibitors of cell-free protein synthesis on ribosomes in these organisms (Hunt et al., 2002). Macrolides are generally used against a wide range of infectious bacteria, including *Haemophilus* sp., *Legionella* sp., *Chlamydia*, and some strains of *Mycobacterium* spp. They specifically stimulate the dissociation of the peptidyl-tRNA (transfer RNA) molecule from the ribosome during elongation of the amino acid chain, inhibiting the synthesis of proteins. Macrolides were pioneered by the introduction of erythromycin, a purely natural molecule produced by *Saccharopolyspora erythraea*. Erythromycin has variable and erratic bioavailability possibly because of its lability to stomach acid. More recent improvements of these compounds to clarithromycin and roxithromycin have led to more acid-stable derivatives with improved bioavailability and extended in vivo half-life (Chu, 1999). Erythromycin has also been extensively used in the livestock industry. In terms of popularity, erythromycin has been displaced by various semisynthetic macrolides, some of which are listed in Table 1.8. These more recent semisynthetic derivatives have been designed to primarily improve intracellular and tissue penetration. They are also more stable and have fewer incidences of gastrointestinal disturbance. Besides zithromax and erythromycin, other macrolides that have been most prescribed recently include Biaxin, Bactroban, and Macrobid (Table 1.8; Fig. 1.7). Like several other macrolides, Biaxin has a moderately high bioavailability (55%), and 40% is excreted as the parent compound through renal excretion.

Tetracyclines are characterized by a conjugated four-ring structure with a carboxylamide functional group and several other ionizable functional groups attached as side chains. Such groups include tricarbonyl methane, phenolic diketone, and dimethyl ammonium. These groups give a molecule its pK_a, ultimately giving the whole molecule a net negative, cationic, or zwitterion charge (Sassman et al., 2005). Their general structure is shown below, and their respective sources as well as molecular weight are summarized in Table 1.9. Tetracyclines are sparingly soluble in water. They are broad-spectrum antibiotics active against gram-positive and gram-negative bacteria, chlaymdiae, mycoplasmas, and rickettsias.

TABLE 1.9 Common Quinolones, Their Molecular Structures, and Chemical Characteristics

Generation	Compound	MW	Composition at Respective Position								$\log K_{ow}$	pK_a
			R1	X	R5	Y	R6	R7	Z	R8		
First	Nalidixic acid	232	C_2H_5	CH	n.a.[a]	C	H	CH_3	N	n.a.	1.18	6.21
	Flumequine	261	(structure)	CH	n.a.	C	F	H	CR1	n.a.	2.56	6.25
	Oxolinic acid	261	C_2H_5	CH	n.a.	C	R7	(structure)	CH	n.a.	1.67	6.3
	Cinoxacin	262	C_2H_5	N	n.a.	N	R7	(structure)	CH	n.a.	1.62	3.07
	Piromidic acid	288	C_2H_5	CH	n.a.	C	H	(structure)	N	n.a.	1.64	6.3
Second	Pipemedic acid	303	C_2H_5	CH	n.a.	N	H	(structure)	N	n.a.	0.12	6.3
	Norfloxacin	319	C_2H_5	CH	n.a.	C	F	(structure)	CH	n.a.	0.93	6.26
	Enoxacin	320	C_2H_5	CH	n.a.	C	F	(structure)	N	n.a.	0.82	5.85
	Ciprofloxacin	331	(structure)	CH	n.a.	C	F	(structure)	CH	n.a.	1.03	6.27
	Desmethyl-danofloxacin	343	(structure)	CH	n.a.	C	F	(structure)	CH	n.a.	1.18	6.27
	Lomefloxacin	351	C_2H_5	CH	n.a.	C	F	(structure)	CF	n.a.	0.99	5.9
	Danofloxacin	357	(structure)	CH	n.a.	C	F	(structure)	CH	n.a.	1.47	6.21
	Enrofloxacin	359	(structure)	CH	n.a.	C	F	(structure)	CH	n.a.	1.78	6.21
	Ofloxacin	361	(structure)	CH	n.a.	C	F	(structure)	CH	n.a.	1.01	5.97

(*Continued*)

TABLE 1.9 *Continued*

Generation	Compound	MW	R1	X	R5	Y	R6	R7	Z	R8	log K_{ow}	pK_a
Third	Sarafloxacin	385	 [structure]	CH	n.a.	C	F	 [structure]	CH	n.a.	2.22	6.15
	Temafloxacin		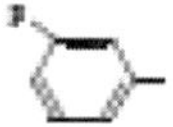 [structure]	CH	n.a.	C	F	 [structure]	CH	n.a.		
	Sparfloxacin		 [structure]	CH	NH$_2$	C	F	 [structure]	C	F		
	Grepafloxacin	423	 [structure]	CH	CH$_3$	C	F	 [structure]	CH	n.a.		
	Levofloxacin		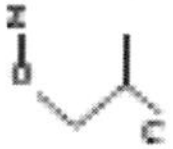 [structure]	CH	n.a.	C	F	 [structure]	C	R1[b]		
	Trovafloxacin		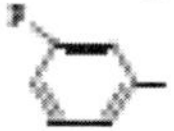 [structure]	CH	n.a.	C	F	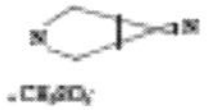 [structure]	N	n.a.		
	Gatifloxacin	402	 [structure]	CH	n.a.	C	F	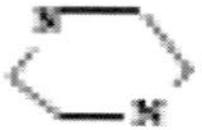 [structure]	C	OCH$_3$		
	Moxifloxacin		 [structure]	CH	n.a.	C	F	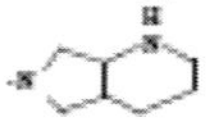 [structure]	C	OCH$_3$		
	Gemifloxacin		 [structure]	CH	n.a.	C	F	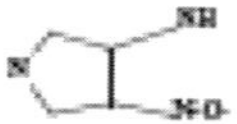 [structure]	N	n.a.		
	Garenoxacin		 [structure]	CH	n.a.	CH		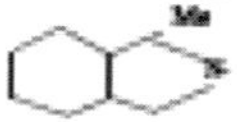 [structure]	C	OCHF$_2$		

Composition at Respective Position

[a]n.a. = not applicable.
[b]R1 and R8 are interconnected to give [structure].

48

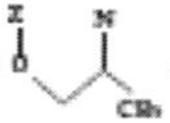

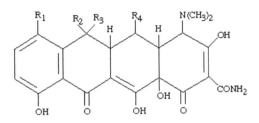

Tetracyclines have been marketed since 1984, and, since discovery, some first-generation tetracyclines such as tetracycline, oxytetracycline, and chlorotetracycline have been extensively used in livestock and aquaculture, besides clinical use by humans. For human use, the second-generation tetracyclines doxycycline and minocycline have been prescribed to a great extent, and indeed prescription of the latter has steadily increased in the United States over the 2003–2005 period (Fig. 1.7). However, each of these tetracyclines is less than 0.5% of all the other 200 most prescribed drugs. Tetracyclines are also prescribed to a good extent in several European countries (Fig. 1.6), with the exception of Italy and Denmark where per capita prescriptions are quite minimal, that is, <25 prescriptions per 1000 inhabitants (Mölstad et al., 2000). They are also widely used in animal husbandry where daily therapeutic doses of 40 mg tetracycline kg^{-1} liveweight are typical (Kühne et al., 2000).

Tetracyclines prevent the growth of bacterial cells by blocking the ribosomes from binding to messenger RNA (mRNA), thus preventing the synthesis of proteins. To reach the ribosomes, they have to be actively pumped into the cell. Loss of the ability for a cell to pump these compounds inside culminates into resistance to tetracycline. However, both the more recently popular minocycline and doxycycline are soluble in lipids compared to their predecessors and are, therefore, still absorbable by cells without the pumping mechanism. To that effect, bacteria are less prone to developing resistance against them. Such lipophilicity also makes both of these drugs highly bioavailable (Table 1.10). The more polar (i.e., hydrophilic) tetracyclines are eliminated through the kidneys (i.e., urine), whereas the more lipohilic ones such as minocycline and doxycycline are eliminated mostly through the biliary (i.e., fecal) route, typically after extensive transformation.

Like macrolides, streptogramins inhibit peptide chain elongation by binding to the 50S ribosome subunit. They have marked effect on gram-positive bacteria including the methacillin-resistant *S. aureus* and other multiresistant gram-positive pathogens. However, they tend to have poor solubility. Streptagramins have an important place in pharma-ecology as Synercid, one of the leading members in this class used by humans to treat vancomycin-resistant enterococci, has a very similar structure with virginiamycin, which is extensively used in animal husbandry. Virginiamycin has been extensively used for more than three decades in poultry production to control clostridial diseases and as a growth promoter.

TABLE 1.10 Tetracyclines, Their Origin, and Characteristics

Compound	Origin	MW	Bioavailability	R1	R2	R3	R4
Tetracycline	*Streptomyces texas*	444.4	60–80% oral; <40% intramuscular	H	CH$_3$	OH	H
Chlortetracycline	*S. aureofasciens*	478.9	30%	Cl	CH$_3$	OH	H
Demeclocycline[a]	*S. aureofasciens*	464.9	?	Cl		OH	
Oxytetracycline	*S. rimosus*	460.4	?	H	CH$_3$	OH	H
Minocycline	Semisynthetic	457.5	100% (oral)	N(CH$_3$)$_2$	H	H	H
Doxycycline	Semisynthetic	444.4	100%	H	CH$_3$	H	OH

[a]Lacks an —OH on the ring with R1.

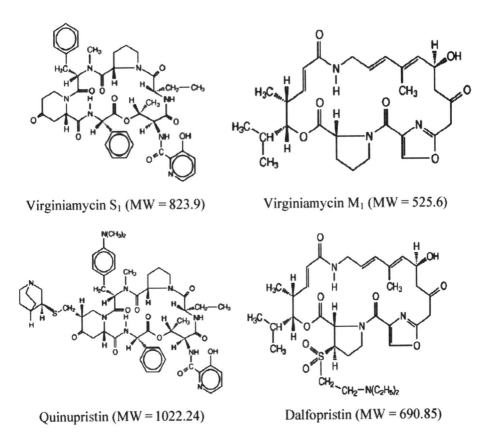

Virginiamycin S_1 (MW = 823.9)   Virginiamycin M_1 (MW = 525.6)

Quinupristin (MW = 1022.24)   Dalfopristin (MW = 690.85)

Virginiamycin is a streptogramin mixture of virginiamycin M_1 and S_1 in proportions of 75 and 5%, respectively. Similarly, Synercid is a mixture of quinupristin and dalfopristin in proportions of 70 : 30 on a weight by weight basis. This mixture is also commonly referred to as streptogramin compounds A and B. As is clear from the structures above, the structure of virginiamycin S_1 and quinupristin are similar. The structures of the highest percentage components of each drug, that is, virginiamycin M_1 and dalfopristin, are even more similar, with the slight difference being the sulfonyl moiety in dalfopristin. The prevalence of *Enterococcus faecium* resistant to quinupristin–dalfopristin in Europe and the United States has been reported (Hayes et al., 2001) and will be discussed in more details in Chapter 4.

Oxazolidinones (Table 1.8) were originally developed as monoamine oxidase inhibitors for the treatment of depression (Moellering, 2003). Their antibiotic activity was realized later when they were shown to be effective against bacteria and fungal infections in plants. From a clinical perspective, their suppression of bone marrow (i.e., myelosuppression) is a major side effect in patients receiving large doses, but this condition is reversible. None of them has been in the top 200 drugs prescribed in the United States lately as they are generally used as a last resort against gram-positive bacteria including methicillin-resistant *S. aureus* (MRSA), methicillin-resistant *S. epidermitis* (MRSE), penicillin-resistant *S. pneumoniae* (PRSP), and vancomycin-resistant enterococci (VRE). They are highly bioavailable (Moellering, 2003).

Aminoglycosides differ considerably in structure but generally contain a cyclohexane ring and amino sugars. In terms of structure and pharmacokinetic properties, aminogylcosides also differ from streptogramins and macrolides, although their mode of action is similar to those of these other classes. Thus, they act by binding to the 30S ribosome, interfering with initiation of protein synthesis. Such interference causes a misreading of mRNA, resulting in a misreading of the genetic message carried by the mRNA. To be effective, they have to be actively pumped into the bacterial cell, across the cytoplasmic membrane against a potential. That process requires energy and oxygen. Thus, they are ineffective against anaerobic bacteria as these thrive in the absence of oxygen, conditions under which the drug will not be pumped into the cells. They are generally used to control infections of gram-negative aerobes and *Mycobacterium* sp. They diffuse through porins in gram-negative bacteria. By contrast, anaerobes are mostly targeted with lincosamides such as clindamycin, chlorophenicol, and lincomycin (Table 1.8).

Most prescribed among aminoglycosides is TobraDex (which is really a combination of tobramycin—the aminoglycoside—and the corticosteroid dexamethasone). Besides TobraDex, other aminoglycosides commonly used include streptomycin, kanamycin, and gentamicin. Aminogylcosides are ototoxic and nephrotoxic to humans. They are excreted by glomerular filtration in the kidneys with the parent compound ending up in the urine. Other classes that block the synthesis of proteins in bacteria includes furantion of which nitrofurantoin (Macrobid) registered 0.2% of the top 200 most prescribed drugs in 2003 and 2005 (Fig. 1.7). Approximately 30–40% of this drug is excreted as the parent compound (Queener and Gutierrez, 2003).

1.2.4.3 Nucleic Acid Synthesis Inhibitors   This mechanism is best exemplified by quinolone antibiotics. The basic structure of all quinolone antibiotics is similar, with an exocylic oxygen at position 4. The core structure is:

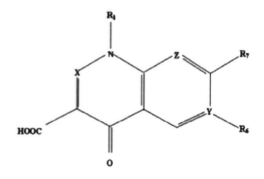

The first quinolone, naladixic acid, which has been marketed since the late 1960s, was discovered as an accidental product of the antimalarial drug chloroquine. It has a low oral adsorption with peak serum levels of less than $0.5 \, \mu g \, mL^{-1}$ (Hunt et al., 2002). Through additional research, it was discovered that adding a floro group at position 6 formed flumequine, which had improved pharmacokinetics. Floro-based modifications have since become routine with quinolones to the extent that this class is more widely referred to as fluoroquinolones or more collectively as 4-quinolones. The range of 4-quinolones and how they relate to this core structure are summarized in Table 1.9. It is apparent that some manipulation has occurred at the C7 and N1 positions since nalidixic acid. The addition of the piperazine ring at the C7 position in compounds 6–15 in Table 1.9 improved efficacy against gram-negative bacteria (Andersson and MacGowan, 2003). The addition of methyl groups to the piperazine ring exemplified by compounds 11–14 increased oral absorption and in vitro activity. The third-generation fluoroquinolones are active against both gram-positive and gram-negative bacteria, and they have a great potential to effectively penetrate phagocytic cells (Applebaum and Hunter, 2000). To increase the solubility and bioavailability of these third-generation flouroquinolones in the body, the C7 position has been designed with a methyl and/or —NH$_2$ moiety (e.g., gemifloxacin) or provided with an azabicyclo group (e.g., garenoxacin, moxifloxacin, and travafloxacin).

The importance and use of 4-quinolone antimicrobial agents is rapidly increasing both in clinical and veterinary practice, where they are widely used against a variety of infectious diseases caused by gram-positive and gram-negative bacteria as well as mycoplasmas. On a global scale, their widespread use is only second to that of penicillins and macrolides, with current annual worldwide sales amounting to about $3 billion (Appelbaum and Hunter, 2000; Emmerson and Jones, 2003). The study of antibiotic use in 13 European countries showed more per capita usage of 4-quinolones in Spain, Portugal, and Italy (Fig. 1.6). The data about usage of these compounds in Denmark were not available and per capita use in Finland, the United Kingdom, and the Netherlands was quite low. In the United States, Levoquin (levoflaxacin) and ciproflaxacin are the most commonly prescribed antibiotics in this class (Fig. 1.7; Table 1.8). Both are quite effective in both gram-positive and gram-negative bacteria. Avelox (monofloxacin) was also widely prescribed in 2003 but its use has declined in the recent past.

Quinolones are currently used to treat urinary tract, systemic, and respiratory infections. Quinolone compounds, particularly those that are fluorinated, are desirable in clinical practice because of their good bioavailability, which enables them to penetrate and concentrate intracellularly, enabling contact with intracellular pathogens such as *Legionella pneumophila, Listeria monocytogenes, Neisseria gonorrhea,* and *Moraxella catarrhalis* (Wise, 2003). Unlike various other classes of antibiotics, all members of this class are synthetic and do not have any naturally occurring counterpart in the environment (Wetzstein et al., 1997). They generally act against microorganisms by selectively inhibiting DNA synthesis, affecting its supercoiled nature (Hawkey, 2003; Brighty and Gootz, 1997). The supercoiling is affected by topoisomerase I (i.e., gyrase) in gram-negative bacteria and topoisomerase IV in gram-positive bacteria. The difference between the two types of targets is that gyrase from gram-positive bacteria is less susceptible to inhibition by 4-quinolones compared to gyrase in gram-negative bacteria. Distorting the supercoiled structure of the DNA impedes the bacterial chromosome from winding, making it too long to fit into the two daughter cells. At the concentration that these compounds are used in clinical practice, they are unable to affect these enzymes in mammalian cells. However, therapeutic dosages of quinolones in general have a whole range of side effects including dizziness, headaches, seizures, shakiness, and changes in vision. They can also damage cartilage, which may lead to permanent damage and lameness, diarrhea, nausea, vomiting, rashes, and itching. Some of the third-generation quinolones have also been associated with cardiac problems, including heart failure (Jjemba and Robertson, 2005). Ciprofloxacin has also been implicated as a mutagen in hospital-derived wastewaters in Switzerland (Hartmann et al., 1998a). Other antibiotics that block DNA replication belong to the class of rifamycins.

1.2.4.4 Antagonism to Metabolic Processes   The antibiotics in this class are primarily structural analogs to metabolic intermediates. In this instance, the analogs compete with the metabolites in metabolic processes; but, because the analogs are not functional in metabolic processes, the metabolic process that is targeted fails. Sulfa drugs are perhaps the oldest form of antibiotics known to modern life. They were introduced in the 1930s. They consist of a benzene ring, a sulfonamide group ($-SO_2NH_2$), and an amine moiety ($-NH_2$):

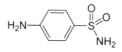

They are amphoteric but generally function as weak organic acids at physiologic pH (pK_a 4.5–7.5) and are quite hydrophilic. They range in solubility from 0.1 to 8 g L^{-1} (Sarmah et al., 2006). However, there are significant differences in the pharmacokinetics between individual members of sulfonamides. Most of them have at least two nitro (or amine) functional groups with one of them (referred to as the N^4) attached to the aromatic ring. The N^4 is easily protonated at very low pH (<2.5) giving the sulfonamides a positive charge under acidic conditions. They are neutral between pH 2.5 and 6 and negatively charged above pH 6. Sulfonamides are the most widely used antibiotics worldwide.

Sulfonamides are used for controlling urinary tract infections, acute and chronic lung infections (norcadiosis), protozoan infections of the nervous system (i.e., toxoplasmosis), and a variety of infections in humans and livestock. Their mode of activity is by inhibiting the multiplication of bacteria by competitively inhibiting para-aminobenzioc acid (PABA) in the folic acid metabolism cycle (O'Neil et al., 2001). More specifically, they block the synthesis of folic acid in bacteria as the drugs are structurally similar to PABA. Folic acid is essential to the synthesis of amino acids and nucleic acids. In bacteria, folic acid is synthesized from PABA

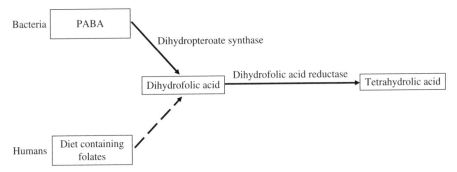

Figure 1.8 Folic acid pathway: Tetrahydrofolic acid synthesis in bacteria and humans. PABA = Paraminobenzoic acid.

compared to humans where folic acid is obtained directly from our dietary intake (Fig. 1.8). Sulfa drugs such as sulfadiazine and sulfamethoxine have also been widely used in livestock and aquaculture settings.

As indicated earlier, sulfonamides are effective in both gram-positive and gram-negative bacteria. Mostly prescribed for humans in the United States, in this class is sulfamethoxazole, mostly in combination with trimethoprim (SMZ-TMP) in a 5 : 1 ratio. Trimethoprim inhibits dihydropholic acid reductase and this, just like sulfonamides, also interferes with the synthesis of folic acid (Fig. 1.8). As a matter of fact, use of the combined SMZ-TMP has been steadily increasing recently as is displayed by the number of prescriptions (Fig. 1.7). Oral doses of sulfonamides are absorbed well and eliminated by the liver and kidney with 20–60% excreted as the parent compound (Queener and Gutierrez, 2003).

1.2.4.5 Antibiotics That Disrupt Membrane Integrity
Biological membranes are assembled in sheetlike profiles and are composed of proteins as well as lipids. The lipid components are primarily phospholipids, glycolipids, and cholesterol or, in the case of fungi, ergosterol. Membranes are very essential to life as they give each cell its individuality by separating it from its environment. They are embedded with specific molecular pumps and gates that serve as highly selective barriers. Such selectivity in turn regulates the molecular and ionic composition of the intracellular medium. In eukaryotic cells such as those of humans and fungi, membranes also form boundaries between organelles such as mitochondria, nucleus, and ribosomes. Membranes also control the flow of information between cells and their surroundings.

Antifungal azole drugs inhibit the synthesis of ergosterol, an important component of fungal cell membranes. Diflucan has been the most prescribed antifungal azole in the United States during 2003–2005 to treat infections that range from cryptococcal meningitis and candidiasis. Antifungal azoles are also routinely used for various skin fungal infections. Ketoconazole has also been prescribed to some extent (Fig. 1.7). High doses of azole antifungal drugs can inhibit the synthesis of testosterone leading to enlarged breast tissue. Elidel is an antifungal agent that is increasingly prescribed as a cream for eczema. Its active ingredient, pimecrolimus, is derived from ascomycin, a natural substance produced by the fungus *Streptomyces hygroscopicus* var. *ascomyceticus*. This active ingredient selectively blocks the production and release of cytokines from T cells, which in turn blocks the inflammation, redness, and itching of the skin that is often associated with eczema. Nystatin, a polyene, is also another widely prescribed antibiotic to target membranes. It is primarily used for systemic fungal infections.

1.2.4.6   Other Antimicrobials   Metronidazole (Flagyl) acts by producing a metabolite under anaerobic conditions, conditions under which it is reduced to a short-lived metabolite that damages DNA. It is also effective against some protozoal infections as these organisms are able to reduce it to its active form intracellularly. Because it affects DNA, metronidazole may be mutagenic or carcinogenic. Approximately 5.4 and 6.1 million prescriptions of metronidazole were authorized in the United States during 2004 and 2005, respectively (Fig. 1.7). The acyclic nucleoside analogs are selective against viral infections. They are activated by viral thymidine kinase, an enzyme that is unique to virus-infected cells. They act by inhibiting the viral DNA polymerase, inhibiting the replication of viruses. For purposes of this discussion, the statistics for valacyclovir (Valtrex) are combined with those for acyclovir (Zovirax) as the former is an oral prodrug form that is rapidly converted to the latter in the body. Both of the drugs in this category have been extensively prescribed, and the percent number of prescriptions have consistently increased in the United States during 2003–2005 (Fig. 1.7). In terms of overall frequency of prescribing, their percentages are somewhat similar to those of tetracycline and doxycycline. Chloroquine is one of the most widely used drugs in the world. It is absorbed from the GIT and particularly metabolized by hepatic microsomal enzymes, generating metabolites such as desethylchloroquine:

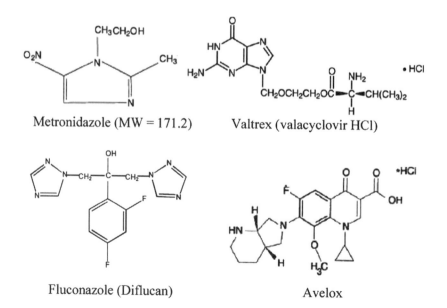

Metronidazole (MW = 171.2)   Valtrex (valacyclovir HCl)

Fluconazole (Diflucan)   Avelox

Antibiotics are obtained without any prescription in some countries. In general, most antimicrobial agents that are taken up by animals for therapeutic or subtherapeutic purposes are poorly absorbed in the gut. Thus, substantial amounts of these, which can in some instances exceed 90%, are excreted as the parent compound (Jjemba, 2002a, 2006; Hirsch et al., 1999). The presence of antibiotics in the environment may be contributing to the increasing incidence of severe nosocomial infections. Furthermore, synthetic antimicrobials particularly pose an unusual advantage from a clinical perspective as they are totally novel to the biosphere and thus should not easily lead to the development of intrinsic resistance mechanisms in the target microorganism. These issues will be fully explored in Chapter 4.

1.2.5 Gastrointestinal Drugs

The gastrointestinal tract (GIT) is a continuous hollow organ extending from the mouth to the anus. It includes the esophagus, stomach, small intestine, large intestine, and the rectum. Its disorders can be structural or neural, leading to the obstruction, slowing, or even acceleration of movement of food through the GIT. Structural disorders include inflammation and ulcerations, impacting the absorption, motility, and secretion of substances through the digestive system. Neural disorders on the other hand manifest themselves through nausea, causing antiperistalsis—resulting in vomiting and the loss of electrolytes. On the other extreme, neural disorders can cause constipation. Both structural and neural disorders can also manifest themselves in abdominal pains or episodes of diarrhea, leading to excessive loss of fluids and electrolytes. Also included among GIT drugs are those that alleviate pancreas and gall bladder disorders as such disorders also affect digestion directly or indirectly.

The popular GIT medications in Table 1.11 provided for 5.3, 4.9, and 4.6% of the top 200 prescriptions in 2003, 2004, and 2005, respectively. Ranitidine and other antacid drugs inhibit histamine receptors type 2 in the gastric system, inhibiting the production of stomach acids. Histamine receptors have also been found in some fish including zebrafish (Peitsaro et al., 2000). Gastroesophageal reflux disease (GERD), popularly known as heartburn, occurs when the gastric acids back up into the esophagus. Compared to the stomach, the esophagus is not as well lined and protected by mucus. The reflux occurs as a result of insufficient closure of the lower esophageal sphincter. Treatments are therefore designed to either reduce the acidity of the stomach contents or to increase the tone of the esophagus. Targeting the stomach contents is justified as they are the ones that come into contact with the esophagus. In the stomach, the acids are secreted by parietal cells for the purpose of breaking down connective tissue in food and killing some bacteria that have been ingested with the food. In terms of constipation, Prevacid, Nexium, Prilosec, and Zantac are very common household names and indeed led the GIT prescription drugs in 2003–2005 (Table 1.11). Most of them act by blocking the enzymes that pump H^+ ions into the secretory side of the parietal cells of the stomach, ultimately reducing the production of gastric acids. For this, they are labeled as antiacids. For gastrointestinal pain, the arachidonic acid that is responsible for pain is converted to prostaglandins, which have many effects such as maintaining cellular

TABLE 1.11   Some of the Most Commonly Prescribed Gastrointestinal Drugs in the Untied States

Product	Total Number (in millions) of Gastrointestinal Prescriptions			Mode of Action
	2003	2004	2005	
Prevacid (lanoprazole)	25.6 (1.2)	23.6 (0.8)	22.2 (1.0)	Block the enzyme that pumps H^+ into the secretory side of the parietal cells of the stomach.
Nexium	21 (1.0)	23.6 (0.8)	22.9 (1.0)	Block the enzyme that pumps H^+ into the secretory side of the parietal cells of the stomach.
Omeprazole (Prilosec)[a]	17 (0.8)	8.4 (0.3)	7.4 (0.3)	Block the enzyme that pumps H^+ into the secretory side of the parietal cells of the stomach.
Protonix (Pantaprazole)	15 (0.7)	18.4 (0.7)	16.4 (0.7)	Block the enzyme that pumps H^+ into the secretory side of the parietal cells of the stomach.
Ranitidine (Zantac)	13.7 (0.6)	29.8 (1.1)	12.8 (0.6)	Antagonizes the receptors that control the secretion of HCl by the parietal cells.
Aciphex (Rabeprazole)	8.1 (0.4)	8.5 (0.3)	7.1 (0.3)	Block the enzyme that pumps H^+ into the secretory side of the parietal cells of the stomach.
Ultracet	5.1 (0.2)	5.6 (0.2)		A pancreatic enzyme that is involved in the digestion of lipids, proteins, and carbohydrates.
Meclizine HCl	4.7 (0.2)	5.6 (0.2)	5.1 (0.2)	Against vomiting and nausea (antiemetic) as it blocks histamine (H_1) receptors.
MiraLAX (GlycoLax, polyethylene glycol)	3.7 (0.2)	3.6 (0.1)	4.4 (0.2)	As an osmotic agent, it enhances the retention of water in the stool alleviating the effects of constipation.
Metoclopramide		6 (0.2)	5.7 (0.2)	Enhances the contractions of the GI smooth muscle, thus accelerating the transiting of fluids to the ileum. No effect on stomach acidity.
Famotidine (Pepcid)		4.3 (0.2)	3.5 (0.2)	Antagonizes the receptors that control the secretion of HCl by the parietal cells.

[a]Prilosec became available over the counter in the United States in 2004 and therefore its sales based on prescription data are an underestimate.

structure in the gut, inflammation, and the like, by the cyclooxyganase enzyme system (COX). As indicated in Section 1.2.3, there are two COX systems, that is, COX-1 and COX-2. The former is preferable as it protects the stomach lining by making good prostaglandins, whereas COX-2 is undesirable as it stimulates the prostaglandins that are responsible for pain and the enzymes that are related to pain. Some of the now infamous COX-2 inhibitors such as Vioxx were developed with the rationale that chronic pain sufferers that take the drug would block the pain without suffering severe consequences of the gut's failure to protect itself from injury.

1.2.6 Antidiabetic Drugs

According to the World Health Organization, there are five different types of diabetes. These are Type 1 diabetes (juvenile-onset or insulin-dependent diabetes), Type 2 diabetes, impaired glucose tolerance, gestational diabetes, and impaired fasting glucose. As the name suggests, Type 1 diabetes mostly attacks at a young age with peak occurrences at 11–13 years but can still occur in adolescents to adults under 30 years. It is thought to emerge from gene–environment interactions and is probably associated with autoimmune processes that lead to the destruction of beta cells. Beta cells are located in the pancreas and are primarily responsible for secreting insulin from the precursor proinsulin. The destruction of beta cells eventually leads to glucose insulin deficiency. The autoimmunity mechanisms are related to cell and cytokine-mediated injury of beta cells. Autoimmunity is itself triggered by chemicals such as alloxan, streptococin, pentamidine, and vacor (Heuther, 2000). Other documented triggers include intake of diets with high levels of nitrosamines and also viral (i.e., mumps, coxsackie, rubella, and CMV) infections. The secretion of insulin is controlled parasympathetically and increasing its levels is promoted when the levels of glucose in the blood are high. Thus, its increased secretion keeps the blood sugars (primarily glucose) in check.

Type 2 diabetes is the most prevalent and manifests itself due to insufficient levels of insulin production. It occurs at a later age ($>$40 years) and is often associated with obesity and hypertension. Even in cases where the blood insulin levels in Type 2 diabetics is normal, the target tissues may be nonresponsive to insulin possibly due to a decrease in the mass of beta cells, abnormal function of beta cells, or due to changes in the insulin receptors. It can also be due to defects in postreceptor activity. Gestational diabetes is associated with a degree of intolerance to glucose during pregnancy due to changes in metabolism that are induced by endogenous and placental hormones. In this instance, a good fraction of the glucose can be steadily transported to the fetus, resulting in a fasting hypoglycemic state of the expecting mother. This condition usually resolves itself postpartum. Impaired glucose tolerance and impaired fasting glucose are basically variations in metabolic stages between diabetes and normal glucose homeostasis.

The bottom line for any of the above disorders is that insulin action is lost. Insulin is the primary hormone that regulates the metabolism of glucose in its conversion to the storage of carbohydrate—glycogen stored in the liver and muscles. Insulin also regulates the metabolism of fats and proteins by directly stimulating the synthesis and storage of lipids within fat cells and promoting the uptake of amino acids, directly stimulating the synthesis of proteins. Loss in the action of insulin culminates in less glucose usage by the muscles and fat cells with more glucose being released into the circulatory system by the liver. Ultimately, the high levels of glucose end up in the urine (glucosuria). Treatment of these conditions normally involves providing insulin from external sources as well as controlling electrolytes and fluid imbalances. Insulin is marketed under different names and the leading ones are listed in Table 1.12. The prescribed insulin products are synthetic or semisynthetic. They are largely similar in their mode of action and the effects they produce but differ in the onset and duration of action (i.e., pharmacokinetics).

The most widely prescribed antidiabetic drug in the United States is Glucophage (metformin). Metformin is also one of the top three most prescribed drugs in the United Kingdom where more than 10 tons were used in 2000 (Bound and Voulvoulis, 2005). It is suspected to act by directly stimulating insulin receptors, increasing the cellular use of glucose, and inhibiting gluconeogenesis. Those increases may also be a result of an inhibition of tyrosine phosphatase (Holland et al., 2004), the enzyme involved in the activation of the insulin receptor. Metformin has a low bioavailability and has a half-life of 4–9 h (DiMicco and Gutierrez, 2003) and most of it is excreted in urine unchanged. The other antidiabetic drugs listed in Table 1.12, notably Humalin, Actos, and Avandia, also stimulate insulin receptors. One of the adverse effects of insulin administration to strictly control glucose is the onset of hypoglycemia (i.e., low blood glucose) with blood glucose levels decreasing to less than 60 mg/dL. Hypoglycemia can manifest itself with a whole range of symptoms including tremors, confusion, fatigue, or coma. Insulin and insulin-like receptors are also fairly widespread in various vertebrates and invertebrates. In fish, these receptors appear to be structurally similar to those of mammals but regulate the metabolism of lipids in a fashion that is the opposite of what is encountered in mammals (Planas et al., 2000). Thus, high levels of insulin in the adipose tissue of trout were associated with a decreased number of insulin receptors, whereas starvation was associated with an increase in insulin receptors. These results show the presence and regulation of specific insulin and insulin-like growth factor receptors in fish, indicating that fish can also be targeted by antidiabetic compounds. Insulin and IGF receptors have also been reported in birds, reptiles, and amphibians (Planas et al., 2000; Méndez et al., 2001).

1.2.7 Diuretics and Electrolytes

Diuretics are compounds that therapeutically affect the mammalian nephral excretion balance, increasing the net excretion of water and solutes. Thus, they modify the excretion of water and the concentration of salts in the body through the kidney so as to ensure a constant volume of body fluids. Their functions influence blood pressure and the actions of several organs. Some of the leading diuretics and electrolytes and the extent of their prescription are summarized in

TABLE 1.12   Antidiabetic Prescriptions in the United States and Their Mode Action

Product	Total Number of Antidiabetic Prescriptions[a]			Mechanism of Action
	2003	2004	2005	
Glucophage (Diabeta, Glyburide, Glucovance, Metformin)	18.4 (1.8)	40.4 (1.4)	49.6 (2.1)	A biguanide that acts by increasing the ability of insulin to bind to peripheral tissues, increasing the uptake of glucose by the muscles and other tissues.
Humulin (Humalog, Insulin, Lantus)	18.4 (0.9)	15.3 (0.5)	11.7 (0.5)	Binds to unique insulin receptors to regulate glucose levels and its metabolism.
Glucotrol (Glipizide)	10.7 (0.5)	14.1 (0.5)	13.1 (0.6)	A second generation sulfonylurea that blocks ATP-sensitive K-channels in the membrane of beta cells, stimulating the release of insulin.
Actos	9.3 (0.4)	9.9 (0.4)	9.7 (0.4)	A thiazolidinedione that activates insulin-responsive genes that are concerned with the metabolism of lipids and carbohydrates.
Avandia	8.8 (0.4)	9.2 (0.3)	10.4 (0.4)	A thiazolidinedione that activates insulin-responsive genes that are concerned with the metabolism of lipids and carbohydrates.
Amaryl (Glimepiride)	6.9 (0.3)	7.2 (0.3)	5.8 (0.2)	A second-generation sulfonylurea. Blocks ATP-sensitive K-channels in the membrane of beta cells, stimulating the release of insulin.

[a]The prescriptions are in millions, whereas the numbers in parentheses are percent number of the respective prescription as a fraction of the total top 200 prescriptions in the United States in the respective year.

Table 1.13. Most diuretics are designed to deal with conditions of edema by reducing the reabsorption of chlorides, sodium, and water as these are the major constituents of edema fluid. Diuretics are grouped into five main categories: loop diuretics, thiazides, potassium-sparing diuretics, osmotic diuretics, and carbonic

TABLE 1.13   Diuretics and Electrolytes Prescriptions in the United States (2003–2005)[a]

	Total Number (in millions) of Diuretics and Electrolytes Prescriptions		
Product	2003	2004	2005
Furosemide (Lasix)	35.1 (1.6)	36.5 (1.3)	34.8 (1.5)
Potassium chloride (Klor-con)	23.7 (1.1)	27.3 (1.0)	27.6 (1.2)
Triamterene (Dyrenium)	22.0 (1.0)	24.5 (0.9)	22.8 (1.0)
Spironolactone (Aldactone)	2.9 (0.1)	7.3 (0.3)	7.0 (0.3)

[a]The numbers in parentheses are percent number of the respective prescription as a fraction of the total top 200 prescriptions in the United States in the respective year.

anhydrase inhibitors. Each of these categories of diuretics mainly acts within a specific portion of the nephron. As the group categorization suggests, loop diuretics primarily act in the loop of Henle, the U-shaped tubules within the kidneys' cortex and medulla where Na^+, K^+, and Cl^- are actively reabsorbed. The most widely prescribed diuretic, furosemide, is a loop duiretic. Thiazides are mostly used to control edema associated with heart disease or hypertension, and, therefore, for convenience those drugs such as hydrochlorothiazide are discussed under hypertensive and cardiovascular drugs (Table 1.5), although in a strict sense they are diuretics. They act by blocking the reabsorption of sodium and chloride ions in the distal tubule by inhibiting the Na^+ and Cl^- symporter. Potassium-sparing diuretics such as Klor-con, triamterene, and spironolactone are designed to reduce the loss of potassium in the urine. The use of loop diuretics or thiazides also indiscriminately leads to the loss of potassium. Thus, potassium-sparing diuretics are often used in combination with loop diuretics or thiazides to counteract this loss of potassium. The usage of other groups of diuretics, that is, carbonic anhydrase inhibitors (e.g., Acetozolamide and Mathazolamide) and osmotic diuretics (e.g., urea) is not as widespread and have not featured in the top 200 most prescribed diuretics in 2003–2005 in the United States.

1.2.8   Thyroid System Medication

Levothyroxine (also marketed as Levoxyl, Levothroid, Synthroid, or L-thyroxine sodium) is the only thyroid system prescription that has recently featured prominently among the top 200 medications in the United States. Thus, 75.1 million, 80.7 million, and 79.4 million levothyroxine prescriptions were dispensed in 2003, 2004, and 2005, respectively. At these rates, levothyroxine by itself comprised 3.5, 2.9, and 3.4% of the total prescriptions in the top 200 most popular drugs in the United States in 2003, 2004, and 2005, respectively, making it the most prescribed drug. Levothyroxine is a synthetic version of the natural hormone thyroxine, which is released by the thyroid gland. The natural hormone is important in the development of fetuses and newborns, playing a role in crucial processes such as the development of tissues and bones. Throughout life, thyroxine also increases the metabolic rate of cells in all tissues and maintains brain function, body temperature, as well as general metabolism. The synthetic thyroid hormones also increase metabolic rates by stimulating the nuclear receptors in skeletal muscles, as well as those in the liver, heart, kidneys, lungs, and intestines. That stimulation increases gluconeogenesis, the process by which glucose is synthesized from noncarbohydrate precursors. Since glucose is the energy currency of the body, all of these thyroxine-driven actions culminate into enhanced synthesis of proteins, cell growth, and differentiation as well as enhanced development of the central nervous system.

Levothyroxine is taken orally or intravenously and is normally prescribed to treat hypothyroidism or to suppress the release of thyroid hormone so as to manage cancerous thyroid nodules (i.e., thyroid cancer) and growth of goiters. Other less frequently prescribed thyroid and parathyroid drugs include methimazole (Tapazole), various iodides, lithioronine (Triostat), and liotrix (Thyrolar).

1.2.9 Respiratory Drugs

Asthma is one of the fastest growing medical conditions, especially among children and adolescents. Its prevalence in the United States has increased more than 60% in the 1990s and mortality from it has more than doubled since the 1970s (Agins et al., 2003). To that effect, most of the respiratory drugs prescribed recently (Table 1.14) are intended to combat asthma. Asthma attacks manifest themselves as a result of three processes: (i) constriction of the bronchioles, (ii) swelling of the bronchial mucosa, and (iii) excessive secretion and accumulation of mucus in the bronchioles. All of these reactions contribute to inflammation, edema, secretion of mucus, and blockage of the airways. The net result of these events is a compounded restriction of the airways leading to shortness of breath, coughing, and wheezing. Some of these drugs, notably Albuterol and Salmeterol, are popular as they relax bronchial smooth muscle (i.e., bronchodilation), decreasing airway resistance to facilitate breathing. The mucus and chemicals contained therein (i.e., histamines and leukotrienes) are primarily secreted from T cells and mast cells located in the epithelial layer of the bronchioles. Leukotrienes include metabolites of arachdonic acid, which as we saw in the case of NSAIDs is a precursor of prostaglandins, which are in turn responsible for pain and inflammation. Thus, some of the drugs used are designed to block the action of leukotrienes. Because some of the drugs used for asthma module nervous system tissue, that is, beta-adrenergic receptor, they are more correctly classified as autonomic nervous system drugs.

In general, second perhaps to cardiovascular and antihypertension drugs, respiratory drugs have been some of the most aggressively marketed through the direct-to-consumer advertisements. Through such marketing campaigns, some of them have rapidly graduated to over-the-counter status within a short time. Also increasingly popular are antihistamines such as Allegra and Zyrtec, which block H_1 receptors, decreasing the discomfort associated with allergic reactions associated with the upper respiratory system during hay fever episodes.

TABLE 1.14   Number of Prescriptions in the United States (2003–2005) Against Respiratory Diseases as a Proportion of the Top 200 Most Prescribed Drugs[a]

Product	Total Number (in millions) of Respiratory Prescriptions			Mode of Action
	2003	2004	2005	
Albeturol sulfate	32.5 (1.5)	41.7 (1.5)	40.0 (1.7)	A selective beta-2 agonist bronchodilator that inhibits the release of mast cell mediators and increases clearance of the mucus membranes.
Salmeterol (including Advair Diskus)	15.9 (0.7)	17.4 (0.6)	18.3 (0.8)	A beta-2 corticosteroid. If in combination (Advair Diskus) a bronchodilator combination that reversibly binds to and stimulates beta-androgenic receptors.
Pulmicort (Rhinocort)	8.0 (0.4)	8.1 (0.3)	3.9 (0.2)	A corticosteroid inhibits the production of anti-inflammatory mediators such as leukotrines, histamines, and prostaglandins.
Triamcinolone (Nasacort AQ)	7.5 (0.3)	11.1 (0.4)	10.4 (0.5)	A corticosteroid inhibits the production of anti-inflammatory mediators such as leukotrines, histamines, and prostaglandins.
Flovent (Flonase)	6.7 (0.3)	20.7 (0.7)	[a]	A corticosteroid inhibits the production of anti-inflammatory mediators such as leukotrines, histamines, and prostaglandins.
Combivent	6.7 (0.3)	6.9 (0.2)	6.0 (0.3)	A combination of Albuterol and Atrovent, with combined mechanisms of both.
Methylprednisolone	3.2 (0.1)	8.7 (0.3)	9.8 (0.4)	A surfactant that improves the exchange of gases across the air–alveolus interface in the lungs.

(*Continued*)

TABLE 1.14 *Continued*

Product	Total Number (in millions) of Respiratory Prescriptions			Mode of Action
	2003	2004	2005	
Atrovent	2.7 (0.1)	[a]	[a]	An anticholinergic bronchodilator that inhibits smooth muscle in the bronchial.
Zyrtec (Citirizine)			4.5 (0.2)	Antihistamine that blocks allergic reactions by preventing fluids from escaping from capillaries.
Allegra (Fexofenadine)			3.8 (0.2)	Antihistamine that blocks allergic reactions by preventing fluids from escaping from capillaries.
Total	49.8 (2.3)	73.0 (2.6)	56.7 (2.4)	

[a]In parentheses are the percentage out of the top 200 most prescribed drugs in the respective year.

1.2.10 Oral Contraceptive and Reproductive Therapeutics

Several nonchemical-based methods to control birth are known and some have been practiced for eternity. In the midnineteenth century, Thomas Malthus's work that related to population growth estimates aroused a lot of interest in the problems of overpopulation. The interest in his work subsequently led to the discovery of the fact that manipulating the levels of reproductive hormones can prevent ovulation. Hormones are chemical cues that are formed in one organ and transported to another organ to stimulate the functions of that second organ. Chemical-based methods, popularly referred to as contraceptive birth control, have gained a lot of popularity among millions of women since the late 1950s when, building on results from reproductive hormone manipulation, Frank Colton invented the first oral contraceptive, which was marketed as Enovid. Colton's invention has since then undergone some modification to improve efficacy and safety, but the key concept and related mechanism of action still remain the same in the sense that the elevated levels of estrogen and/or progesterone in the system tricks the body to believe that pregnancy has already occurred, preventing ovulation and implantation of the ovum into the uterine wall (endometrium). Both estrogen and progestin are steroids, the former influences a number of female reproductive tissues, whereas the latter specifically affects the lining of the uterus. They are synthesized in the ovaries and primarily promote the development of the reproductive system and secondary sexual characteristics. Estrogen also promotes the development and thickening of the endometrium during the first half of the menstrual cycle. Such enlargement provides a suitable environment for successful fertilization and implantation of the ovum. Several estrogens are known and some of them are natural (e.g., estriol, estrone, and estradiol), whereas most of the others are synthetic. In males, the endogenous production of estrogens stands at about 20, 40, and 20 $\mu g/$ day, respectively (Webb et al., 2005), with endogenous levels in prepubescent boys estimated at 6 $\mu g/day$. Under WHO guidelines acceptable daily intake of estradiol in food additives is 50 ng/kg (Webb et al., 2005; http://jecfa.ilsi.org), and Hartmann et al. (1998b) estimate that a normal dietary intake of steroid of 0.1 $\mu g/$ day occurs on a regular basis. Chemical structures of a few from each category are shown in Table 1.15. Common to all of them are at least four fused aromatic rings. Also shown is testosterone, another reproductive hormone, which will be discussed shortly. It is noticeable that the natural estrogens consitently have slightly higher solubility compared to the synthetic ones. By comparison, the synthetic estrogens have higher octanol–water partition coefficents (log K_{ow}). Estrogens are also generally nonvolatile (as is reflected by their very low vapor pressure) and highly lipophilic substances that can adsorb to solids in environmental matrices.

Presently, oral contraceptives are primarily comprised of estrogen combined with progestin. In some instances, the contraceptive is composed of only the latter. Oral contraceptives work by effectively suppressing ovulation and changing the cervix lining, making it difficult for sperm to penetrate the uterus. They also change the conformity of the endometrium, disabling implantation even if fertilization occurs. To be even more effective, they are typically taken cyclically for 3 weeks followed by 1 week of taking a placebo over the 28-day estrus cycle. Some contraceptives have recently been designed for use at a lower frequency or as implants. The exact sequences of consumption, series of events, and the various modifications to this basic routine are quite eloquently described in various reproductive biology books and family planning literature. It is important to note that presently there are more than 20 brand names of birth control pills and possibly more than twice as many formulations on the market. All of them have some element of estrogen and/or progestin as the active ingredient.

The total number of reproductive drugs comprised 91.9, 79.4, and 57.2 million prescriptions in 2003, 2004, and 2005, respectively. As summarized in Figure 1.9, the conjugated estrogen marketed as Premarin has dominated most of the market followed by Ortho-Tri-Cyclen. More recently, though, Ortho-Evra and Yasmin have become equally or even slightly more popular. Note also from these data that Microgestin FE, Trivora-28, Ortho-Novum, Kariva, Low-Ogestrel, and Apri have dropped off the most prescribed 200 drugs in the United States since 2003, whereas Estradiol and Tri-Sprintec (ethinyl estradiol) have filled that vacuum. However, as noted above, the active ingredients are similar with slight compositions marketed under different names. Ethinyl estradiol consumption is estimated at 16 kg/ year for Italy (Zuccato et al., 2001) and 50 kg/year for Germany (Ternes, 2001). At the time these data were compiled, estrogen was especially also used in postmenopausal women as part of their hormone replacement therapy (HRT) to counter the loss of endogenous estrogen at the onset of menopause, which is also associated with osteoporosis (bone loss). Other less widespread uses include stemming the effects of inadequate synthesis of endogenous estrogen during puberty so as to enhance

TABLE 1.15 Structure aand Physicochemical Characteristics of Various Compounds Associated with the Reproductive System

Nature	Compound	Structure	MW	Solubility (mg/L at 20°C)	log K_{ow}	pK_a	Vapor Pressure (mm Hg)
Natural estrogens	Estrone (E1) CAS 51-16-7		270.4	13	3.43		2.3×10^{-10}
	Estradiol (17β-estradiol; E2; CAS 50-28-2)		272.4	13	3.94	10.23^a	2.3×10^{-10}
	Estriol (E3; CAS 50-27-1)		288.4	13	2.81	10.05^a	6.7×10^{-15}

(*Continued*)

TABLE 1.15 *Continued*

Nature	Compound	Structure	MW	Solubility (mg/L at 20°C)	log K_{ow}	pK_a	Vapor Pressure (mm Hg)
Synthetic estrogens	Ethinyl estradiol (steroidal; EE2; CAS 57-63-6)		296.4	4.8	4.15	10.21[a]	4.5×10^{-11}
	Mestranol (steroidal; MeEE2; CAS 72-33-3)		310.4	0.3	4.67		7.5×10^{-10}
	Dienestrol (nonsteroidal; CAS 84-17-3)		266.3				
	Testosterone (CAS 58-22-0)		288.4	18–25	2.9		

[a]pK_a values obtained from Yamamoto et al. (2003).

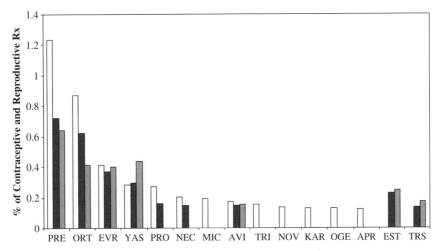

Figure 1.9 Common leading reproductive drugs prescribed in the United States in 2003 (□), 2004 (■), and 2005 (▨) as a percentage of the total top 200 most prescribed drugs in that country. Based on a report of the top 200 prescriptions in those three years, these reproductive drugs comprised 91.9, 79.4, and 57.2 million prescriptions in 2003, 2004, and 2005, respectively. PRE = Premarin, ORT = Ortho-tri-cyclen, EVR = Ortho-Evra, YAS = Yasmin 28, PRO = Prempro, NEC = Necon, MIC = Microgestin FE, AVI = Aviane, TRI = Trivora-28, NOV = Ortho-Novum, KAR = Kariva, OGE = Low-Ogestrel, APR = Apri, EST = Estradiol, and TRS = Tri-Sprintec (Ethinly estradiol).

normal growth and sexual development. There is also some use of these hormones in the treatment of breast cancer, but none of these other side uses reach the quantities consumed for birth control purposes. Estrogenic compounds are also frequently administered to livestock as growth promoters (Khanal et al., 2006), although the amounts used for this purpose are not easily quantifiable.

Under normal usage, estrogens have been associated with overreactions in certain reproductive tissues such as an increased risk of endometrial cancer. There is also an association with anorexia, nausea and vomiting, depression, malaise, atherosclerosis, and a rare but increased risk of myocardial infarction (DiMicco and Gutierrez, 2003). However, the risk for some of these conditions is reduced if estrogen is combined with progestins. These risks have been verified by comparing women who use oral contraceptives with those who do not.

Although not among the top 200 most prescribed, testosterone is also another important sex/reproductive hormone that needs to be discussed at this point. Testosterone is a steroid hormone that is secreted in both males and females although the former secrete about 20–30 times the amount compared to the latter. It is associated with energy and immunity. From a therapeutic perspective, it is administered to treat males that have little or no natural testosterone production (i.e., hypogonadism). However, its usage has gone well beyond those uses, mostly through abusive situations, under what is popularly referred to as performance-enhancing drug use. By nature of such abusive usage, therefore, solid statistics of the amounts dispensed are rare and/or greatly unreliable. Needless to mention, usage under these conditions is reported to give mixed results. Testosterone and other related androgens such as trenbolone acetate are also administered to livestock as growth-promoting agents (Lange et al., 2002) in amounts that are not well quantified.

Estrogen drugs mimic the effects of estrogen in the target tissues by entering cells and binding to cytoplasmic estrogen receptors, forming a complex that in turn interacts with DNA, altering the expression of specific genes. Different proteins are involved in different tissues and therefore the nature of the estrogen–receptor complex may vary. This observation signifies the fact that the estrogen–receptor complex can elicit different expressions on the genes in different tissues. Synthetic steroids such as 17α-ethinyl estradiol (EE2; CAS 57-63-6) and 17α-methyltesterone (MT; CAS 58-18-4) are potent receptor agonists used in reproductive biology. Typical physiological plasma concentrations in women are around 3 nM (Harnagea-Theophilus and Miller, 1998). EE2 is the main component of contraceptive pills. Under normal circumstances, premenopausal women excrete 10–100 μg estrogens/day, but at postmenopause the excreted levels drop off to less than 10 μg/day (Ingerslev and Halling-Sørensen, 2003). Pregnant women excrete up to 30 mg estrogens/day but average values are around 250 μg/day. By comparison, men naturally excrete 2–25 g estrogens/day. Natural estrogens are also excreted by farm animals (cattle, poultry, pigs, etc.), with the quantities varying between animal species and type of animal production system. Women using contraceptive pills are believed to excrete the whole daily dose of 25–50 μg. Approximately 30% of ethinyl estradiol is excreted as conjugates (Schowanek and Webb, 2005). They also have the capacity to disrupt the endocrine system by interfering with the synthesis, secretion, binding action, and elimination of natural hormones in the body that are responsible for reproduction, homeostasis, behavior, and development of individuals. Thus, these hormones have been influenced in an array of effects on various organisms in the environment (see Chapter 4). However, it should be emphasized that a number of natural and synthetic compounds can also act as endocrine disruptors by primarily:

1. Blocking natural estrogen (i.e., estrogenic compounds)
2. Mimicking natural testosterone (i.e., androgenic compounds)
3. Impacting the thyroid (i.e., thyroidal compounds)

1.2.11 Biophosphonates and Other Skeletal Ailment Drugs

The replacement of old bone tissue with new tissue is a natural phenomenon in healthy humans leading to an increase in bone mass. However, with further aging, reduced bone mass naturally becomes more predominant, with the rate at which bone loss occurs far exceeds that at which replacement occurs. The loss of bone tissue and its associated elevated incidence of osteoporosis tend to increase after menopause. A number of pharmaceuticals have been developed to slow this

TABLE 1.16   Number of Prescriptions for Bone Loss Drugs in the United States (2003–2005) as a Proportion of the Top 200 Most Prescribed Drugs[a]

Drug	Number of Prescriptions (in millions) During Year		
	2003	2004	2005
Fosamax (alendronate)	20.6 (1.0)	21 (0.74)	17.9 (0.77)
E-Vista (raloxifene HCl)	7.6 (0.36)	7.1 (0.25)	5.2 (0.23)
Actonel (risedronate)	7.0 (0.33)	9.7 (0.35)	9.7 (0.42)
Miacalcin (calcitonin)	2.9 (0.14)	[b]	[b]

[a]Numbers in parentheses are the percentages of those prescriptions as a fraction of the top 200 prescribed in the United States in that year.
[b]Not among the top 200 most prescribed in that year.

process. Among them are biophosphonates, which are primarily synthetic inorganic pyrophosphates that are designed to block the removal of calcium from bones by suppressing osteoclasts, the cells that remove calcium from bones to meet metabolic needs. Calcium removal ultimately results in bone loss and its associated malignancies such as osteoporosis (a decreased bone mass that increases the risk of bone fractures), hypercalcemia (unusually high concentrations of calcium in the serum), and Paget's disease (osteitis deformans). Most used among these types of drugs are Fosamax (alendronate), E-Vista (raloxifene HCl), and Actonel (residronate) (Table 1.16). It is clear that the number of prescriptions for each of these drugs has remained about the same in the United States over the 2003–2005 period with the exception of Miacalcin whose prescriptions were reduced below the radar screen among the top 200 drugs. Other biophosphonates that are not as equally popular, based on recent prescription data, include pamidronate, etidronate, and tiludronate:

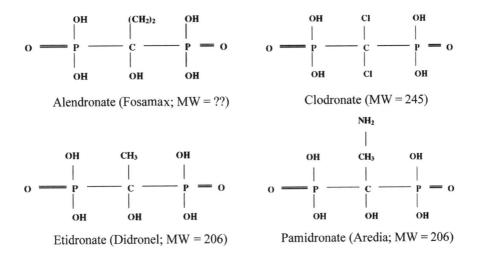

Alendronate (Fosamax; MW = ??)   Clodronate (MW = 245)

Etidronate (Didronel; MW = 206)   Pamidronate (Aredia; MW = 206)

In general, biophosphonates have a low bioavailability in the body and can actually persist in the body for months after therapy. For Actonel, radiolabeling and assaying in experimental animals (rats and dogs) dosed intravenously show that approximately 60% of the drug is distributed into bones with the rest excreted in urine and feces (U.S. FDA, 2006). Additional studies also show that of the fraction that is absorbed, more than 85% is ultimately excreted in urine over several days. Similarly, radiolabeling showed that approximately 50% of an intravenous dose of Fosamax is excreted in the urine within 72 h but also terminal half-life in humans can be as long as 10 years (Merck, 2005). This latter observation suggests continued release of the drug from the skeleton. Miacalcin (calcitonin) is not a biophosphonate but rather a polypeptide hormone that is secreted by parafollicular cells of the thyroid gland in mammals and by the ultimobranchial glands of fish and birds. Its synthetic form has 32 amino acids (Novartis, 2002).

Irrespective of the class, the anti-bone-loss drugs reduce the number of osteoclasts or simply block the ability of osteoclasts to attach to bones. These actions result in a slowing of bone resorption, allowing the retention of calcium in the bones. Just like Miacalcin, Evista, which is the second most popular drug for combating bone reabsorption, is also not a biosphosphonate but rather classified as a selective estrogen receptor modulator (SERM). By distinction, it is also neither an estrogen nor a hormone. Evista exerts its effect by binding to the estrogen receptor in a manner that is different from that of estrogens themselves in selective tissues. It interacts with such receptors, causing an estrogen-like increase in the accumulation of calcium in the bones, an effect that is similar to that realized with biophosphonates.

1.2.12 Steroids

Steroids are broadly characterized by a carbon skeleton with four fused rings. There are several kinds of steroids among which are cholesterol, which was discussed under the antihypertension section (Section 1.2.1), as well as estrogen, progesterone, and testosterone, discussed under the reproductive medications (Section 1.2.10). Of primary focus in this section are other steroids used for various therapeutic purposes to primarily treat endocrine and sensory system deficiencies. Steroid drugs act by primarily turning genes on or off as they enter target cells and bind to specific receptors in the cytoplasm. Prednisone is one of the most widely prescribed steroids for the endocrine system (Table 1.17). For this and other corticosteroids, following such binding on the receptors, the complex enters the nucleus and binds to selected DNA sites known as glucorticoid response elements (GREs) inhibiting the transcription of specific mRNAs. Such steroids are administered to counter cases of adrenal insufficiency in individuals who are unable to produce normal amounts of glucorticoids. Its use weakens the immune system (i.e., immunosuppressant), and to that effect it is also commonly used in autoimmune diseases, inflammation (including inflammation of bronchioles typical with asthma and GIT

TABLE 1.17 Most Prescribed Steroidal, Hematologic, and Nutritional Pharmaceuticals within the United States (2003–2005)[a]

Product	Total Number (in millions) of Steroid, Hematology, and Nutritional Prescriptions			Category and Mode of Action
	2003	2004	2005	
Steroids				
Prednisone (Deltasone)	18.4 (0.9)	22.5 (0.8)	23.0 (1.0)	An endocrine system glucocorticoid.
Xalatan (Latanoprost)	6.9 (0.3)	7.1 (0.3)	6.9 (0.3)	A sensory system steroid.
Alphagan (Brimonidine)	3.0 (0.1)	b	b	A sensory system steroid.
Proscar (Propecia)	2.8 (0.1)	b	b	An anabolic steroid that acts as a 5-α-reductase inhibitor responsible for the biotransformation of testosterone to dihydrotestosterone in the skin and prostate gland.
Hematologic				
Coumadin (Warfarin)	20.1 (0.9)	24.3 (0.9)	22.6 (1.0)	An anticoagulant that acts by preventing the synthesis of active vitamin K, a necessary cofactor for synthesizing active clotting factors.
Ferrous sulfate	b	4.2 (0.1)	b	Provides iron, an essential nutrient in the transport of oxygen, the formation of bone marrow, and an important component of the liver and spleen.
Nutritional				
Folic acid	4.7 (0.2)	8.6 (0.3)	8.5 (0.4)	Essential for the formation of red blood cells in the bone marrow as well as the synthesis of purines and pyramidines, the two bases that are the building blocks of DNA and RNA.
Phentermine (Adipex-P)	b	b	3.2 (0.1)	An anorexiant that suppresses appetite by acting on the hypothalamus, leading to an increased release of catecholamines, ultimately reducing the appetite.

[a]The numbers in parentheses are percent number of the respective prescription as a fraction of the total top 200 prescriptions in the United States in the respective year.
[b]Not among the top 200 most prescribed in the United States at that time.

73

inflammations), and in transplant patients to avoid rejection of the transplanted tissue or organs. Prednisone is also used orally or topically for dermatological disorders.

Both Xalatan and Alphagan have also been substantially used, although the usage of the latter was not among the top 200 most widely prescribed medications in the United States in 2004 and 2005 (Table 1.17). Both of these steroids directly affect the sensory system and are specifically used in optometry. Alphagan is an α_2-adrenegic agonist that decreases the production of aqueous humor in the eye, subsequently decreasing intraocular pressure. High intraocular pressure (i.e., >20 mm Hg) signifies a high risk of developing glaucoma, which is a common cause of visual impairment and blindness. The high intraocular pressure develops as a result of the degeneration of the anterior chamber in the eye, reducing the uptake of aqueous humor back into Schlemm's canal and causing stagnation of this liquid in the anterior chamber. Such elevated pressure in this region can ultimately damage the optic nerve. Remedial actions aim at decreasing the production of the fluid (e.g., with Alphagan) and/or improving its outflow by relieving the inflammation (e.g., with Xalatan). Usage of these medications has been predominant in 2003–2005 as there is an estimated 3 million cases of glaucoma in the United States (Queener and Gutierrez, 2003).

Propecia is another steroid that has been prescribed widely, at least in 2003. It is an anabolic steroid designed to inhibit the biotransformation of testosterone to dihydro-testosterone in target tissues such as the skin and prostate gland. This inhibition reduces the stimulation of the target sites, preventing overgrowth and enlargement of the prostate. Propecia has also found some use as a hair regrowth drug to guard against baldness, a purpose for which individuals have to apply it continuously. This application introduces it directly into the water (sullage) as compared to oral applications for prostate cancer. Under oral application, it is biotransformed by hepatic metabolism to less active metabolites. By their nature, 5α-reductase drugs such as Propecia are especially potent to the developing (male) fetus as the dihydro-testosterone that they inhibit is necessary for the normal development of male gene-talia. Thus, exposure of this and similar drugs to pregnant women can directly pose a high risk to the progeny.

1.2.13 Hematologic Drugs

The most prescribed drugs for blood-related disorders are also summarized in Table 1.17. Blood clots can occur in the veins at extremities (especially after a long period of immobility). Those clots can subsequently become lodged in the blood vessels, the atria, the heart valves, and within the lungs, causing embolism and shortness of breath. Under severe conditions, the clots can also obstruct the flow of blood to the brain, leading to a stroke and paralysis. Coumadin (warfarin; see chemical structure below) is one of the main hemato-logical compounds that has been widely prescribed to counter these conditions

(Table 1.17). It has a 100% bioavailability, with more than 90% excreted as the parent compound.

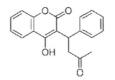

Chemical Structure of Warfarin (MW = 308.3; CAS 81-81-2)

Twenty to 24 million prescriptions of Coumadin were dispensed in the United States during 2003–2005. Coumadin is an anticoagulant that acts by preventing the synthesis of active vitamin K, a necessary cofactor for synthesizing active clotting factors. Thus, preventing the synthesis of active vitamin K indirectly inhibits the formation of active clotting factors, notably factor II (prothrombin), VII, IX, and X in the liver. Ironically, coumarins, of which Coumadin is a member, are also used as rodenticides.

Similarly, iron is an essential mineral in hemoglobin that is required to transport oxygen. It is also an essential nutrient in bone marrow, liver, and spleen. Iron deficiency, for example, in cases of anemia, is therefore routinely supplemented with ferrous sulfate or a variety of other iron-based compounds. If its supplementation is not urgent, it can also be replenished by adapting iron-rich diets. Prescribed iron supplements feature in the top 200 most prescribed drugs in the United States only in 2004 (Table 1.17).

1.2.14 Nutritional Drugs

Only folic acid (vitamin B_9, folate) and phentermine (Adipex-P) were the major nutritional drugs that were extensively prescribed during 2003–2005 in the United States (Table 1.17). The former is essential for the formation of red blood cells in the bone marrow and also for the synthesis of purines and pyramidines, the two bases that are the building blocks of DNA and RNA. Phentermine is an antiobesity drug and has been increasingly prescribed to meet the needs of an increasingly obese population. Obesity in developed countries has, at least in part, been attributed to the continuous shift from a manual-labor-dominated to a more sedentary life-style. In a majority of cases, obesity is also compounded by the improper balance between caloric intake and caloric needs as diets have steadily shifted in the last century from carbohydrates to those that are mostly high in fats. This contention explains the nutritional nature of obesity problems, although there are indications that some obesity issues are due to a genetic predisposition. Phentermine, the leading prescription to address problems of obesity, is an anorexiant that suppresses appetite. The drug seems to work by acting on the hypothalamus, leading to an increased release of catecholamines, ultimately reducing the appetite. Other drugs in this same anorexiant category include Didrex, Sanorex (mazindol), Bontril (phendimetrazine), Meridia (sibutramine), and

Depletite (diethylpropion), but these have not been prescribed to the same extent in recent years.

1.2.15   Triptans

Triptans are commonly used to relieve pain particularly due to extreme headaches—including migraines. Triptans stimulate vascular smooth muscle by acting on serotonin 1B and 1D receptors, leading to constriction of the blood vessels surrounding the brain. This sequence of events is thought to reduce the release of neuropeptides from the sensory neurons. Several triptans are on the market, including rizatriptan, sumatriptan (Imitrex), frovatriptan, and almotriptan. Among this category of medications, only sumatriptan was prescribed most often in the United States with its prescription contributing only 0.2% of the total prescriptions in these popular drugs. This amounts to about 4.5 million, 5.8 million, and 4.6 million prescriptions dispensed in 2003, 2004, and 2005, respectively. Oral doses are known to act against migraine pain within about 10–30 min, whereas nasal sprays seem to have an even faster effect. These faster response statistics may partly explain its highly preferred status for these kinds of pain where instant relief is the desired outcome.

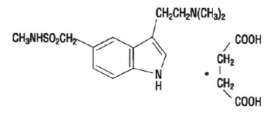

Chemical Structure of Sumatriptan (MW = 295.4)

Studies in rabbits and mice showed that oral and intravenous doses of 100 mg/kg/day and 2 mg/kg/day of sumatriptan, respectively, to be lethal to embryos, whereas dosages as low as 2 mg sumatriptan/kg/day over a 1-month period led to corneal opacities and defects in the cornea epithelium in dogs (GSK, 2006). No controlled studies have been conducted to investigate this aspect in humans.

1.2.16   Anesthetics

Anesthetics block the capacity of sensory transmission impacting the perception and response to stimulation due to pain. As a matter of fact, anesthesia literally means "without sensation." Anesthetics generally effect their action by reversibly blocking the transfer of nerve fiber membrane-based electrical impulse voltage channels to the brain. Some of them are gaseous, whereas most of them are liquids or volatile liquids. The gaseous and volatile types are administered by inhalation and are mainly excreted through the lungs. The intravenously applied opiod, Duragesic (fentanyl; Sublimaze), is the only anesthetic reported among the top 200 most prescribed drugs in the United States. Its prescription contributed only about 0.2% each year among the whole range of popular prescription drugs in that country. Other anesthetics that are not used to the same extent include ketamine (which is chemically related to the illicit hallucinogen phencyclidine—PCP) and the alkyl phenol propofol, all of which are highly lipophilic. Propofol is an anesthetic that is frequently used in a number of other countries. It is metabolized within a few hours to 4-hydroxyprofol or glucuronated, but the extent to which this occurs greatly varies between individuals (Court et al., 2001). Approximately 90% is excreted unchanged (Guitton et al., 1997).

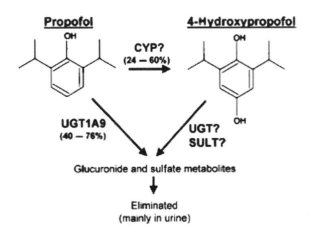

The preceding structure shows the biotransformation pathway for propofol in humans (adapted from Court et al., 2001).

1.2.17 Antineoplast and Immunosuppressants

Antineoplasts are also referred to as cytostatic agents. They are primarily used in cancer therapy. Many of them are carcinogenic, teratogenic, and mutagenic. Some of the common antineoplasts and immunospressants include ifosamide, cyclophosphamide, epirubium, bleomycin, 5-fluorouracil, mitoxantron, carboplatinum, and cis-platinum. They are clearly diverse in composition, molecular weight, and structure (see structures below). Their carcinogenicity, mutagenicity, and terogenicity have been documented (Skov et al., 1990; Hirose et al., 2005).

Antineoplasts (anticancer drugs) such as cyclophosphamide have been widely used since the 1950s for breast cancer, ovarian cancer, lung cancer, rheumatoid arthritis, and malignant lymphoma. These drugs are cytostatic, interacting with cell proliferation. In general, antineoplasts inhibit enzymes and prevent the action of enzymes such as DNA polymerase that are required for DNA synthesis. Others are very chemically similar to the bases that comprise DNA (i.e., purines and pyrimidines), and, when they are incorporated in the DNA during its synthesis, these decoys make the DNA unstable and nonfunctional. Approximately 3.8–4 million prescriptions of methotrexate (Folex), the leading antineoplast, have been dispensed in the United States each year during 2003–2005. It is antagonistic to folic acid, specifi-cally inhibiting the folate dehydrogenase enzyme, which is instrumental in synthesiz-ing the two DNA bases—purines and pyrimidines. This occurrence ultimately terminates the synthesis of DNA without causing any harm to other cells that are not trying to make DNA. To be effective, such cell-cycle-specific drugs have to be taken in repeated doses so as to knock out the dividing (cancerous) cells at the most sensitive stage. Methotrexate also has immunosuppressive actions. Other antineoplastic drugs include cytarabine (Cytosar), fluorouracil (Adrucil), paclitaxel (Taxol), and mercaptopurine (Purinethol):

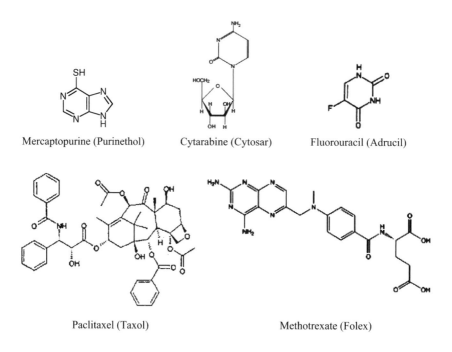

Mercaptopurine (Purinethol)   Cytarabine (Cytosar)   Fluorouracil (Adrucil)

Paclitaxel (Taxol)   Methotrexate (Folex)

It goes without saying that many of these drugs have a high mutagenic and carcinogenic potential that can impact not only mammalian cells but also those of other organisms in the environment, enhancing mutation frequencies.

1.3 CONCLUSION

This chapter has mainly focused on the categories of compounds used to address various medical conditions, their mode of action, and side effects. Until recently, research on the use of PPCPs has mainly focused on their beneficial and adverse effects on the end users—both humans and livestock. The mode of action of these compounds in these target organisms has mostly been well studied, with little regard to whether similar targets exist in lower vertebrates and invertebrates.

Furthermore, the known side effects of these compounds on the target organisms have not been considered to extend to nontarget organisms. It is very obvious that PPCPs are indispensable and their general use is essential to our well-being. It is also apparent that we cannot continue to ignore their presence in the environment without making a full assessment of their effects as some of them have similar targets in other organisms. To that effect, a number of adverse effects that are attributable to PPCPs are increasingly reported, particularly in the so-called lower organisms such as bacteria, fungi, invertebrates, plants, fish, birds, and amphibians, which reportedly have limited sentience. Some of these effects will be extensively examined in Chapter 4.

To address the issues of PPCPs in the environment, we need a profound understanding of the process of how they get into the environment and characterize how the ecosystem that receives those emissions is impacted. We have blindly embraced every PPCP that comes on the market without seriously considering its direct and indirect effects on the ecosystem. We need a better understanding of the risks associated with the use of PPCPs and possibly minimize or even substitute the most risky ones with alternative cousins that may have co-equal or even better efficacy but with less potency in the environment. To attain that, the medical/health care community has to join the growing list of multidisciplinary stakeholders to address issues of PPCPs in the environment. The occurrence of PPCPs and their persistence, impacts, and remedial actions are discussed in subsequent chapters.

2

DETECTION AND OCCURRENCE OF PPCPs IN THE ENVIRONMENT

Pharmaceutical and personal care products have been characterized as "new" or "emerging" contaminants in the environment. However, PPCPs have been around for several decades now. To that effect, a more accurate characterization is the fact that our attention to their presence in the environment is new or just emerging. Interest in their presence in the environment is also directly or indirectly stimulated by the fact that they are produced in increasingly large quantities. Furthermore, their use and diversity is also steadily increasing every year. Current accurate statistics about what proportions of PPCPs enter the environment through these numerous channels in various countries are not readily available. However, it appears that most of the introductions into the environment are directly or indirectly through two main routes, that is, disposal and excretion. A recent survey by Bound and Voulvoulis (2005) in the United Kingdom revealed that almost everyone interviewed (98%) had some type of pharmaceutical compound in their household, with only half of the respondents routinely finishing their dosage, leaving none to be disposed. Pharmaceuticals are disposed of for various reasons including discontinuation due to intolerance to their side effects, changes in dosages, and the medications reaching their expiration date. In most instances, we have been trained to get rid of expired or unwanted medications by "disposing of them in a manner that children cannot get access to them." In practical terms, this usually involves flushing them down the toilet or putting them into the household trash. In the former instance, they end up in the sewer, whereas in the latter instance, they end up in the landfill. This form of disposal is also codified in medical institutions, including hospitals, nursing homes, and pharmacies where unwanted drugs have to be destroyed in the presence of at least two witnesses so as to avoid misuse and abuse. The requirement of that codified medical disposal procedure at institutions usually necessitates flushing such drugs down the drain too. Other typical pathways through which PPCPs enter into the environment are summarized in Figure 2.1 and will be examined in various sections of this chapter. It is important to note that PPCPs that end up in the environment are primarily from both veterinary and human therapeutic uses. As alluded to in Chapter 1, the compounds from veterinary source can be a result of therapeutic or subtherapeutic purposes. Other compounds, particularly personal care products, also enter the environment through routine practices such as showers, swimming, and conducting laundry operations.

Under regular therapeutic and subtherapeutic use in humans and livestock, some compounds are metabolized in the body before they are excreted. Drug metabolism will be discussed in Chapter 3, but where excretion is the mode of entry into the environment, the excreted metabolites are often not well characterized. From the discussion in Chapter 1, it is apparent that a good fraction of the respective medical compounds that are administered orally or intravenously can be excreted in their parent form, ending up in the sewer system. Excretion as a driver of the occurrence of PPCPs in the environment will be discussed further in Section 2.3. Personal care products are mostly applied topically but several pharmaceuticals (e.g., antimicrobial creams) are also used topically. Topically applied PPCPs can be introduced directly into the environment through routine practices such as swimming and bathing, substantially contributing to PPCP loads into the environment. Topically applied PPCPs can be absorbed into the body and later excreted. To that effect, a variety of PPCPs have been detected in streams, rivers, sediments, dust, manure, biosolids, and soils as detailed in Section 2.1. Biosolids refer to solid, semisolid, and liquid residues generated during the treatment of domestic sewage in treatment works. Thus, PPCPs have been used in some instances as tracers and indicators of human activity and impact on the environment (Möller et al., 2000; Glassmeyer et al., 2005).

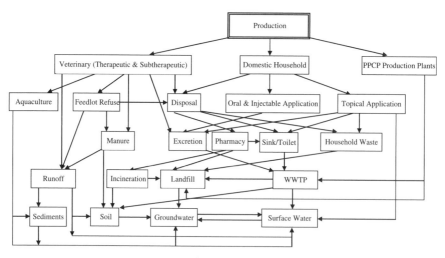

Figure 2.1

The statistics about actual usage and discharge of PPCPs in the environment are rare or at most murky due to privacy issues and competitive marketing tactics. However, as discussed in Chapter 1, prescription trends show that usage greatly varies not only between countries but also, in some instances, between different regions of the same country. Thus, what PPCP is readily detectable in one locality or country may not be as good a marker (or surrogate) across the board as it may not be as widely prescribed in another country. For example, clofibrate is widely used in several European countries and its metabolite, clofibric acid, is very persistent in the aquatic environment (Zweiner and Frimmel, 2003). Clofibric acid is thus frequently detected in European waters, serving as a good surrogate in that region. By contrast, clofibrate is not as popular in the United States and does not appear among the most prescribed drugs highlighted in Chapter 1. Thus, its use as a tracer for others or indicator of human activity in U.S. waterways would not be justifiable. Prescription volumes can be used to select candidate drugs to use as tracers for environmental studies. Using this argument, the most prescribed pharmaceutical compounds are more likely to be frequently encountered in the environment. By the same argument, the most sold (in terms of volume) personal care products are more likely to serve as reliable markers of other personal care products in the environment. Furthermore, selection of a compound to study or monitor in the environment is guided by the perquisite that the compound be detectable in wastewater effluents as the effluents are often diluted severalfolds after discharge. As a guideline, Sedlak et al. (2004) recommend focusing on compounds that have relative concentrations that far exceed their limits of quantification (LOQ) in the environmental matrix under consideration, that is,

$$DR_{median} = \frac{[Effluent]_{median}}{[LOQ]}$$

where $[Effluent]_{median}$ is the median concentration in the effluent from full-scale municipal waste treatment and DR_{median} is the median detection ratio. Following these guidelines and sometimes instinct, various research groups used analytical methods that have revealed a plethora of chemicals in the environment. Such findings have raised the awareness of PPCPs in the environment including streams, lakes, rivers, biosolids, and estuaries. However, what is often not highlighted from these findings is the reality that even though the current methods can enable one to find what one is looking for, the chemicals that are not targeted in the analysis can elude detection. Section 2.1 is devoted to the detection of PPCPs in the environment so as to give us a better understanding of the process and how the concentrations that have been published are determined.

2.1 DETECTION OF PPCPs IN THE ENVIRONMENT

Determining the occurrence of a chemical in the environment should be viewed as an integral process that involves several steps including sampling, storage, sample preparation, and ultimately detection of the chemical(s) of interest. Each of these steps has profound effects on the whole analytical and detection process of PPCPs in the environment. Thus, sampling has to be conducted in a manner that ensures that the sample is representative and homogenous. This may involve taking the sample at different times or from different locations within the same general locale. A decision has to be made as to whether one will aim at obtaining a single or composite sample. Whatever sampling procedure is adapted has to be justifiable and minimize problems such as degradation or sorption of the compound(s) of interest onto sampling equipment. Oftentimes, it is not possible to analyze the sample instantly, and, indeed in most instances, there is a lag period (hours to several days) between sample collection and analysis. Thus, once obtained, the sample has to be stored in a manner that maintains its integrity. Thus, the container it is stored in has to be clean so as to avoid introduction of extraneous material and must be of sound composition so as to avoid unwanted reactions of the analyte within the container. For example, some analytes can sorb on plastic surfaces, and collecting the sample in such a container will hamper recovery of the analyte right from the onset. Similarly, some analytes can be photosensitive, raising the need to collect and store the sample in an opaque container, shielding the sample from direct light. It is imperative that some losses may occur if the sample is not stored at low or even subzero temperatures so as to slow down degradation processes.

Environmental matrices are quite diverse, making sample preparation equally important. The general procedure for analyzing PPCPs from solid and liquid matrices are provided in a schematic (Fig. 2.2). Preparation steps specifically involve extracting the analyte from different matrices, removing interfering substances (i.e., cleanup) and in some instances concentrating the sample. Extraction is a critical part of the analytical process as, considering the low concentrations in which PPCPs typically occur, interference from other compounds (both organic and inorganic) can limit recovery unless sophisticated cleanup processes are adapted at the onset. Matrices such as wastewater, biosolids, and soils usually contain high levels of organic material and suspended solids. The initial preparation of sample from such matrices usually involves filtering or centrifuging the sample or conducting a more rigorous extraction process such as ultrasonic microwave-assisted or accelerated solvent extraction. Accelerated solvent extraction was used by Golet et al. (2002b) with a 50 mM phosphoric acid–acetonitrile (1:1) mixture at 100°C and 100 bar for 1–1.5 h. The solvent used during this process depends on the compounds of interest. It may also be necessary to include a filtration step to remove suspended material. Liquid samples also typically have to undergo some pretreatment such as filtration. Filtration is usually achieved using glass filters of pore size 0.22–1.2 μm.

To enhance recovery of the analyte, a typical approach is to acidify the environmental sample (pH $\approx$ 2) prior to extraction. However, that process can greatly reduce the extractability of compounds that have a high pK_a value (i.e., $pK_a \geq 8$). For example, the extraction of sulfanilamide ($pK_a = 10.43$; MW = 172.21) from spiked water on acidification was reduced to a meager 32% in a mixture containing sulfadiazine ($pK_a = 6.4$; MW = 250.28), sulfathiazole ($pK_a = 7.2$; MW = 255.32), sulfamerazine ($pK_a = 7$; MW = 322.43), sulfamethazine ($pK_a = 7.4$; MW = 278.33), sulfamethoxypyridazine ($pK_a = 6.7$; MW = 280.31), sulfadimethoxine ($pK_a = 6.2$; MW = 310.33), and sulfaquinoxaline ($pK_a = 5.5$; MW = 300.34)

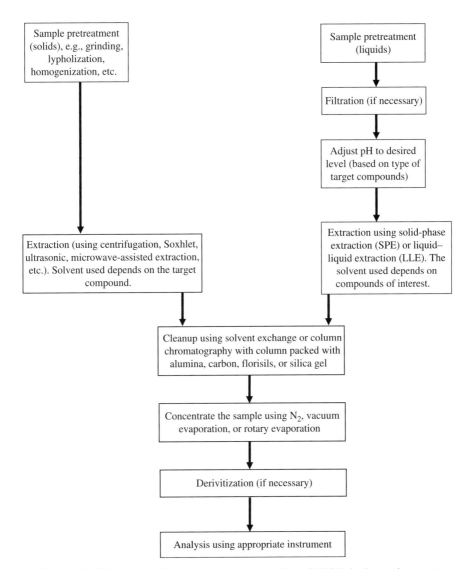

Figure 2.2 Schematic of the preparation and detection of PPCPs in the environment.

(Malintan and Mohd, 2006). The amphoteric nature of these compounds and the wide range of pK_a values displayed by these sulfonamides explain why collectively extracting this group can be challenging. The conditions that allow wholesale extraction of some groups of compounds may not be easily attainable. Thus, a buffer is typically used as the mobile phase for such compounds using high-performance liquid chromatography (HPLC), although this strategy has apparent disadvantages as the salts in the buffer can crystallize within the column over time, reducing its separation capacity (Malintan and Mohd, 2006). Extracting the compounds under alkaline conditions enhanced the recovery of sulfanilamide but reduced the recoveries of all the other sulfonamides. The fact that the extraction of one of these sulfonamides is only optimized at a totally different pH displays the compilations in relying on a single set of extraction conditions as representative for all members within each class of PPCPs. To "trap" acid components in the sample, the sample has to be acidified to $pH < 2$, passed through a conditioned column such as an RP C18 solid-phase extraction column, and then eluted with a volatile solvent. Neutral compounds are, on the other hand, extracted by adjusting the sample to pH 7–8 before running the sample through the extraction column, which has been conditioned with acetone, methanol, or distilled water. Basic compounds are extracted by initially adjusting the sample to $pH \geq 12$ with EDTA and KOH.

The compounds can then be cleaned using solvent exchange or column chromatography. With the latter, a cartridge such as octadecyl (C18) bonded silica, alumina, or florisil is used. Other commercially available cartridges used for this purpose include SDB-XC disks, Amberlite XAD2, and Isolut ENV+. Obviously, the choice of cartridge depends on a number of factors including the nature of the matrix, compound of interest, and price of the cartridges. On the other hand, use of HPLC–MS (mass spectrometry) often requires cleaning up the extracted sample to minimize interference from organic matter. Natural organic matter in the concentrated extracts can interfere with the detection of PPCPs (and other organic contaminants) in environmental samples. If not cleaned out of the sample, natural organic matter can produce interfering peaks. By comparison, LC–MS–MS and gas chromatography (GC)–MS–MS are less susceptible to interference from natural organic matter. In some instances, it may be necessary to concentrate the sample through evaporation so as to improve sensitivity and detection of the compound of interest. Concentration is achieved by drying with N2, evaporating under a vacuum, or by rotary evaporation. When GC–MS or GC–MS–MS is used, derivatization of the compound is often necessary before injecting the sample into the GC. Derivatizing improves the stability of the compounds and the sensitivity as well as precision of the GC–MS and GC–MS–MS. Derivatizing involves resuspending the residues of the concentrated sample into a solvent such as pentafluorobenzyl bromide, heptafluorobutyric anhydride, and triethylamine (Blau and Halket, 1993; Lunn and Hellwig, 1998; Ingerslev and Halling-Sørensen, 2003; Lerch and Zinn, 2003). The sample is then evaporated to dryness and redissolved in a volatile solvent such as propanol. The volatile mixture is then injected into the GC for analysis. Derivatization also works effectively for compounds that are thermolabile, although it is prone to considerable interference from organic matter from environmental matrices such as soils, wastewater, and sludge. However, GC-based methods still mostly have detection limits that are still higher than the concentrations present in matrices such as water.

Finally, actual detection of the analyte has to be done by choosing a cost-effective and reliable method. In general, the functional groups of many of the PPCPs can render their extraction from water and measurement with GC and HPLC difficult. Furthermore, the environmental matrices such as soil and sediments also impose significant interference in the analysis of these compounds. To that effect, no single method or piece of equipment can measure all the compounds of concern. Individual laboratories typically have to optimize the methods.

2.1.1 Detection Using Instrumentation

The detection methods have to be sensitive. Drug metabolites often have a high polarity of the conjugates relative to the parent compound (Webb et al., 2005), and investigations about the occurrence and fate of PPCPs in the environment have been greatly driven by advances in the analysis of residues, particularly resulting from major advances in chemical analytical methods that are able to detect more polar compounds even at trace concentrations without derivatization. The instrument-based chemical analytical methods are based on the principles of chromatography—primarily GC or LC. With gas chromatography, the compounds to be analyzed are vaporized and eluted in a stream of gas (the mobile phase) through a column. The mobile phase acts only as a carrier gas and does not have any significant interaction with the analytes. The analytes, which are typically volatile, are normally dissolved in a liquid solvent and principally become separated as they partition between the liquid stationary phase and the gaseous mobile phase. For enhanced analysis and detection, various studies have strongly recommended that GC be combined with mass spectrometry (GC–MS) so as to exploit the benefits of a powerful separation technique and the structural information provided by mass spectrometry (Ingerslev and Halling-Sørensen, 2003; Xia et al., 2005). Selectivity and analytical power are even more enhanced when GC is combined with a tandem mass spectrometry (i.e., GC–MS–MS), popularly referred to as triple quadrupole. In that instance, the first spectrometer isolates precursor ions, which undergo further fragmentation, yielding product ions and neutral fragments. The product ions are then further analyzed by the second spectrophotometer. This strategy significantly reduces the interference from the matrix.

The effect of any interference can be elucidated by including some internal standards (Koester et al., 2003). Whether with LC–MS–MS discussed below or GC–MS–MS techniques, tandem mass spectrometry is often used to control interference from organic matter that is present in environmental samples. GC–MS may not be very suitable for compounds that have a low volatility. To improve on its detection, the samples have to be derivatized. However, this increases the sample preparation time and can limit the number of samples to assay. By comparison, LC–MS is less labor intensive compared to GC–MS since no derivatization is necessary. It is also quite specific and sensitive. High-performance liquid chromatography is performed by separating the analytes in a liquid medium through a chromatographic column. HPLC is advantageous compared to GC as glucuronic and sulfuric conjugates can be detected without derivertizing the sample. Similar to GC, selectivity of the analytes is also enhanced if LC is combined with MS or MS–MS. The equipment for LC–MS is more expensive than GC–MS, and its usage can be limited particularly in less developed countries. Other techniques used include HPLC coupled with fluorescent detection (Golet et al., 2001, 2002a, 2002b) and immunochemical analysis (Huang and Sedlak, 2001).

2.1.2 Detection Using Bioassays

It is important to note that pollutants rarely affect a single factor or parameter. To that effect, chemical-based criteria by themselves cannot sufficiently predict the ecological intensity and biological end points. This realization has led to the development of biomarkers as economical rapid early warning indicators of environmental contamination. The issue of biological end points and biomarkers will be explored further when we discuss ecotoxicity (Chapter 4), but in the present chapter we focus the discussion on the general aspects of bioassay analyses. They are conducted in vivo or in vitro and are typically short-term screening tests that evaluate the ability of a chemical to elicit biological activity. They involve the determination of the presence and/or strength of a substance by comparing its effect on a test organism versus that of a standard preparation. They rely on biological systems, including whole organisms (plants, animals, microorganisms), individual cell lines, or biological processes. Since PPCPs can occur in the environment in mixtures and/or together with other contaminants, it may be difficult to pinpoint the observed effects on whole organisms or individual cell lines and attribute those effects on a particular PPCP. However, the use of bioassays in this regard can be quite beneficial under controlled conditions (e.g., under laboratory conditions) where a known compound and possibly its concentration are known. For example, Microtox and other indicator organisms such as *Daphnia* spp. have been used to detect the presence and effects of various contaminants, including PPCPs. Lower detection limits are increasingly being reported using bioassay-based methods, and bioassays are in some instances preferred compared to classical chemical analytical methods because they are fast, easy, convenient, cost effective, highly sensitive, and robust.

Increasingly used for detecting and/or investigating the effects of PPCPs in the environment is the immunoassay technique (Aga and Thurman, 1997; Aga et al., 2003). Immunoassays involve the use of antibodies that recognize specific interactions with a homologous antigen (the analyte). They are based on the capability of antibodies to specifically recognize and form stable complexes with antigens. Immuno-based techniques benefit from the fact that the immune system is very complex and has an important feature of the ability to distinguish itself from foreign materials forming antibodies, that is, proteins that specifically and non-covalently bind with chemical molecules. Thus, various PPCPs evoke the body to generate antibodies that can be measured. Antibodies are the critical component in all immunoassays, significantly contributing to specificity and sensitivity. This concept has been used in medicine for several decades to study the pharmacokinetics and pharmacodynamics of various drugs (Laurie et al., 1989; McCann et al., 2002). Under these types of environments, immunoassay studies can give very good correlation with conventional instrument-based techniques (Fig. 2.3). Immunoassay techniques have also been used extensively in environmental studies to detect herbicides in environmental samples (Brady et al., 1995; Hennion, 1998), with the assay deriving from the observation that the distribution of the analyte between the antibody–antigen (i.e., chemical) complex and free form is quantitatively or qualitatively related to the concentration of the analyte in that environment. Immunoassays are also becoming increasingly used in detecting PPCPs in the environment. In most instances, the antibodies are immobilized on a solid surface and detection is enhanced by labeling with fluorescence, radioisotope, enzymes, or chemiluminence. When based on enzymes, it is referred to as enzyme-linked immunosorbent assays

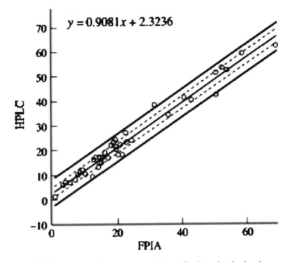

Figure 2.3 Relationship between the concentration of teicoplanin in the serum of patients as determined using high-performance liquid chromatography (HPLC) versus fluorescence polarization immunoassay (FPIA) ($r^2 = 0.974$). (From McCann et al., 2002.)

(ELISA). Sample preparation prior to immunoassay is usually minimal for water samples but is often needed for solid matrices such as soils and sediments. Such preparations can include some form of extraction followed by subsequent dilution before immunoanalysis. Diluting especially minimizes the effects of interference from the matrix. To extract the antibiotics from a set of hog manure samples, Kumar et al. (2004) added 40% H_2SO_4 and some disodium ethylenediaminetetracetate (Na_2EDTA). The manure slurry was then incubated on a rotary shaker for 4 h after which it was filtered using 0.7-μm glass fiber filters. It is important to note that in the ELISA analysis, the absorbance is inversely related to the concentration of the target compound.

Analysis using immunochemistry shows that tylosin was absent in one of the manure samples (sample 3), whereas low levels of this antibiotic were detected in sample 2 using ELISA but not detected with LC–MS. Slightly higher quantities of both antibiotics were generally also detected in the other manure samples using ELISA compared to LC–MS, in most instances suggesting that the immunoassay method was detecting some of the transformed products and partially decayed intermediates as well. Note in Table 2.1 that the ultratrace concentration of tylosin in manure 2 was detectable by ELISA but not by LC–MS, indicating that ELISA can have a low detection limit compared to LC–MS. Similarly, the concentrations of 17β-estradiol in secondary effluents that had undergone divergent treatment methods (i.e., activated sludge and chlorination versus trickling filters and UV disinfection) were in the nanogram per liter range, concentrations that were comparable to those of GC–MS–MS in wastewater matrices (Table 2.1). Kumar et al. (2004) obtained 100% recoveries of both tylosin and tetracycline when each of these antibiotics was spiked in lake waters, runoff samples, nanopure water, and soil saturation extracts. Furthermore, the tetracycline ELISA kit used in that study was highly specific for tetracycline and chlorotetracycline but not for any of the other forms of tetracyclines (i.e., oxtetracycline, demeclocycline, and doxycycline). During that research, recoveries of the steroid hormone 17β-estradiol from various wastewaters, the Colorado and Sacramento rivers, as well as an engineered wetland ranged between 51 and 117% using ELISA. Possible sources of losses during immunoassay analysis include the chemical adsorbing to glassware and/or colloidal material.

TABLE 2.1   Comparison of Concentrations of Various Pharmaceutical Compounds Using ELISA and LC–MS or GC–MS–MS

Compound	Matrix	Concentration		% Difference	Reference[a]
		ELISA	LC–MS or GC–MS–MS		
Chlortetracycline (μg/L)	Manure 1	7931	5230	34.1	Kumar et al. (2004)
	Manure 2	0	0	NA[b]	Kumar et al. (2004)
	Manure 3	5146	4310	16.2	Kumar et al. (2004)
	Manure 4	4698	3540	24.6	Kumar et al. (2004)
Tylosin (μg/L)	Manure 1	4032	3780	6.3	Kumar et al. (2004)
	Manure 2	6	0	100	Kumar et al. (2004)
	Manure 3	0	0	NA	Kumar et al. (2004)
	Manure 4	3304	3690	−11.7	Kumar et al. (2004)
17β-estradiol (E2; ng/L)	WWTP A (secondary effluent; activated sludge and chlorine disinfection)	3.68	3.9	0	Huang and Sedlak (2001)
	WWTP B (secondary effluent; trickling filter and UV disinfection)	0.2	0.27	0	Huang and Sedlak (2001)

[a]The determinations by Kumar et al. (2004) were determined using LC–MS, whereas those by Huang and Sedlak (2001) were determined using LC–MS–MS.
[b]NA: Not applicable.

Immunochemical approaches are cheaper, readily adaptable, rapid, portable, and reduce the need for expensive analytical equipment. They can also be used to simultaneously assay a large number of samples over a short period of time. One of the major factors that still limits the use of this technique in the detection of a wider range of PPCPs in the environment is the lack of suitable antibodies sensitive to most PPCPs that occur in the environment. Furthermore, immunoassay accuracy can be susceptible to cross reactions and other effects from the matrix, giving false positives in some instances (Huang and Sedlak, 2001). Thus, it is recommended that immunoassay analytical results be validated with GC- or LC-based methods.

On the whole, programs that are designed to monitor the presence of PPCPs in the environment still face several major challenges. For example, different compounds have different properties that normally require the use of different methods or sample preparation, pretreatment, and/or measurement conditions. Furthermore, the metabolites and transformation products may still be biologically active but not detectable using the same methods or treatment conditions as the parent compound. Decisions on which method to use, for example, instrumental versus immunoassay, can be guided by factors such as sensitivity, variability, selectivity, and price. Some general guidelines with these four factors in mind were compiled by Ingerslev and Halling-Sørensen (2003) and are presented in Figure 2.4. Although initially published specifically with regard to estrogenic compounds in the environment, they seem to be generally applicable to a variety of other PPCPs. Based on those guidelines, sensitivity is highest with the quadrupole, that is, GC–MS–MS and LC–MS–MS methods, but immunochemical techniques are also satisfactorily sensitive. Depending on the compound(s) of interest, sensitivity may fall below the threshold with GC–MS and LC–MS, and becomes even lower with GC–FID and LC–fluorescence. Sensitivity is least with LC–UV techniques. Based on variability as the primary criterion, immunochemical techniques can be most variable compared to all the chemical techniques. Immunochemical techniques may also fall below the threshold, particularly in complex environmental samples, which almost always contain more than one compound. However, they are cheap compared to the highly selective and less variable techniques. By comparison, the quadrupole techniques show slow variability, whereas GC–MS and LC–MS show variability that may not be acceptable in sewage effluents but may be acceptable in other aquatic matrices. Selectivity and cost are the other factors to consider. Selectivity problems with GC–MS and LC–MS can be severe but these can be manageable. By comparison, the selectivity problems with immunochemical techniques—GC–FID, GC–EC, LC–fluorescence, and LC–UV—can be severe and unmanageable. In terms of cost, immunochemical techniques trounce all of the others, making them ideal for massive screening projects that, because of high variability and cross reactivity, always have to be validated with chemical-based methods.

As with any analytical determination procedure, both quality assurance and quality control issues have to be built into the detection of PPCPs in the environment. For example, analysis of polycyclic musks (PCM) can be easily skewed during analysis as the compound might be present in laboratory soaps and creams, thus coating laboratory glassware used for the analysis. Therefore, programs that are designed to research

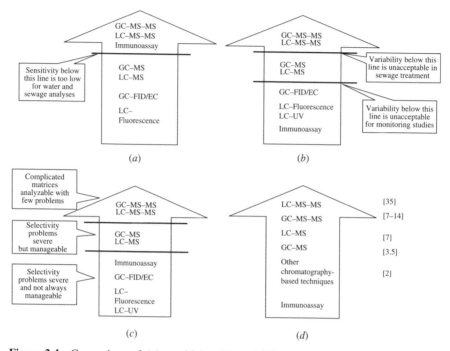

Figure 2.4   Comparison of (*a*) sensitivity, (*b*) variability, (*c*) selectivity, and (*d*) pricing between various chemical and immunological analyses for the presence of PPCPs in the environment. FID = flame ionization detector and EC = electrochemical detection. Note that GC–MS–MS can have mass detectors such as triple quadrupole and ion trap with ionization from EI = electron ionization or CI = chemical ionization, whereas LC–MS–MS with ionization from ESI = electrospray ionization, APCI = atmospheric pressure chemical ionization, or APPI = atmospheric pressure photoionization. (Adapted from Ingerslev and Halling-Sørensen, 2003.)

or monitor these compounds in environmental samples have to maintain very rigorous quality control procedures so as not to inadvertently introduce PCM contaminants in the samples being analyzed. Appropriate controls with blank samples and spike samples that are processed in a similar fashion as the environmental samples using the same glassware, solvents, and filters have to be part of the quality control procedure.

2.2   OCCURRENCE OF PPCPs IN VARIOUS ENVIRONMENTS

More than 100 pharmaceutically active compounds of various prescription classes have been detected in water, livestock-derived manure, sewage, sediments, potable water, biosolids, and soil. The compounds detected belong to a wide range of classes of PPCPs, including antibiotics, antiphlogistics, antiepileptics, beta blockers, lipid regulators, vasodilators, and sympathomimetics. Most of the evidence for the presence of these compounds in the environment has been accumulated in aquatic systems. Occurrence in different matrices is also discussed.

2.2.1 Aquatic Systems

Water is essential for life and covers approximately 70% of Earth's surface (Butcher et al., 1992). It is constantly used and reused in a cyclic fashion of drinking water–wastewater–receiving water. Such cycling exposes the water to various contaminants such as pathogens, pesticides, fertilizers, herbicides, radioactive waste, pharmaceuticals, as well as personal care products. Of primary focus in this book are PPCP contaminants in the water and other environments. The aquatic ecosystem is of particular importance because this is where most PPCPs, and other contaminants in general, are deposited. Such contamination has been documented worldwide (Kolpin et al., 2002; Ternes, 1998; Golet et al., 2001, 2002a, 2002b; Boyd et al., 2001, 2003; Ternes et al., 1999a; Atkinson et al., 2003). As is discernible from Figure 2.1, there are several points of entry of these compounds into the aquatic system. The excreted compounds as well as those that are disposed of in a variety of other forms end up in some water matrices including runoff (e.g., from farms), landfill leachate, and sewage treatment plants. Water is also by itself an important medium for transporting these and other pollutants in the environment. The frequency of occurrence of PPCPs and their concentrations also largely differ between different aquatic systems as will be evidenced in the discussion below.

2.2.1.1 *Presence of PPCPs in Wastewater*   Many wastewater treatment plants (WWTPs) still have a combined sewer that collects both domestic sewage and stormwater. Thus, during the rainy season, the volumes of water to be treated far exceed the treatment capacity, which leads to the discharge of untreated effluents. Similarly, many WWTPs use only primary treatment technologies, relying on the local hydrological conditions to dilute the sewage discharges to "safe" levels. Such dilutions have been somewhat studied with a focus on pathogens and conventional pollutants, with a lot less emphasis on PPCPs. Where conducted, secondary treatment to generate activated sludge reduces organic content through metabolism of the existing suspended solids, whereas tertiary treatment may be necessary to reduce nutrient loads. To ameliorate the levels of contamination, water treatment schemes have been developed by various entities that range from treatment at a single household level to entire municipalities. Measures to treat and remediate PPCPs from wastewater will be discussed in Chapter 5. The presence of pharmaceutical compounds in sewage and wastewater, particularly estrogenic hormones, was hypothesized in the early 1960s by Stumm-Zollinger and Fair (1965). However, at that time the techniques to analyze the presence of most of these compounds were not yet in place. An emerging pattern shows that PPCPs occur in all municipal treated sewage, particularly if the system does not have some of the more advanced treatment technologies such as activated carbon or some form of membrane filtration. As a matter of fact, more data on the occurrence of PPCPs in wastewater have been accumulated compared to other aquatic systems and the frequency of such reports is increasing. A few of the compounds typically reported and the concentrations encountered in wastewater are displayed in Figure 2.5. Based on this figure, it appears that the fragrance ingredients HHCB and AHTN seem to be the most frequently detected. In some instances,

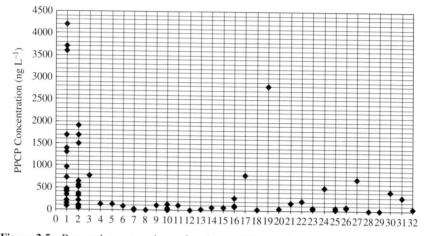

Figure 2.5 Reported concentrations of various PPCPs in Wastewater effluents by several research groups. On the x axis are respective PPCPs that are primarily cosmetics (1 = HHCB, 2 = AHTN, 3 = acetophenone, 4 = camphor, 5 = isoborneol, 6 = skatol, 7 = celestolide, i.e., AHMI, 8 = Phantolide, i.e., AHMI), the lotion ingredient (9 = methyl salicylate), two disinfectants (10 = triclosan and 11 = trilocarban), antihypertensive (12 = dehydronifedipine, 13 = diltiazem, 14 = bezafibrate, and 15 = gemfibrozil), analgesics and anti-inflammatories (16 = naproxen, 17 = ibuprofen, 18 = codeine), antimicrobials (19 = chlortetracycline, 20 = erythromycin, 21 = novobiocin, 22 = oxytetracycline, 23 = sulfamethaxazole, 24 = thiabendazole, 25 = trimethoprim), anxiolytic sedative (26 = carbamazepine), antidiabetic (27 = metaformin), reproductive (28 = 17β estradiol, 29 = 17α-ethinyl estradiol), GIT (30 = cimetidine, 31 = ranitidine), and respiratory (32 = Albuterol). The concentrations were compiled from Boyd et al. (2003), Gagné et al. (2006), Glassmeyer et al. (2005), Halden and Paull (2005), Huang and Sedlak (2001), Ricking et al. (2003), and Ternes et al. (2003).

HHCB and AHTN have been detected in concentrations as high as 4200 and 1900 ng L^{-1}, respectively. Fragrances are used topically, and after routine use they are released into sewage and wastewater, and they have been detected in influents and effluents, seawater, as well as in marine biota (Ricking et al., 2003; Berset et al., 2004). In surveys conducted in Canada and Sweden, the concentrations of HHCB, AHTN, ADBI, and AHMI in the effluents was moderately to strongly correlated with the population size served by the respective treatment plants ($r^2 = 0.796$, 0.822, 0.717, and 0.514, respectively). However, nitro musks (i.e., musk ketones and musk xylene) were not detected in any of the samples (Table 2.2) despite the fact that both have a low K_{ow} compared to polycyclic musk fragrances. The lower K_{ow} suggests that they are less likely to partition in the sediments and more likely to be in the liquid column. Similarly, among the polycyclic musk fragrances, Cashmeran (DPMI) and Transeolide (ATTI) were also not detected. The absence of these compounds in the water during that survey can be explained by several possibilities:

1. The method used was inadequate for their extraction.
2. They are not widely used in those areas.

TABLE 2.2   Polycylic Musk Fragrances and Nitro Musks Detected in Various Waters in Canada and Sweden

Location	Population Served	Type of Environment	Polycyclic Musk Fragrances (ng L^{-1})							Nitro Musk (ng L^{-1})	
			Galaxolide (HHCB)	Tonalide (AHTN)	Celestolide (ADBI)	Phentolide (AHMI)	Cashmerun (DPMI)	Traseolide (ATTI)		Musk Xylene (MX)	Musk Ketone (MK)
Sweden											
Enköping (Uppsula)	21,000	Household and industry effluents	336	90	7	4	<1	<1		<1	<1
Skene	17,280	Effluents of STP through trickling filter. Activated sludge with nitrogen removal combined with Al precipitation.	218	42	3	2	<1	<1		<1	<1
Gässlösa	79,000	Same type of effluent treatment as that at Skene.	423	104	6	5	<1	<1		<1	<1
Nolhaga		Same type of effluent treatment as that at Skene.	157	42	2	2	<1	<1		<1	<1
Ljusne	2,596	Effluents of activated sludge combined with Al precipitation.	407	77	8	3	<1	<1		<1	<1

(*Continued*)

TABLE 2.2 *Continued*

Location	Population Served	Type of Environment	Polycyclic Musk Fragrances (ng L^{-1})						Nitro Musk (ng L^{-1})	
			Galaxolide (HHCB)	Tonalide (AHTN)	Celestolide (ADBI)	Phentolide (AHMI)	Cashmerun (DPMI)	Traseolide (ATTI)	Musk Xylene (MX)	Musk Ketone (MK)
Canada										
Bedford (in city of Miramichi)	350,000	After secondary treatment	1300	520	19	6	<1	<1	<1	<1
Lancaster STP (St. John; New Brunswick)	90,000	After secondary treatment	205	110	4	2	<1	<1	<1	<1
Straw Marsh ATP (Nova Scotia)	20,000	Secondary conventional STP with preliminary treatment, settling and finally UV treatment	480	220	7	2	<1	<1	<1	<1

Source: Based on data from Ricking et al. (2003).

3. They are effectively removed by the treatment system in place.

4. They are quite degradable compared to HHCB, AHTN, ADBI, and AHMI.

Concentrations of HHCB and AHTN as well as two other fragrances (i.e., benzophenone and ethyl citrate) were detected in all of the wastewater effluents (i.e., 100% frequency) from 10 locations in the United States in a comprehensive study that detected 110 chemical analytes (Glassmeyer et al., 2005). Quite recently, both HHCB and AHTN have also been detected in Lake Michigan waters with HHCB occurring at higher concentrations than AHTN (Peck and Hornbuckle, 2004). These authors speculate that the high concentrations of HHCB in these waters is possible because this compound is manufactured and used in larger quantities compared to AHTN. As indicated in Chapter 1, current production of AHTN in the United States stands at more than 4500 tons/year (Kannan et al., 2005). Numerous reports cited by Balk and Ford (1999) also show the presence of these compounds in surface water, sludge, sediments, and in animal lipid tissue. Available data show occurrence to be greatly related to the vicinity of sewage treatment outflow. They are not readily degradable and can in fact form more polar metabolites, such as HHCB-lactone, which is even more persistent (Balk and Ford, 1999). Where elimination during sludge and wastewater treatment has been reported, such elimination is mostly due to sorption of the compound to the colloidal material that is present rather than due to degradation. Acetophenone, another fragrance, has been detected in wastewater at high concentrations of 780 ng L^{-1} but with less frequency. Transformation products of polycyclic musk fragrances have also been detected in the environment, which serves as evidence of some degradability. However, most of their disappearance may be due to sorption onto environmental matrices. Other personal care products such as the detergent metabolites nonylphenol and alkylphenol polyethoxylates have also been detected in wastewater at high concentrations (Ferguson et al., 2000; Montgomery-Brown and Reinhard, 2003). Nonylphenol is also a component of fire retardants. Also detected are EDTA (Bedworth and Sedlak, 1999; Nowack, 2002) and the antimicrobial compounds, triclosan and trilocarban (McAvoy et al., 2002). Triclosan was also detected with 100% frequency in the comprehensive study conducted by Glassmeyer et al. (2005) cited earlier.

Most of the other compounds, notably antihypertension drugs, analgesics, antimicrobial agents, sedatives, reproductive hormones, and respiratory drugs that have been detected in wastewater are in concentrations that are less than 200 ng L^{-1}. Notable exceptions to this generalization are the analgesic ibuprofen and the antibiotics chlortetracycline as well as thiabandazole (Fig. 2.5). Both 17β-estradiol and 17α-ethinyl-estradiol were detected in four wastewater system effluents at concentrations that ranged between 0.2–4.05 and 0.13–2.42 ng L^{-1}, respectively, over two consecutive years (Huang and Sedlak, 2001). Although they are at such low concentrations in wastewater, the presence of reproductive hormones is a source of concern (see Chapter 4) as these are known endocrine disruptors. Relatively high concentrations of the antidiabetic drug metformin (700 ng L^{-1}) and the GIT drugs cimetidine and ranitidine (Zantac) have also been reported, albeit, less frequently.

A wide range of antibiotics have been detected at low concentrations (micrograms/grams or micrograms/liter) in sewage (Hirsch et al., 1999; Kühn and Müller, 2000) and wastewater. Thus, Golet et al. (2001, 2002a) detected fluoroquinolone in tertiary wastewater effluents at concentrations of 249–405 ng ciprofloxacin/L and 45–120 ng norfloxacin/L. Similarly, median concentrations of 0.02 μg ciprofloxacin/L were detected in 26% of the streams in the United States surveyed by Kolpin et al. (2002). The concentrations in hospital and nursing home effluents tend to be higher (Kümmerer and Helmers, 2000; Brown, 2004) as such facilities are hotbeds for heavy usage of pharmaceuticals by virtue of their occupants. Typical concentrations of various antibiotics from two hospital facilities versus residential areas encountered in New Mexico are shown in Figure 2.6. Thus, effluents from these facilities can have antibiotic concentrations that are in the same order of magnitude as the minimum inhibitory concentrations (MICs) for sensitive bacterial pathogens. However, hospitals and long-term care facilities contribute an estimated volume of only about 1% of the municipal sewage in most areas (Kümmerer and Henninger, 2003).

Occurrence of some PPCPs in the environment is largely dependent on seasonality. For example, in temperate regions, some macrolides such as erythromycin, clarithromycin, and roxithromycin, which are commonly used for respiratory tract infections, are more prevalent in sewage effluents during winter (McArdell et al., 2003). This time also coincides with a high incidence of respiratory infections and therefore increased use of those drugs. On another note, despite a high usage of β-lactams such as cephalosporins and penicillins compared to any other class of antibiotics, this class has only occasionally been detected in the wastewater possibly

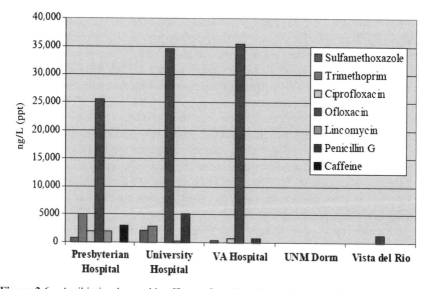

Figure 2.6   Antibiotics detected in effluents from three hospitals in New Mexico compared to a couple of residential sites in the same general locality. (From: Brown, 2004.)

because they are easily hydrolyzed particularly in the basic pH range, which is typically encountered in STP effluents (Al-Ahmed et al., 1999). In contrast, β-lactamase genes have been amplified from environmental samples quite frequently (Kümmerer, 2004a; Schwartz et al., 2003). The rare detection of β-lactams in the environment may also be attributed to their high propensity to sorb onto organic matrices such as sewage, sediments, and soil colloids, mechanisms that may make them unavailable with the extraction methods currently in use. Some of them may also be easily photodegraded, although photolysis may hold limited practical value in turbid wastewater and sludge due to the high turbidity associated with such waters. Sulfonamides have also been detected in wastewater. For example, a wide range of sulfonamides were detected from swine farms in all three provinces in Malaysia tested by Malintan and Mohd (2006).

Polar compounds such as fibrates, ibuprofen, and carbamazepine are relatively easier to detect in environmental samples, and their occurrence has therefore been more widely reported. Thus, several beta blockers such as acebuterol, metoprolol, nadolol, oxprenolol, bisoprolol, propranolol, and betaxolol have been detected in sewage effluents but at low concentrations (Andreozzi et al., 2003; Miège et al., 2007). The lipid regulator clofibrate is not encountered in the environment. Rather, its metabolite clofibric acid has been widely reported in sewage. In some instances, its concentrations are as high as $270 \mu g \, L^{-1}$ (Heberer and Adam, 2005). Ternes (1998) reported daily loads of $60 \, g/day$ in the influent of some effluents in Germany, subsequently getting reduced to about $40 \, g/day$ during routine treatment of the wastewater. It can be fairly persistent under both oxic and anoxic conditions (Zwiener et al., 2000; Zwiener and Frimmel, 2003). Atorvastatin (Lipitor) has also been detected in sewage effluents (Miao and Metcalf, 2003). Other lipid regulators detected in sewage effluents include bezafibrate, fenofibric acid, and gemfibrozil (Heberer and Feldmann, 2004).

Fluoxentine has been detected in water and municipal STP effluents (Kolpin et al., 2002; Metcalfe et al., 2003; Brooks et al., 2003, 2005). The muscle relaxant diazepam (a tranquilizer) detected in low concentrations (i.e., $40 \, ng \, L^{-1}$) in STP effluents in Germany (Ternes et al., 2001) and at $0.7–1.2 \, ng \, L^{-1}$ in the Po River in Italy (Zuccato et al., 2000). Also frequently detected among the antiepileptic drugs is carbamazepine (Ternes, 1998; Andreozzi et al., 2002; Tixier et al., 2003). It has very low removal rates in STPs. For example, Ternes (1998) reported removal of only 7% in wastewater treatment plants in Germany. In a recent survey by Metcalfe et al. (2004), all 26 samples collected contained detectable levels of carbamazepine, making it the most frequently detected drug in Canadian waters. Other epileptic drugs detected in water include primidone (Drewes et al., 2002, 2003).

Chemotherapeutic agents (i.e., anticancer drugs) such as ifosfamide have also been detected (Steger-Hartmann et al., 1996; Kümmerer et al., 1997b). Occurrence of these in any substantial amount is primarily interesting as they are inherently mutagenic, carcinogenic, and/or embroyotoxic. Iodinated X-ray contrast (AOI) media such as iopamidol and iopromide have been reported in concentrations as high as $10 \mu g \, L^{-1}$. They have a molecular weight range of 700–900 and a half-life in the body of approximately $2 \, h$ (Kümmerer, 2004b). In general, the concentrations of estrogen encountered in the environment range 44–490 ng L^{-1} in sewage influents (Ingerslev and Halling-Sørensen, 2003). Desbrow et al. (1998) also detected estrogenic substances in sewage effluents in the United Kingdom. Some fraction of estrogens is excreted by both humans and animals as conjugates. According to Lange et al. (2002), farm animals in the United States excrete 49 mg of estrogenic and 4.4 mg of androgenic hormones per annum. Thus, their analysis in environmental samples needs a deconjugation step or selection of a method that is able to directly analyze conjugates.

To transition our discussion to surface water, the work by Glassmeyer et al. (2005) is worth revisiting. That research group documented the presence of PPCPs in the immediate upstream, wastewater effluent, and downstream at 10 treatment plants located in widely divergent geographic regions of the United States. The plants served populations ranging from 27,000 to 1.5 million people. The number of compounds detected at each sampling location and the total concentration of PPCPs that were detected are summarized in Figure 2.7. Those results consistently show the

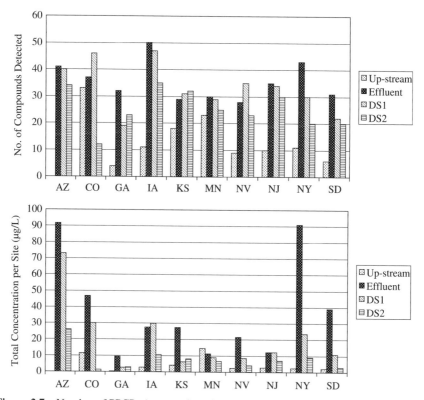

Figure 2.7 Number of PPCPs (top panel) and total concentration of analytes (bottom panel) found at 10 wastewater treatment locations within the United States. Samples were taken from an immediate up-stream location, the wastewater effluent, and two points, DS1 and DS2 downstream of the plant. DS2 was always further downstream compared to DS1. The plant in Arizona did not have any up-stream point as the stream was entirely a result of the treatment plant discharge. (Based on data from Glassmeyer et al., 2005.)

presence of some PPCPs in the upstream, most possibly as a result of human activity upstream. It is also noticeable that both the number and the concentration of PPCPs increased in the effluent, indicating introduction of these compounds in those environs from wastewater. Furthermore, both the number and total concentration of PPCPs (and related indicators such as caffeine and cholesterol) decreased in most instances in the waters sampled further away from the plants. It is also worth noting that the total concentration of PPCPs detected in the effluent were highest in the plants in Arizona and New York, which served 419,000 and 65,000 inhabitants, respectively, compared to the plant in Colorado, for example, which served an even larger population of 1.5 million people. These differences may be accounted for by several reasons, including differences in PPCP removal efficiencies, differences in PPCP use and disposal patterns, as well as differences in household water consumption. It is also apparent from these data that the concentrations of PPCPs generally decrease with distance away from the wastewater discharge points, possibly as a result of dilution, degradation, and sorption. We will now examine the occurrence of these compounds in surface water.

2.2.1.2 Presence of PPCPs in Surface Water Surface water in rivers, streams, and lakes provides water for many of our water supply systems. Also included among surface water is seawater, although this reservoir can adequately supplement our water supplies only after aggressive desalination and treatment. Most surface water originates directly from precipitation. However, substantial amounts of surface water are also derived from wastewater and farm runoff. Surface water reservoirs are also liable to contamination by natural and human activities in their surroundings. For example, WWTP effluents and raw sewage is discharged directly into surface water. Other forms of discharges include those used in aquaculture and other farm operations (Fig. 2.1). Thus, the quality of surface water in a stream, for example, will largely depend on the activity of the users upstream and land-use patterns in the watershed.

Extensive surveys of U.S. streams by Kolpin et al. (2004) showed an increased occurrence of pharmaceutical compounds in low-flow conditions. This observation suggests a correlation between water quality with seasonal conditions as river flow often changes with seasons. Some of the pharmaceuticals reported in surface water and typical concentrations are presented in Figure 2.8. It is apparent from this figure that most of these compounds are similar to those frequently encountered in wastewater as presented in Figure 2.4. However, the concentrations in surface water are typically lower than those encountered in wastewater and sewage, clearly demonstrating the effects of dilution. Most frequently encountered in surface water is pentobarbital, in some instances its concentration being as high as $5400\,\text{ng L}^{-1}$. Other reported compounds at concentrations as high as $1000\,\text{ng L}^{-1}$ in surface water include the fragrance acetophenone, the antibiotic sulfamethaxazole, the sedatives carbamazepine, primidone, butalbital, secobarbital, and phenobarbital, and the antihypertensive drug bezafibrate. As a matter of fact, from these data, the antihypertensive drug bezafibrate can occur at one of the highest concentrations (i.e., $3 \times 10^5\,\text{ng L}^{-1}$; Ternes et al., 2002) in surface water. Present in surface water but at low concentrations (i.e., $\leq 10\,\text{ng L}^{-1}$) are acetaminophen, clofibric

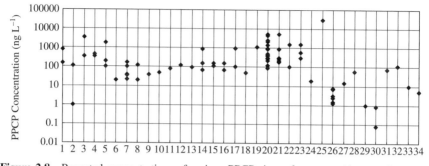

Figure 2.8 Reported concentrations of various PPCPs in surface water (rivers and streams) by several research groups. On the *x* axis are the cosmetic (1 = acetophenone), analgesics and anti-inflammatory drugs (2 = acetaminophen, 3 = acetylsalicylic acid, 4 = diclofenac, 5 = ibuprofen, 6 = indomethacin, 7 = naproxen), antimicrobials (8 = ciprofloxacin, 9 = doxycycline, 10 = enfloxacin, 11 = erythromycin, 12 = norfloxacin, 13 = ofloxacin, 14 = sulfamethaxazole, 15 = tetracycline, 16 = trimethoprim), anxiolytic sedatives (17 = carbamazepine, 18 = fluoxentine, 19 = primidone, 20 = pentobarbital, 21 = butalbital, 22 = secobarbital, 23 = phenobarbital), antihypertensives (24 = atorvastatin, 25 = bezafibrate, 26 = clorofibric acid, 27 = lovastatin, 28 = provastatin, 29 = simvastatin), reproductive (30 = 17β-estradial, 31 = mestranol, 32 = testosterone), GIT (33 = ranitidine), and an antineoplast (34 = cyclophosphamide). The concentrations were compiled from Boyd et al. (2003), Buser et al. (1998a), Heberer (2002a), Huang and Sedlak (2001), Kolpin et al. (2002), Metcalfe et al. (2004), Peschka et al. (2006), and Ternes et al. (2002).

acid, 17β-estradiol, and ranitidine. A recent study by Atkinson et al. (2003) showed comparatively higher concentrations of estrone in surface waters with increasing proximity to sewage treatment plants (Fig. 2.9). The fact that the data presented by Atkinson et al. (2003) were derived from divergent regions of the world strongly reinforces the idea that the occurrence of these hormones (and possibly PPCPs in general) in the environment is a fairly widespread phenomenon around the globe.

Diclofenac has been identified as one of the major pharmaceuticals in the waters in Berlin (Germany) at median concentrations of $0.47\,\mu g\,L^{-1}$ (Ternes et al., 2002; Ferrari et al., 2003). Other analgesics such as naproxen, phenazone, codeine, indometacin, fenoprofen, ketoprofen, phenylbutazone, and propyphenazone have also been detected in sewage and surface water (Heberer and Adam, 2005; Boyd et al., 2005), although data about their occurrence in the environment are not extensive. The data presented in Figure 2.5 are not exhaustive but are rather intended to show the great diversity in concentrations of some of the key PPCPs in wastewater.

Tylosin, which is exclusively used in livestock, was also encountered at $90\,ng\,L^{-1}$, whereas another macrolide erythromycin was most prevalent among those detected and was at concentrations as high as $300\,ng\,L^{-1}$ in surface water in Germany (Christian et al., 2003). Erythromycin is typically applied dermally, which results in a lower absorption and metabolism, a scenario that can enable most of the applied dose to end up in the sewer during normal activities such as showering or swimming. Calamari et al. (2003) detected a range of compounds including the antihypertensive drugs such as atenolol, clofibric acid, furosemide, and benzafibrate; the GIT drugs such as ranitidine (Zantac); and the sedative diazepam (Valium) in the Po River in Italy.

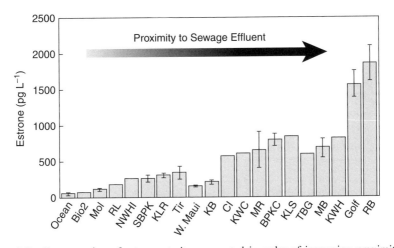

Figure 2.9 Concentration of estrone at sites presented in order of increasing proximity to sewage effluents. The open ocean (ocean) data set is composed of 7 sites near Florida, Hawaii, French Polynesia, and Marianas Islands. Biosphere 2 (Bio2) is an enclosed reef meso-cosm in Arizona, Molokai (Mol) is an uninhabited fringing coral reef near Hawaii, Rangiroa Lagoon (RL) is a large atoll in French Polynesia, South Big Pine (SBPK) is an offshore patch reef and seagrass meadow near Florida, Key Largo Reef (KLR) is also off the Florida coast. Tirian (Tir) is a fringing reef near a town on Marianas Islands, Kaneohe Bay (KB) is a coral reef lagoon adjacent to a city in Hawaii, Coconut Island (CI) is a lagoon within Kaneohe Bay, Moorea Resort (MR) is an indoor reef in French Polynesia, Maalaca Bay (MB) site is near a set of condominiums in Hawaii, Golf course (Golf) pond is located in the center of a golfing facility that received irrigation water, and Rahoboth Bay (RB) is an estuarine bay in Delaware with a sewage diffuser outflow. NWHI = NW Hawaii Islands, KWC = Key West Channel, BPKC = Big Pine Key Channel, KLS = Key Large Shore, and KWH = Key West Harbor. (Redrawn from Atkinson et al., 2003.)

Halden and Paull (2004) found triclosan and triclocarbon in a variety of types of water (i.e., finished drinking water, groundwater, urban streams, and sewage slurries) in the greater Baltimore area. The concentrations they encountered varied by a five-fold order of magnitude but were also positively correlated by a linear empirical model:

$$\log_{10} C_{TCC} = 0.9491$$

$$\log_{10} C_{TCC} \quad R^2 = 0.9882$$

where C_{TCC} in the concentration of triclocarbon and C_{TCS} in the concentration of triclosan.

The R^2 value of 0.9882 (almost unity) was indicative of the one-to-one correspondence of the fate of these two compounds. Both compounds are widely used in soaps and other personal care products on a regular basis. This finding shows the possibility of using information about the presence of one of these compounds to predict the fate of the other despite the fact that both chemicals belong to different groups but have similar uses and sources. Triclosan was also reported in the upper Detroit River by Hua et al. (2006).

Surfactants such as SDS and LAS have also been reported in the environment, causing serious pollution problems in rivers and lakes (Cserháti et al., 2002; Zoller, 1993; Poiger et al., 2004). Their concentrations in aquatic environments have long been associated with diurnal discharge of sewage into surface water (Kantin et al., 1981). Sunscreen products such as EHMC are released into the environment through swimming, bathing, and in laundry operations. In the temperate regions, their use and release into the environment is mostly during summer. Some of them may have estrogenic effects (Schlumpf et al., 2002).

2.2.1.3 *Presence of PPCPs in Groundwater*

2.2.1.3 Presence of PPCPs in Groundwater   Water naturally moves down-gradient toward the lowest point. Groundwater results from such movement infiltrating and percolating to the water table through voids in the soil and cracks in consolidated rocks, forming an aquifer. An aquifer is any porous water-bearing geologic formation. Aquifers are either confined or unconfined. In the latter types of aquifers, the upper surface of the saturated zone (i.e., water table) rises and declines freely. By contrast, confined aquifers have a characteristic permeable layer that is sandwiched between two less permeable upper and lower layers. The less permeable layer can be comprised of clay or consolidated rock. In most instances, water in the confined aquifer is under pressure from the sandwiching layers.

Groundwater is an extremely important source of water and supplies almost half of the water needs in the United States, especially in the rural areas where 95% of the population depends on wells (AWWA, 1995). Even larger proportions of the population's water needs in some other countries are derived from groundwater. For example, Denmark derives as much as 98% of its drinking water from groundwater sources (McKay, 1998). In most instances, groundwater does not undergo any form of treatment prior to its use for domestic purposes. The quality of groundwater is primarily affected by incidences of recharge of loaded surface water and leaks from sewer systems and landfills. PPCPs leach into the groundwater through various pathways including landfills, soil, and sediments (Fig. 2.1). The occurrence and transport of PPCPs in groundwater poses a threat to this source of drinking water. It is relevant for all compounds that are highly soluble and have a low propensity to sorb to soil colloids. In many jurisdictions worldwide, it is still perfectly legal to discard expired and/or unwanted pharmaceuticals in household wastewater that ultimately ends up in landfills. Many of these landfills are still open dumps that do not have protective liners and leachate collection systems. Thus, the pollutants in the unlined landfills end up in the groundwater. A recent report by Schneider et al. (2004) showed the presence of a variety of pharmaceutical compounds in two landfills in Germany (Table 2.3). Most abundant among the analgesics and anti-inflammatory drugs in those two landfill leachate were ibuprofen, prophenazone, and phenazone. Anxiolytic sedatives and antipsychotic drugs, particularly primidone, carbamazepine, and diazepam, were also detected in concentrations that are even much higher than what has typically been encountered in wastewater and surface water. Antineoplastics, antihypertensive drugs, and X-ray contrast media were also

TABLE 2.3   Median Concentrations of Various Pharmaceutical Compounds in Leachate from Two Landfills in Stuttgart (Germany)

Common Use	Compound	Concentration (ng L^{-1})	
		Landfill A	Landfill B
Analgesic and anti-inflammatory	Ibuprofen	9362	4894
	Propylphenazone	9173	2455
	Phenazone	5507	1761
	Dimethylaminophenazone	4764	2668
	Diclofenac	3190	1183
	Ketoprofen	697	438
	Piroxicam	481	931
	Naproxen	445	288
	Dihydrocodeine	101	14
	Indomethacine	17	141
Anxiolytic sedatives and antipsychotics	Primidone	5011	2002
	Carbamazepine	1415	202
	Diazepam	453	192
	Valproic acid	205	122
Antineoplastic	Cyclophosphamide	192	97
	Ifosfamide	42	32
Antihypertensives	Clofibric acid	2658	2879
	Pentoxifyline	2875	1116
	Bezafibrate	1353	2773
	Atenolol	44	34
	Metoprolol	31	24
	Propanolol	10	10
X-ray contrast media	Iopamidol	2485	2944
	Amidotrizoic acid	242	n.d.[a]
	Iopromide	199	236
	Iomeprol	92	42

[a]n.d. = Not detected.
Source: Schneider et al. (2004).

present in the landfill leachate. The presence of X-ray contrast media is worth noting as they are not household waste but rather used in hospital settings. Their presence in the landfill indicates a high possibility of such wastes having originated directly from hospitals as part of "medical waste." Eckles et al. (1993) detected pentobarbital at concentrations of $1 \mu g L^{-1}$ in groundwater from a landfill in Florida. A study by Holm et al. (1995) documented the presence of various sulfonamides in landfill leachate at a facility in Grindsted, Denmark (Fig. 2.10). The concentrations of these compounds generally decreased with the increasing depth of the wells. Concentrations also generally decreased as one moves further away from the landfill. This trend suggests some level of degradation or adsorption of some PPCPs in the groundwater. Estrogens and the androgenic compound testosterone have also been detected in groundwater (Finlay-Moore et al., 2000; Peterson et al., 2000) with the

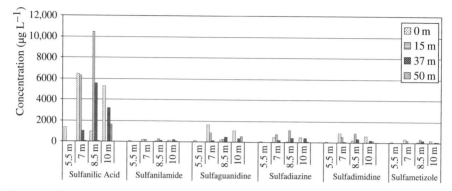

Figure 2.10   Distribution of sulfonamides in groundwater as a function of distance from a landfill in Grindsted, Denmark. (Based on data from Holm et al., 1995.)

latter investigating team using them as indicators of animal waste contamination of karst aquifers.

In general, the detection of pharmaceutical compounds in landfills is not entirely surprising. For example, two-thirds (63.2%) of the population in the United Kingdom discard their pharmaceuticals in household waste, which ultimately ends up in land-fills (Bound and Voulvoulis, 2005). Those authors also found that only 21.8% dispose of their unwanted medications by returning them to the pharmacy, whereas 11.5% flush them down the toilet. Bound and Voulvoulis (2005) found that only about 18% of respondents in their survey finished their antibiotic prescriptions, with 71% emptying antibiotic leftovers in the household trash and 4% disposing them into the toilet. Dumping pharmaceuticals into the trash was also the leading method of choice for disposing of painkillers (70% of respondents), antihistamines (75% of the respondents), antiepileptic drugs (100% of the respondents), beta block-ers (67% of the respondents), hormones (75% of the respondents), as well as lipid regulators and antidepressants (67% of the respondents for each drug category). During those interviews, the respondents were given an opportunity to list/enumerate all of the disposal practices they use for each category of medication. Naturally, most individuals had used more than one disposal practice. A similar study in the United States found an even higher percentage of respondents (35.4%) disposing their unwanted or expired medication down the toilet and 54% disposing them into the garbage (Kuspis and Krenzelok, 1996). A more recent study by Seehusen and Edwards (2006) found that more than 36% of people found it acceptable to flush medications down the toilet and 21% believed it acceptable to rinse them down the sink. Similar levels of disposal of pharmaceutical compounds in household trash have also been reported in Canada (COMPAS, 2002).

In some instances, the sludge generated from wastewater is landfilled. Under such circumstances, the remaining PPCPs in the sludge are also introduced into the land-fill. Alternatively, the sludge may be applied to agricultural fields, introducing whatever organic and inorganic contaminants that may be present into the soil (see Section 2.2.3). Not all of the PPCPs detectable in landfill leachate directly enter other compartments of the environment as some landfills have leachate collection systems in which the collected leachate is subjected to treatment processes that are similar to those in wastewater treatment plants before releasing into surface water (Barlaz, 1996). Unless the leachate is treated, the pharmaceuticals deposited into the garbage bypass transformation by the body and wastewater treatment systems. Of course, where the landfill leachate leaks into the groundwater before collection, leachate treatment would be of minimal consequence to the quality of the groundwater.

Other sources of PPCP contamination to groundwater can originate from farms, leaking septic tanks, and lagoons. For instance, Campagnolo et al. (2002) detected several types of antibiotics including macrolides, tetracycline, sulfonamides, and β-lactams in groundwater samples collected from sites that were in proximity of a swine farm.

2.2.1.4 Presence of PPCPs in Potable Water By comparison to other aquatic systems, PPCPs have been reported least frequently in treated potable water. However, the possibility of having PPCPs in drinking water has aroused a lot of interest (Aherne and Briggs, 1989; Raloff, 1998, 2000; Zuccato et al., 2000; Kuch and Ballschmitter, 2001; Campbell, 2007; Collier, 2007). Typical PPCP concentrations that have been reported in treated drinking water are displayed in Figure 2.11 from which it is noted that the types of PPCPs reported are less diverse and the concentrations lower. It is possible that far beyond what are indicated in the figure occur in this matrix but at concentrations that are below the current method detection limits. The PPCPs that have been detected in such waters (Fig. 2.11) are mostly below 100 ng L^{-1}, although some high concentrations of the antihypertensive clofribic acid, and sedatives, particularly carbamazepine and diazepam, have also been encountered. Ternes and Hirsch (2000) detected iodinated contrast media in drinking

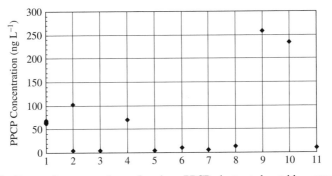

Figure 2.11 Reported concentrations of various PPCPs in treated potable water by several research groups. So far reported are the analgesic (1 = naproxen), antihypertensive (2 = clofibric acid, 3 = dehydronifedipine, 4 = gemfibrozil), reproductive (5 = ethinyl estradiol, 6 = norethindrone), antineoplasts (7 = metyhotrexate, 8 = bleomycin), sedatives (9 = carbamazepine, 10 = diazepam), and antimicrobials (11 = penicillin). Concentrations compiled from Boyd et al. (2003) and Collier (2007).

water at concentrations of $0.07 \, \mu g \, L^{-1}$, as compared to $3 \, \mu g \, L^{-1}$ in STP effluents. However, it should be emphasized that to the best of my knowledge, there are no statutory regulations in any potable water facility to specifically monitor for these compounds. It looks like no municipality is mandated to routinely test its water supplies for the presence of pharmaceutical and personal care products at their trace concentrations. For example, the European Union under the Drinking Water Directive (Council Directive 98/93/EC) requires monitoring 48 parameters, none of which relate to PPCPs. The U.S. EPA (2003) has an evolving Drinking Water Contaminant Candidate List (CCL) that includes chemicals that are being considered for possible drinking water standards. Making the list is primarily based on two criteria, that is, whether (i) the contaminant adversely affects public health and (ii) it is known or is substantially likely to occur in public water systems at levels and frequencies that threaten public health. Recently, the U.S. EPA updated its guidelines specifying 170 parameters for drinking water (U.S. EPA, 2006), none of which include PPCPs.

Absence of such monitoring may, in part, be due to the lack of systematic and affordable testing methods. It is also possible that their occurrence is only sporadic, being mostly removed during the water treatment process. Whatever the case might be, where PPCPs are detected in treated potable water, their presence raises concerns about the intentional use of such water (Snyder et al., 2003). The PPCPs detectable in drinking water tend to have a high chemical stability, low biodegradability, of low sorption coefficients, and a high solubility. Recently, California is also contemplating some form of legislation to mandate monitoring potable water for PPCPs and other endocrine disrupting compounds in general. However, such regulations have been bogged down in ironing out compromises on what standard methods to use to satisfactorily quantify the low concentrations that are present. Several methods have been used by different laboratories, and they all do not seem to agree. In general, most of them involve some form of extraction procedure that is followed by detection using either instrumentation or a bioassay technique, as described in Section 2.2.

Salicylic acid, a widely used over-the-counter (OTC) drug is frequently encountered in the environment (Table 2.4). It has also been detected in the plasma of virtually all persons tested in Scotland, including people who were not taking the salicylic-containing drugs (Table 2.4). Blacklock et al. (2001) suspect the salicylic

TABLE 2.4   Concentrations of Salicylic Acid (SA) in Three Groups of Individuals

Parameter	Patients Taking Aspirin[a]	Nonvegetarians	Vegetarians
Number of people tested	14	39	37
Average SA concentration ($\mu mol \, L^{-1}$)	10.03	0.07	0.11[b]
Range of SA concentration ($\mu mol \, L^{-1}$)	0.23–25.4	0.02–0.2	0.04–2.47

[a]Patients were consuming 75 mg aspirin per day.
[b]SA concentration in the plasma of vegetarians was significantly greater than that in nonvegetarians.
Source: Blacklock et al. (2001).

acid in the serum of the individuals that were not taking aspirin to be attributed to fruits and vegetables, but it is impossible to rule out some of it originating from other sources such as potable water.

2.2.2 Occurrence of PPCPs in Sediments

In aquaculture, about 70% of the therapeutic and subtherapeutic agents administered are released into the environment, ending up in sediments and surrounding soil where they remain stable (Fig. 2.1). Sediments can also receive PPCPs from runoff. They can in fact be a large sink for PPCPs compared to the overlaying water column. Almost exclusively applied to aquaculture waters are antibiotics. For example, concentrations as high as 285 mg oxytetracycline kg^{-1} sediments have been reported in sediments of a fish farm after application of this compound (Samuelsen et al., 1992). When introduced as part of the feed, the antibiotics settle directly into the sediments from the fish feed. Some of the antibiotics are also excreted, ending up in the sediments. PPCPs can bind to sediments (Drewes et al., 2002) or be remobilized on perturbation. PPCPs in sediments are often ignored and not routinely considered in aquatic ecotoxicological studies. However, their concentrations may be greatly reduced in the aqueous phase posing no risk to pelagic organisms but remain detrimental to sediment dwellers. As a matter of fact, sediments in aquaculture systems tend to have very high concentrations of antibiotics, which can even inhibit the growth of bacteria. Thus, antibiotic-resistant bacteria in sediments is often used as a sensitive indicator of past antibiotic use for such environments (Kümmerer, 2004a).

Besides aquaculture settings, other PPCPs can also accumulate in sediments. For example, several reports have documented the accumulation of SDS and LAS (Quiroga et al., 1989; Prokvová et al., 1997; Marahesi et al., 1994). The fragrances HHCB and AHTN, previously discussed in wastewater, have also been detected in sediments at concentrations that are higher than those in the surrounding water column, which has important ramifications for sediment dwellers.

2.2.3 Occurrence of PPCPs in Soil

Pharmaceutical and personal care products can be introduced into soils through various entry points, including runoff, dust, as well as manure and biosolids when these are applied on arable land (Fig. 2.1). As a matter of fact, the application of manure and biosolids may be the two most significant sources of PPCPs in soils. More than 6.5 million tons (dry weight basis) of sewage sludge are generated in the European Union, of which 37% is utilized for agriculture (Smith, 1996). Similarly, 5.6 million tons (dry weight basis) of biosolids are generated in the United States per annum of which 61% is used as a soil conditioner and 20% is disposed of in landfills (NRC, 2002). These forms of excreta are an important source of nutrients (e.g., nitrogen, phosphorus, and organic matter) on arable land. They also enhance soil physical properties and plant growth. Biosolids are also used for landscaping, land reclamation, and for revegetation programs. For livestock, such excreta come in different forms, depending on the animal species and the production methods practiced. It can be in the form of a liquid slurry (feces mixed with urine) or solid bed-yard manure (feces and straw). Finlay-Moore et al. (2000) detected 17β-estradiol and testosterone in soils and runoff from grasslands that had been amended with broiler litter, and their presence in sewage has been detected for more than three decades (Kirk et al., 2002). Hamscher et al. (2002) detected a variety of antibiotics in 10 out of 12 biosolids-treated soils. A closer examination of the data from Hamscher et al. (2002) indicated that the concentrations of some antibiotics in the soil, notably tetracycline and chlortetracycline, were increasing with soil depth possibly due to the additional release of bound antibiotic residues in the form of 4-epi-tetracycline, a metabolite of tetracycline. Similarly, Christian et al. (2003) detected sulfadimidine in soil 7 months after liquid manure application. Aga et al. (2003) and Boxall et al. (2005) also detected antibiotics in soils on fields where animal slurry had been applied, whereas Boxall et al. (2002) found sulfona-mide sulfochloropyridazine at microgram per liter concentrations in soil pore water in soils that received manure. Sulfonamides are widely used in veterinary medicine and excreted as either the parent compound or its conjugates. The conjugates are, in the manure tank, converted back into the parent compound (Boxall et al., 2002). Similarly, concentrations in the range of $9–12\,\mu g$ tetracycline kg^{-1} soil were also detected in soils treated with manure (Kühne et al., 2000). Although not exten-sive, these data also suggest that, similar to aquatic environments, pharmaceutical compounds occur in soil at low concentrations that are likely to be bioavailable to nontarget organisms.

The PPCPs become highly concentrated in sludge (Golet et al., 2002b). The antiepileptic drug carbamazepine has been detected in soils where it is reportedly readily mobile and is not readily attenuated (Heberer and Mechlisnki, 2003; Kühn and Müller, 2000). Detergent metabolites nonylphenol and alkylphenol polyethoxy-lates have been detected in aquifer soils (Montgomery-Brown et al., 2003). A recent survey by Kinney et al. (2006) found a minimum of 30 and a maximum of 45 organic contaminants in any biosolids product generated from wastewater treatment plant from seven different states in the United States. The biosolids had been processed through several production methods and the wastewater had been derived from differing population demographics.

In the analyses by Hamscher et al. (2002), increasing concentrations of tetra-cycline in soil with depth were noted, clearly suggesting the high mobility of this compound in soil as most of it was being leached into the soil profile. Antibiotics occur naturally and are a main factor in the ecology of microorganisms. Under these natural conditions, the concentrations of antibiotics are very low and their effects are confined to microenvironments that are within the vicinity of the organ-isms that produce them. That natural phenomenon is changed when antibiotics are of a synthetic or semisynthetic nature. Strongly hydrophobic compounds tend to sorb and accumulate in soils and sediments, whereas strongly hydrophilic compounds tend to be quite mobile, easily finding their way into groundwater, surface runoff, and drainage water. Organic compounds with a low octanol/water partition coefficient (K_{ow}) tend to be highly mobile in soil and groundwater. The detection of PPCPs in soil can be problematic compared to aquatic environments due to interference of

TABLE 2.5 Antibiotic Residues in Dust Particles Obtained from a Piggery over a 20-Year Period (1981–2000)

Sampling Year	Antibiotic Concentration (mg kg^{-1})						Sum (mg kg^{-1})
	Oxytetracycline	Tetracycline	Chlortetracycline	Tylosin	Chloramphenicol	Sulfamethazine	
1981	1.10	—	—	0.42	—	1.85	3.37
1982	0.18	—	—	0.09	—	0.06	0.33
1983	—	0.19	2.12	5.65	—	2.90	10.86
1984	—	—	—	—	—	—	—
1985	—	—	—	—	—	—	—
1986	—	—	—	12.18	—	0.32	12.50
1987	—	—	—	8.72	—	0.39	9.11
1988	—	—	—	0.72	—	0.43	1.15
1989	—	—	—	0.45	1.96	0.34	2.75
1990	—	—	—	0.14	—	0.09	0.23
1991	0.43	—	0.32	0.26	9.07	0.41	10.49
1992	—	—	—	0.35	5.49	0.05	5.89
1993	—	0.19	—	0.10	—	0.12	0.41
1994	—	0.23	—	0.37	—	0.12	0.72
1995	—	0.37	0.52	0.29	—	—	1.18
1996	0.29	5.18	—	0.55	—	0.16	6.18
1997	—	0.47	—	0.16	—	—	0.63
1998	—	0.50	—	0.20	—	—	0.70
1999	—	0.61	—	—	—	—	0.61
2000	—	0.19	—	—	—	—	0.19
Frequency (%)	20	45	15	80	15	65	

Source: Hamscher et al. (2003).

111

organic matter. Schlüsener et al. (2003a, 2003b) successfully used pressurized liquid extraction combined with HPLC–MS to determine various antibiotics from soil and liquid manure, respectively.

2.2.4 Aerial Environments

The use of pharmaceuticals in the livestock industry has more recently led to the identification of another route through which these compounds get into the environment. Hamscher et al. (2003) tested dust samples collected from a pig farm over a 20-year period (1981–2000). During each of those years, the dust samples were collected over 14–30 days in a standardized metal frame with an effective sampling surface area of 3002 cm^2. The sampling device was mounted at a typical breathing height for humans. Some results from that long-term study show that all of the six antibiotics tested were detected at concentrations ranging between 0.05 and 12.18 mg kg^{-1} dust (Table 2.5). Tylosin was detected more frequently than all the others, whereas chlortetracycline and chloramphenicol were least frequent. Tylosin was also detected in the highest concentrations in 1986, whereas chloramphenicol was detected in high concentrations in 1991 and 1992. Sulfamethazine occurred at a high frequency of 65% during the two decades of sampling. None of the antibiotics was detected in the dust in 1984 and 1985. Based on these six antibiotics, the total antibiotic load in the dust ranged between 0.19 mg kg^{-1} in 2000 and 12.5 mg kg^{-1} dust in 1986. The antibiotics in the dust are believed to have originated from animal feed mixed with veterinary drugs for therapeutic and prophylactic purposes. The feed is usually in pellets or powder form, and its aerosols are released during handling and feeding. Some antibiotic-containing dust is also released from the dried manure particles. The risks associated with inhalation of such pharmaceutical-laden particles have not yet been assessed.

2.3 EXCRETION AS A DRIVER OF PHARMACEUTICAL OCCURRENCE IN THE ENVIRONMENT

Table 2.6 summarizes information about the fractions of various pharmaceutical compounds that are excreted. However, the exact proportions can vary between individuals, age groups, and physiological status. Some of the drugs are conjugated, forming glucuronic acid or sulfate conjugates. They can also be hydroxylated prior to releasing in the feces. The conjugates are generally biologically inactive but may, depending on various factors, be easily cleaved, becoming reactivated. Except for specific dumping situations such as through production facilities, the concentration and range of PPCPs tends to be comparatively higher in sewage and livestock manure than in the adjoining water, and sediments due to the dilution effects and natural attenuation (i.e., sorption, degradation, volatilization) get into the environment. This trend also gives evidence to the fact that excretion is the major route by which these compounds are introduced in the environment (Fig. 2.1). These excreted proportions presented in Table 2.6 have been documented

TABLE 2.6 Categorization of Various Commonly Used Pharmaceutical Compounds Based on the Proportions of the Parent Compound Excreted in Clinical Setting[a]

	Proportion of the Parent Compound Excreted[a]				
Common Use	Low (≤5%)	Moderately Low (6–39%)	Relatively High (40–69%)	High (≥70%)	
Skeletal ailments	Clodronate [0.18–19% (Ylitalo et al., 1999)], pamidronate [0.01–0.18% (Hylastrup et al., 1993)]		Actonel [40% (U.S. FDA, 2006)], Alendronate [40–60% (Porras et al., 1999; Merck, 2005)]		
Antimicrobial	Acylovir [0.42% (Yamashita et al., 1993)], chloramphenicol [5–10% (Hirsch et al., 1999)]	Ampicillin [30–60% (Hirsch et al., 1999)], clindamycin [4% in feces & 10% in urine (Martindale, 1999)], nitrofurantoin [30–40% (Queener and Gutierrez, 2003)], **sulfamethoxazole** [15% (Ternes, 1998)]	Biaxin [40% (Queener and Gutierrez, 2003)], dicloxacillin [50% (Vinge et al., 1997)], didanosine [50% (Martindale, 1999)], ethambutol [50–80% (Breda et al., 1999)], fluconazole [62–80% (Ripa et al., 1993)], metronidazole [40% (Kümmerer et al., 2000)], minocycline [60% (Hirsch et al., 1999)], **norfloxacin** [30% (Martindale, 1999)], **trimethoprim** [60% (Hirsch et al., 1999)], valaciclovir [57–65% (Burnette and Demiranda, 1994)], zithromax [50% (Queener and Gutierrez, 2003)]	Amoxacillin [80–90% (Hirsch et al., 1999)], **ciprofloxacin** [83.7% (Volmer et al., 1997)], **doxycycline** [70% (Hirsch et al., 1999)], cephalexin [80% (Martindale, 1999)], flucytosine [90% (Martindale, 1999)], genaconazole [76–87% (Mojaverian et al., 1994)], **tetracycline** [80–90% (Kühne et al., 2000)]	

(Continued)

113

TABLE 2.6 *Continued*

	Proportion of the Parent Compound Excreted[a]			
Common Use	Low ($\leq$5%)	Moderately Low (6–39%)	Relatively High (40–69%)	High ($\geq$70%)
Analgesics and anti-inflammatory	**Aspirin** [2–30% (Martindale, 1999), **ibuprofen** [1–8% (Ternes, 1998)], **paracetamol** [<5 (Martindale, 1999)]	**Diclofenac** [15% (Ternes et al., 1999)]		Morphine [71.6% (Zakowski et al., 1993)]
Respiratory		Cromoglycate [25% (Yoshimi et al., 1992)]	Dexamethasone [65% (Martindale, 1999)],	
Sedative and antipsychotic	**Carbamazepine** [1–2% (Ternes, 1998)], **diazepam** [1% (Smith-Kielland et al., 2001)], **fluoxentine** 2.5–11% (Altamura et al., 1994)], naltrexone [<1% (Martindale, 1999)], sertraline [0.2–50% (Kobayashi et al., 1999; Warrington et al., 1992)]	**Phenobarbitone** [25% (Martindale, 1999)], Paxil [36% (Hiemke and Härtter, 2000)], primidone [15–40% (Gutierrez and Queener, 2003)]	Atropine [50% (Alimelkkila et al., 1993)], Celexa [50% (Hensiek and Trimble, 2002)],	Baclofen 70–85% (Gutierrez and Queener, 2003)]
Cardiovascular and antihypertensive drugs	**Atorvastatin** [5% (Prueksaritanont et al., 2002)], labetalol [1.61% (Yeleswaram et al., 1993)], quinapril [2.6% (Swartz et al., 1990)], verapamil [<4% (Martindale, 1999)]	Enalapril [36% (Lo et al., 2000)], hydrochlorothiazide [24% (Martindale, 1999)], procainamide [30–70% (Martindale, 1999)], quinidine [20–50% (Gutierrez and Queener, 2003)], ramipril [25.2%	**Bezafibrate** [45% (Martindale, 1999)], clonidine [40–60% (Martindale, 1999)], digoxin [50–70% (Martindale, 1999)], **furosemide** [40% (Bindschedler et al., 1997)],	

Category			
Dopamirgics		(Meyer et al., 1995)], **simvastatin** [10–15% (Martindale, 1999)] Nimodipine [32% (Parnetti, 1995)]	Plavix [50% (Wallis and Sciacca, 2003)]
Reproductive		**Ethinylestradiol** [30% (Schwanek and Webb, 2005)	**Testosterone** [60% (Schumann, 1991)]
Antineoplasts	**Cyclophosphamide** [5–15% (Joquevail et al., 1998)], idarubicin [5% (Stewart et al., 1991)], Fluorouracil [7–20% (SICOR, 2005)], **methotrexate** [<5% (Widemann and Adamson, 2006)]		
Anesthetics		Lidocaine [<10% (Martindale, 1999)]	Propofol [≈90% (Guitton et al., 1997)]
Gastrointestinal		Domperidone [10% (Martindale, 1999)], **ranitinide** [30–70% (Martindale, 1999)]	
Triptans			Rizatriptan [89–94% (Vyas et al., 2000)]
Hematologic			Coumadin [>90% (Queener and Gutierrez, 2003)]

[a] In boldface are some of the compounds for which occurrence in the environment has been reported (see Section 2.2). The numbers in brackets indicate the proportions that are excreted and the pertinent reference.

Source: Modified from Jjemba (2006).

115

mostly under clinical conditions during the research and development of each of these drugs. The proportions excreted have been arbitrarily categorized as low ($\leq 5\%$), moderately low (6–39%), relatively high (40–69%), or high ($\geq 70\%$) (Jjemba, 2006). It is apparent that this categorization is not clear-cut as some compounds that have the proportions excreted are reported as a range that overlaps across these arbitrary categories. Based on that loose categorization, however, it appears that only low proportions of most of the biophosphonates and other skeletal drugs are excreted, the only exception to this observation being alendronate (Fosamax). By comparison, a relatively high to high proportion of most antimicrobial drugs is excreted as parent compound with the exception of acylovir, chloramphenicol, ampicillin, clindamycin, and sulfamethoxazole (and possibly a variety of other sulfa-related antimicrobials). Note that very low proportions of acylovir and chloramphenicol are excreted. Low to moderately low proportions of most analgesic and anti-inflammatory drugs, notable of which are the household companions aspirin, ibuprofen, and paracetamol, are also excreted. Low to moderately low proportions of the most widely used sedatives, cardiovascular drugs, and GIT drugs are also excreted. Relatively high or even high proportions of reproductive drugs and triptans are excreted. This categorization also reveals that the pharmaceutical compounds that have been detected in the environment belong to all of the four excretion categories. Thus, even compounds that have only low (i.e., $\leq 5\%$) proportions of the parent compound excreted have also been encountered in the environment. Thus, although excretion may be the main mode of entry of PPCPs into the environment, persistence of the respective compounds can also be significant. Persistence and degradability are examined in Chapter 3. It is worth noting that the ones that have not yet been reported to occur in the environment may be present but have not been looked for. It is also possible that the ones that have not yet been detected are not easily detectable using the (chemical) methods on which we currently rely.

The quantities of individual pharmaceutical compounds that are present in the environment are probably not large in themselves, but their continuous use and our failure to effectively remove them during routine treatment of sewage as well as the absence of treatment of manure prior to disposal or use may lead to sustained concentrations and promote subtle effects on nontarget organisms (Jjemba and Robertson, 2003). In extreme cases, some of these effects may be irreversible. The mere detection of (or lack thereof) the PPCPs in the environment may not suffice for assessing the associated risk. Pharmaceuticals may be the most investigated compounds in terms of pharmacokinetics, pharmacodynamics, and by inference toxicity during their routine development for clinical efficacy. Noticeably absent during that development stage is rigorous research about their ecotoxicity. The PPCPs in STP (or their metabolites) may be rapidly degraded, mineralized, or persist, binding to sludge and sediments and ultimately leaching into the groundwater. It is important to note that the accumulating compounds in the environment may attain a higher concentration in organisms than in the respective environments.

3

ECOPHARMACOKINETICS AND ECOPHARMACODYNAMICS

The development of a pharmaceutical compound prior to clinical use is a protracted process that involves a huge investment of time and money. A typical screening cascade during drug development includes the following sequence: primary screening (in vitro) → secondary functional screening (in vitro) → drug metabolism and pharmacokinetics (in vitro and in vivo) → pharmacodynamic model (in vivo) → disease model (in vivo) → development and marketing. In vitro systems are experimental systems that use biological materials other than intact animals. Such systems include primary cell cultures, tissue slices, and established cell lines. Noticeably absent from the screening cascade outlined above is the determination of the effects of pharmaceutical compounds on other organisms besides the organisms for which the compounds in question are designed. On the whole, the general public already appreciates the fact that our individual actions and activities can contribute to the pollution of our environment. However, that appreciation and recognition is still mostly limited to activities that are related to industrial and agricultural processes. It has not popularly been extended to our activities that are associated with pharmaceutical and personal care products, products that we use on a regular basis. That recognition has also not been widely extended to the way that we dispose of these compounds once they have outlived their usefulness. Part of that disconnect has probably emerged from our training, which tends to confine us into seemingly disconnected disciplines. To that effect, a majority of us remain fully uninformed of the intricacies of health care delivery, whereas those that are involved in providing health care are not fully informed of the concerns and fate

Pharma-Ecology: The Occurrence and Fate of Pharmaceuticals and Personal Care Products in the Environment. By P.K. Jjemba
Copyright © 2008 John Wiley & Sons, Inc.

of the therapeutic products they market or prescribe to address noble tasks in their normal practice, once those compounds get into the environment. This chapter focuses on the dynamics and fate of PPCPs in our bodies and in the environment. More specifically, it is intended to devote some attention to the similarities between pharmacokinetics and pharmacodynamics of PPCPs in clinical and environmental settings by primarily looking at the sorption kinetics, mobility (distribution), and degradability of these compounds in the environment versus what is known about these same processes in our bodies.

3.1 OVERVIEW OF PHARMACOKINETICS AND PHARMACODYNAMICS

Pharmacodynamics is a branch of pharmacology that deals with the reactions between drugs and living systems, whereas pharmacokinetics characterizes the interactions of a drug and the body in terms of its absorption, distribution metabolism, and excretion. As a parallel, the Gaia hypothesis conceptualizes Earth as the largest living organism in the solar system (Lovelock, 1988). This hypothesis is quite opposed to conventional science that depicts Earth as an inert rock upon which plants and animals thrive. One does not have to fully subscribe to this Greek-mythology-based hypothesis to realize the fact that PPCPs are subjected to distribution, absorption, metabolism (i.e., biodegradation), and mobility (equivalent to excretion in living organisms) in the environment, making the environment behave in a manner that is somewhat analogous to how living organisms behave when they are undergoing pharmaceutically- based therapy (Jjemba, 2006). Pharmacokinetics and pharmacodynamics in the environment have been referred to as enviropharmacokinetics and enviropharmacodynamics (Jjemba and Robertson, 2005) or ecopharmacokinetics and ecopharmacodynamics, respectively.

In clinical settings, pharmaceutical compounds are administered intravascularly or extravascularly. In the former instance, they are administered directly into the bloodstream, that is, intravenously or intervascularly, whereas the latter application includes administering the drug through the oral cavity, subcutaneous, rectally, sublingual, buccally, dermally, intramuscularly, or into the pulmonary system. Drug design often starts by identifying the biological target and from a practical perspective, a high affinity of the compound to the biological target, that is, target protein (pK_i), is an important consideration in the rapid screening of promising candidate compounds. The drugs are also designed to have a high stability and, thus, less degradability in the body. All extravascularly administered drugs must be absorbed so as to enter the blood (circulatory) system. Thus, on administration, the concentration of the compound in the blood increases dramatically but ultimately declines as some of it is eliminated or absorbed. Absorption in this instance is the process by which the unchanged drug moves from a site of administration to the site of measurement within the body, whereas elimination is the irreversible loss of the drug from the site of measurement. Elimination occurs by metabolism (i.e., biotransformation), exhalation, or excretion. Some of the excreted fractions are in the form of metabolites. In the body, a mass balance of the drug can be constructed from the model outlined by Rowland and Tozer [1995; Eq. (3.1)]:

Dose =Amount of drug at absorption site + Amount of drug in the body

+ Amount of drug excreted + Amount of drug metabolites in the body

+ Amount of metabolite eliminated   (3.1)

The dynamics between all of the five components in the above model are summarized in Figure 3.1. As the drug reaches the systemic circulation, its concentration rises to a maximum (C_{max}), indicating the trend that absorption is occurring more rapidly than distribution and elimination. C_{max} is attained after a duration (T_{max}). To that effect, the rate of change of drug in the body is the difference between the rate of absorption and that of elimination [Eq. (3.2)], and after C_{max} is attained the rate at which the drug is eliminated exceeds the rate of its absorption. The rate of drug change in the body will have a negative value, signifying an overall decline in the concentration of the drug in the body:

Rate of drug change in the body = Rate of absorption

− Rate of elimination   (3.2)

Studies of drug absorption, distribution and elimination comprise what is referred to as *pharmacokinetics*. By contrast, the concentration of a pharmaceutical compound at the site(s) of action in relation to the magnitude of its effect(s) is referred to as *pharmacodynamics*. Both pharmacokinetics and pharmacodynamics have their roots in physiology, chemical kinetics, biochemistry, and pharmacology. They seek to provide a mathematical basis of the absorption, distribution, metabolisms, and

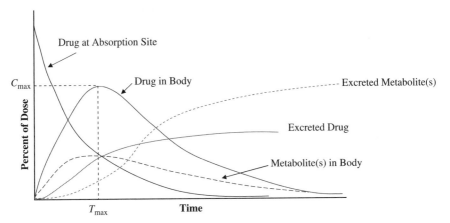

Figure 3.1 Fraction of drug and its metabolites in each model compartment over time. (Adapted from Rowland and Tozer, 1995.)

excretion of drugs. Put more simply, they define what the body does to the compound or drug. They are most commonly assessed by measuring the concentration of the drug in the bloodstream or in the serum. Thus, blood (or serum) is periodically sampled after the drug has been administered and the concentrations plotted against time.

Once administered into the body, the drug is absorbed and distributed into the vascular system. In instances where the pharmaceutical compound is placed away from the site of action, the drug has to "move" in the blood to the site of action. During this movement, it is also distributed to other tissues including those that eliminate it from the body, notably the kidneys, liver, GIT, and lungs (Fig. 3.2). The elimination of a pharmaceutical compound normally requires its conversion to water-soluble compounds through metabolism and subsequent excretion via urine and feces. The kidney is the primary site of elimination (excretion) of the parent compound, whereas the liver is a primary site of excretion of the metabolites. However, some fraction of the parent compound can also be secreted into the bile. When administered orally, all pharmaceutical compounds will go through the liver before they reach the circulatory system. Besides processing foreign substances including drugs into nontoxic waste, the liver is also responsible for the synthesis of glucose for nourishing the cells, storage of excess sugars as glycogen, circulating bile, which breaks down nutrients into usable forms, synthesis of blood clotting factors, as well as metabolizing proteins and fatty acids. Transiting through this vital organ, therefore, subjects the compounds to hepatic elimination. In pharmacokinetic jargon, the elimination by the liver through the hepatic system is referred to as first-pass metabolism. Very specifically, drug metabolism in medicinal chemistry refers to the chemical alterations of the drug in vivo. However, as displayed in Figure 3.2, materials can move back and forth between the liver and the GIT, subjecting the pharmaceutical compounds to an enterohepatic cycle. The enterohepatic cycle for the fraction that is not eliminated is analogous to a situation where water is pumped from one reservoir to another and then drained back into the original reservoir.

Clearance is an important concept of pharmacokinetics that describes the rate at which a chemical is eliminated from the body. The total clearance of a drug from the body is calculated from

$$CL = \frac{F \times \text{Dose}}{\text{AUC}} \tag{3.3}$$

with F being the fraction of the dose that is absorbed, and AUC is the total area under the curve (AUC). Pharmaceutical compounds that undergo significant first-pass elimination have a high clearance rate since

$$CL_H = \frac{Q_H(C_A - C_V)}{C_A} \tag{3.4}$$

where CL_H is the clearance by the hepatic vein, Q_H is the quantity flowing in the hepatic veins. Thus, the overall efficiency of removal of a chemical from the body

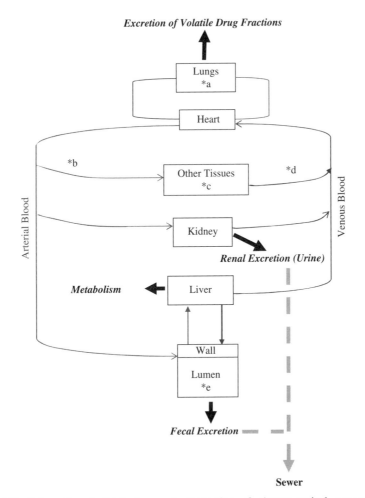

Figure 3.2 Schematic of distribution and elimination of pharmaceutical compound on normal administration. Compound is distributed through the (a) lungs, (b) arteries, (c) other tissues (e.g., muscles, subcutenous tissues), (d) veins, and (e) gastrointestinal tract (i.e., oral). Notice the enteroheptic cycle where recirculation occurs between the liver and the GIT with most of the drug being excreted in the bile and is released into the gall bladder, transits into the small intestine, and is absorbed into the circulatory system.

can be characterized by clearance, and high clearance values indicate efficient and generally rapid removal rates and vice versa. Clearance is an exceedingly important concept in clinical pharmacology and a question that comes to mind then would be whether one can use total clearance values to predict the removal of a PPCP, for example, during a particular sewage treatment process. In the body, the detoxification process is facilitated by the special cells in the liver (i.e., hepatocytes) by converting the fat-soluble molecules, including drugs into water-soluble molecules that are excreted. Some pharmaceuticals are more toxic to the liver than others. Some metabolites are also toxic to this organ. As a matter of fact, effects on the liver is usually one of the main causes for withdrawing pharmaceuticals off the market (see Chapter 4). The detoxified wastes are subsequently excreted into the gut through the bile ducts, making the GIT another important site through which pharmaceutical compounds are eliminated. Although metabolism mostly occurs in the liver, it also occurs at a number of other sites in the body, including the gastrointestinal tract, skin, kidneys, lungs, eyes, nasal mucosa, brain, heart, pancreas, testis, spleen, plasma, ovary, placenta, erythrocytes, and lymphocytes. Loss through the lungs is primarily by volatilization.

A portion of the pharmaceutical can be excreted via the kidney, that is, renal excretion. Thus, renal excretion (F_e) can range between 0 and 1 with the latter representing a situation in which renal excretion is the only route of elimination. Renal excretion can be estimated from

$$F_e = \frac{\text{Total drug excreted unchanged}}{\text{Dose}} \tag{3.5}$$

By definition, the fraction that enters the circulatory system is eliminated by extrarenal mechanisms (usually metabolism by the liver and other tissues) and is derived by the difference from renal excretion that is, $1 - F_e$. The excretory organs are able to eliminate polar compounds such as tetracycline and tylosin more efficiently than compounds that are highly soluble in lipids (i.e., lipophilic) such as metronidazole, erythromycin, clindamycin, and trimethoporin. Thus, the highly lipophilic compounds will not be eliminated until they are metabolized to more polar intermediates.

Naturally, drug elimination can involve a combination of the above-listed processes, and elimination by the individual organs is additive, summing up to total clearance. For each organ, elimination is computed by considering the product of quantity (Q) of blood flowing into that organ and the concentration of the drug in that blood (C_A) minus the product of the blood that is flowing out and the concentration of the drug in that blood (QC_V) as depicted in Eq. (3.6).

Rate of extraction from a particular organ $= QC_A - QC_V = Q(C_A - C_V)$ (3.6)

A more informative relationship derived from Eq. (3.6) is the extraction ratio (E), that is,

$$E = \frac{\text{Rate of extraction}}{\text{Rate of presentation}} \times \frac{Q(C_A - C_V)}{QC_A} = \frac{C_A - C_V}{C_A} \tag{3.7}$$

When there is no extraction from the organ, the extraction ratio will be zero (i.e., $E = 0$) and when there is complete extraction (i.e., elimination), $C_V = 0$ and $E = 1$. In general, pharmaceuticals with an extraction ratio of 0.7 or greater are classified as high-extraction-ratio drugs, whereas those with a ratio of 0.3 or lower are classified as low-extraction-ratio drugs under clinical settings (Jeffrey, 2002). Drugs with ratios of 0.3–0.7 are classified as having a moderate extraction ratio.

Pharmaceuticals are designed to be resistant to metabolic degradation and are mostly lipophilic, a trait that gives them a high propensity to be absorbed effectively by organisms. The hydrogen bonding of a drug also correlates well with its absorption by the human intestine. Thus, instead of metabolism toward total degradation (i.e., mineralization) PPCPs are generally metabolized in the body to more polar and more water-soluble derivatives that have a reduced pharmacological activity than the parent compound. This is as opposed to what we want to happen to them once they get into the environment, where total mineralization is the desired outcome. To that effect, many therapeutic agents are only partially metabolized in the body before they are excreted, with the rest of the parent compounds and/or their metabolites being excreted. Irrespective of the extent of metabolism, that process changes the pharmaceutical and physicochemical properties of the compound. Drug metabolism in the body occurs through two main pathways or phases, rightfully classified as phase I or phase II. The first phase results in the modification of the active compound through oxidation, hydrolysis, alkylation, or reduction (i.e., dealkylation), introducing functional groups such as a hydroxyl (—OH), carboxyl (—COOH), amine (—NH$_2$), or thiol (—SH) to the drug molecule. These modifications are facilitated by several enzymes, most common of which are the enzymes of the cytochrome P450 (i.e., CYP450). The second phase involves further modification of phase I metabolites by conjugation, glucuronation, and sulfatation, typically increasing the water solubility of the molecule and, by default, also enhancing the excretion of the compounds due to the formed water-soluble metabolites (La Du et al., 1979). Glucuronation is the most common form of conjugation as glucose is abundant in biological systems. It basically involves the condensation of the pharmaceutical compound (or its biotransformed product) with D-glucuronic acid. Most prone to gluconation are phenols, alcohols, amines, carboxylic acids, although some thiols and steroids may also undergo glucuronation. It generally diminishes the pharmacological and biological activity of the pharmaceutical compound in question. However, on excretion, some glucoronides can be deconjugated once they get into the environment (Huang and Sedlak, 2001) primarily by glucuronidase enzymes that are present in fecal coliform, although the mechanisms and conditions that favor such deconjugation are not entirely clear. Because coliforms are almost always present in wastewater, deconjugation of pharmaceuticals occurs almost immediately once the glucuronide conjugates enter the wastewater treatment system, a scenario that has major ecological implications for the activity of PPCPs in the environment.

A lot of research about the metabolism of PPCPs in clinical practice has been conducted under the auspices of pharmacology. As alluded to above, drug metabolism in the body results in the breakdown of the compound into metabolites that are chemically simpler, just like what chemical and biological processes attain in the environment. When such breakdown occurs in the environment, it is typically referred to as degradation. Initial steps in the degradation process in the environment also involves modification of the active sites (similar to phase I processes in our bodies) but phase II does not occur. Similar to pharmacokinetics in clinical practice, we can determine the concentration of a pharmaceutical compound in a particular environment (see Chapter 2). We can also establish the duration it takes for the compound to be reduced (i.e., degraded) to acceptable levels. It is possible that some of the vast wealth of data holds cues about the potential fate and ecotoxicity of PPCPs in the environment. For example, it is assumed that a drug that is highly metabolized in animal systems is also subject to extensive degradation in the environment (Calamari et al., 2003). It is also assumed that the metabolites are not more toxic or recalcitrant than the parent compound.

Desorption of chemicals in environmental matrices typically occurs in two phases, that is, a rapid followed by a slow phase. That desorption pattern assumes first-order kinetics and is very similar to that displayed by the drug at the absorption site in the body presented in Figure 3.1. Most chemicals are assumed to be eliminated by first-order kinetics, and this kinetics applies when the rate of elimination of the compound at any one time is proportional to the amount of chemical in the body at that same time. Under first-order kinetics, the percentage of the total dose that is eliminated per unit time is independent of the dose. However, in practice, binding, transport, and biotransformation can affect the rate at which compounds are eliminated from a particular system. In our bodies the binding will occur on proteins (i.e., protein binding), and transport will be facilitated by active forces (i.e., active transport). All of these actions cause the elimination of compound to be anything but proportional to the dose and may instead impose zero-order kinetics. Mobility of the drug in the body is mostly reliant on the circulatory system and its intricate but ever-present cardio rhythm. By comparison, the mobility of PPCPs in the environment is largely dependent on the aqueous flow patterns and is counteracted by physical and chemical factors such as sorption, presence of inorganic compounds (including heavy metals), pH, and permeability of the respective environmental matrices. Despite these similarities between pharmacokinetics in the body and the environment (i.e., ecopharmacokinetics), there are also some apparent differences. For one, the excretion of PPCPs from the body occurs after the compounds and their chiral derivatives have been screened through a series of membranes that vary in permeability and retention potential. Once excreted, the compounds completely leave the body. By contrast, their mobility in the environment only displaces the compounds from one locality to another. As will be discussed in Section 3.3, mobility and degradation in the environment is dependent on various environmental factors such as temperature, organic matter (OM) content, chemical properties, and the pH that is prevailing. Some of these factors are going to be quite divergent from one environment to another, unlike the case in our bodies, setting the stage for unlimited complexity in analyzing the fate of PPCPs in the environment.

It is apparent from the schematic in Figure 3.2 that most of the losses of drugs from the body are due to renal and fecal excretion. Some small fractions can also be lost through the lungs as a result of volatilization. Whatever is lost through renal and fecal excretion from humans is likely to end up in the sewer system, whereas most excretions from livestock end up in manure and farm runoff. Assuming that the concentration of the compound in the bloodstream is in equilibrium with that in the tissues (i.e., the site of action), the minimum effective concentration (MEC) represents the minimum concentration of the drug that is required at the receptors to produce the desired effect. Other pharmacokinetic parameters include bioavailability, clearance, volume of distribution, and half-life. A brief description of each of these parameters is warranted and the equations that are associated with each are inescapable.

When the pharmaceutical compound is administered intravenously (i.e., IV administration), 100% of that compound will be available to the systemic circulation. However, pharmaceuticals are not always administered intravenously and to get a sense of their availability when administered in other ways, the term bioavailability (F) has been developed. Thus, under clinical pharmacology, bioavailability reflects the fraction of the dose that is absorbed into the systemic circulation in an intact form (i.e., prior to metabolism). Pharmacokinetic data after IV administration is used as the reference from which to compare extravascular administration and the bioavailability for various compounds range between 0 and 1. Complete absorption is demonstrated when $F = 1$. Bioavailability is proportional to the total area under the curve (AUC) and from a practical perspective, absolute bioavailability is computed by comparing the AUC after oral application (i.e., AUC_{po}) with that after intravenous application (i.e., AUC_{iv}) with some correction for any differences in dosage:

$$\text{Absolute bioavailability} = \frac{AUC_{po}}{AUC_{iv}} \times \frac{Dose^*_{iv}}{Dose_{po}} \qquad (3.8)$$

The area under the curve (Fig. 3.3) can be estimated using the trapezoidal rule whereby

$$AUC = \text{Average concentration over the time interval} \times \text{Time interval} \qquad (3.9)$$

The AUC from Figure 3.3 can be calculated as shown in Table 3.1. However, in a number of instances, the decrease in concentration is not linear but rather exponential. A more accurate method for calculating AUC is to use a log trapezoidal rule. Two consecutive observations on the exponential curve $C_{(t_i)}$ and $C_{(t_i+1)}$ at times t_i and $t_{(i+1)}$ are related to each other by

$$C_{(t_i+1)} = C_{(t_i)}e^{-k_i \, \Delta t_i} \qquad (3.10)$$

where k_i is the rate constant that permits an exponential decline in concentration from $C_{(t_i)}$ and $C_{(t_i+1)}$ in the interval $t_{(i+1)} - t_i$, that is, Δt_i. The constant (k_i) can be calculated by transforming the above equation logarithmically to

$$\ln C_{(t_i+1)} = \frac{\ln \ C_{ti}}{k_i \ \Delta t_i} \qquad (3.11)$$

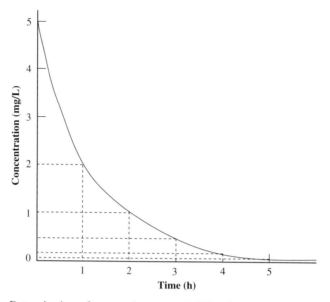

Figure 3.3 Determination of area under curve (AUC) of a compound in the body (see Table 3.1 for calculations).

After rearranging,

$$k_i = \frac{\ln\left[C_{ti}/C_{(ti+1)}\right]}{\Delta t_i} \tag{3.12}$$

Since

$$\text{AUC} = \frac{C_{ti} - C_{(ti+1)}}{k_i} \tag{3.13}$$

TABLE 3.1   Calculation of the AUC Values from Prevailing Concentrations at Different Intervals

Time (h)	Concentration (mg L^{-1})	Time Interval	Average Concentration (mg L^{-1})	Area (mg-h L^{-1})
0	5	—	—	—
1	2	1	$\dfrac{(5+2)1}{2} = 3.5$	3.5
2	1	1	$\dfrac{(2+1)1}{2} = 1.5$	1.5
3	0.45	1	$\dfrac{(1+0.45)1}{2} = 0.725$	0.725
4	0.175	1	$\dfrac{(0.45+0.175)1}{2} = 0.3125$	0.3125
5	0.03	1	$\dfrac{(0.175+0.03)1}{2} = 0.1025$	0.1025
			Total area	6.14

Substituting for k_i in Eq. (3.13) leads to Eq. (3.14):

$$\text{AUC} = \frac{(C_{ti} - C_{(ti+1)}) \, \Delta t_i}{\ln [C_{ti}/C_{(ti+1)}]} \tag{3.14}$$

The total amount of drug eliminated is equal to the clearance time as a product of the AUC (i.e., clearance $\times$ AUC). This product is also co-equal to the amount that is absorbed (i.e., $F \times$ Dose) as is presented in Eq. (3.15):

$$F \times \text{Dose} = \text{Clearance} \times \text{AUC} \tag{3.15}$$

Thus, if the dose, clearance, and AUC are known, bioavailability can be determined from the relationship shown in Eq. (3.16):

$$F = \frac{\text{Clearance} \times \text{AUC}}{\text{Dose}} \tag{3.16}$$

An orally bioavailable compound will have to be stable in the stomach and survive stomach acids, diffuse across the gut wall, survive first-pass metabolism in the liver, partition between aqueous and lipid phases at the target site and (in the case of antimicrobial agents) penetrate the microbial cell wall, inhibit the enzyme, and finally be metabolized and excreted from the body. Meeting all these challenges requires superb physicochemical properties, diffusion capability, and intricate interaction with specific existing proteins such as hydroxylation by cytochrome P450 metabolic enzymes. Based on animals models, absorption and subsequent bioavailability of pharmaceutical compounds has thus been linked to physicochemical characteristics such as intrinsic solubility of the molecule, connectivity of the molecule, electronic nature (i.e., electron affinity, number of aromatic rings, dielectric and conformational energy), the shape as well as size of the molecule (Turner et al., 2004). From such studies, it has been shown that molecular size generally limits the absorption of pharmaceutical compounds through membranes, the bioavailability of the compound changing in a cubic fashion with molar reflectivity, which is based on molar mass and density. However, such extensive analysis has yet to be applied in environmental studies to understand the bioavailability of PPCPs in such settings.

3.1.1 Sorption and Bioavailability of PPCPs in the Environment

Most of what is known about bioavailability of organic compounds in the environment is based on other organic contaminants rather than PPCPs. From those reports, the bioavailability of compounds is affected by a number of ill-defined, often uncharacterized processes, with the only evidence for changes in bioavailability being their declining rate of biodegradation. There is evidence from other organic compounds that the compound can move into microsites within the solid matrix (e.g., sewage, soil, or sediments), which reduces the degradation process because of the physical separation between the degraders and the compound. This process has been characterized as "aging" but should not be perceived as an instance that depicts a change in the identity but rather a change in the behavior of the compound (Alexander, 1994). Thus, the aged PPCP molecule retains its original (parent) form or excreted metabolite status but cannot be completely removed from the environmental matrix by even the most rigorous extraction process with polar or non-polar solvents. However, a series of sequential extraction processes may recover incremental quantities of the compound, confirming its presence beyond what is bio-available. Such an approach clearly shows that whatever fraction of the compound is remaining in the environmental matrix is weathered or aged. Thus, sequential extraction approaches show that as time goes by, the availability of organic compounds to biological systems usually becomes diminished (Hatzinger and Alexander, 1995; Alexander, 2000; Tang et al., 2002) due to the compounds diffusing into micropores where they remains largely inaccessible. For example, the organic fraction of soils contains nanopores of 0.3–1.0 nm diameter in which chemicals can become localized and remain unavailable to potential degraders (Alexander, 2000; Malekani et al., 1997; Xing and Pignatello, 1997).

Organic matter plays an important role in the retention of organic pollutants in the environment. By this token, the mere presence and certainly increases in organic matter in any particular environment decrease the bioavailability of the compounds, attenuating their toxicity. It is important to realize that organic matter is not only a major player in soils but also in sewage and livestock dung, both of which are primary routes of PPCP introduction into the environment. For example, the addition of more and more biosolids to a soil decreased the degradation of ^{14}C-testosterone (Fig. 3.4) due to increasing amounts of the testosterone sorbed to the organic matter fraction. Thus, an increased addition of biosolids to the soil led to a higher percentage of the ^{14}C-label in the extracted fraction corresponding to a decreased amount of $^{14}CO_2$ recovered as the degradation product. Additions of 50% biosolids to the soil severely inhibited the degradation of testosterone. The extent to which adsorption occurs largely depends on the properties of the environmental matrix in which the compound is embedded. More studies about sorption of organic compounds in general (though less specifically for PPCPs) have been done with soils, and from those it is apparent that sorption depends on the organic matter content, pH, octanol–water partition coefficient, solubility, the type and amount of clays, as well as ion exchange capacity. Adsorption is also dependent on both hydrophobic and electrostatic interactions of the compounds with microorganisms and particulate material. Some of the information from those studies can be relatively applicable to the sorption of PPCP, and the role of some of these physicochemical factors in determining the fate of PPCPs in the environment will be discussed in Section 3.3.

In general, pharmaceutical compounds have a very high propensity to sorb to particles (Tolls, 2001; Boxall et al., 2003). However, sorption varies considerably in different soils based on the cation bridging at clay surfaces, hydrogen bonding, cation exchange capacity, and surface complexation. A few studies have specifically been conducted on the sorbing and complexing of PPCPs with clay minerals such as montimorillonite, kaolinite, vermiculite, bentonite, and illite in the terrestrial

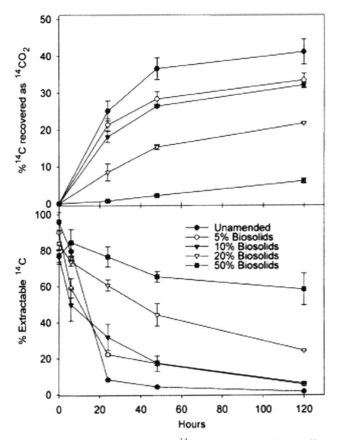

Figure 3.4 Dissipation and mineralization of ^{14}C-testosterone in loam soil amended with increasing quantities of liquid municipal biosolids (i.e., 5–50% v/w). Comparisons are made with the unamended soil. Data are based on three replicates SD. (Adapted from Jacobsen et al., 2005.)

environment (Bewick, 1978; Ghosal and Mukherjee, 1970; Sithole and Guy, 1987a, 1987b; Pinck et al., 1961a, 1961b). Sorption in soils is even more enhanced by various types of clay (Bakal and Stoskopf, 2001; Nowara et al., 1997), soil organic matter content (Marengo et al., 1997; Thiele-Bruhn, 2003), as well as ionic strength, humic substances, and pH (Sithole and Guy, 1987a). Thus, sorption of the compounds occurs instantly under very acidic conditions (pH 3). Bentonite and montimorillonite have expanding lattices that enable a greater exchange surface compared to illite and kaolinite. Sithole and Guy (1987b) used the dodecyltrimethylammonium forms of bentonite to reduce the surface area accessible to tetracycline. They obtained a Langmuir type of adsorption isotherm that suggested that sorption was occurring at limited sites. Thiele-Bruhn et al. (2004) indicate that the sorption of PPCPs onto clay and soil matrices is controlled by the ionic functional groups of the compound.

Besides reducing bioavailability, adsorption and complexation may also cause a loss of biological activity. That possibility is derived from the work presented by Lunestad and Goksøyr (1990) whereby oxytetracycline, used in aquaculture settings, formed complexes with magnesium and calcium ions present in seawater, rendering the antibiotic ineffective. However, not all PPCPs display this loss in bioactivity on sorption. Thus, according to various groups (Ingerslev and Halling-Sørensen, 2000; Thiele, 2000, Halling-Sørensen et al., 2003), such sorption and fixation may only reduce the antimicrobial activity but not completely eliminate it. The differences observed in regard to this aspect may be due to:

1. Differences between the composition of the sediments used
2. Differences in the compounds themselves
3. Differences in the bacterial populations in the specific experiments

It is also recognized that the sorption of hydrophobic organic compounds onto dissolved organic matter can significantly increase the solubility of the compounds under aqueous conditions. The importance of solubility in determining the fate of PPCPs will be discussed in Section 3.3.2. Where available, the sorption data can be fitted to the Freundlich equation [Eq. (3.17)] to derive sorption parameters for the matrix:

$$q = k_F C^n \tag{3.17}$$

where q = amount of compound sorbed in microgram per unit weight of matrix, C = equilibrium concentration of compound (micrograms/milliliter), k_F is the Freundlich distribution coefficient, and n is the Freundlich constant (i.e., the slope for the adsorption–desorption relationship). The k_F is often related to the affinity of the sorbent and thus reflects the degree of sorption. A logarithmic form of Eq. (3.17) is

$$\log q = \log k_F + n \log C \tag{3.18}$$

If the adsorption–desorption relationship is linear, then n is equal to 1 and Eq. (3.18) would be the same as

$$q = K_d C \tag{3.19}$$

where K_d is the distribution coefficient (milligrams/liter). The K_d can be estimated from K_{ow}, the latter being estimated as the ratio between the concentration of the compound to the sorbent (C_s) to its concentration in solution (C_{aq}). Thus, K_d values (also known as distribution coefficient) are an indicator of the sorption and mobility of the compound. In essence, compounds with high K_d values are loosely bound to environmental matrices and more readily transported in the terrestrial environment, impacting surface and groundwater.

The K_d can in turn be used to compute K_{oc}, that is, the distribution coefficient normalized to the percent organic carbon (%OC) present in the sorbent [Eq. (3.20)]:

$$\log \ K_d = \%OC \times 0.41 \times \log \ K_{ow} \tag{3.20}$$

However, further research to predict the sorption and subsequent degradability in the environment is still needed as the K_d values that are calculated can, in some instances, greatly underestimate the sorption of pharmaceutical compounds in various environments (Stuer-Lauridsen et al., 2000; Tolls, 2001; Jjemba, 2004). Such discrepancies may be attributed to differences in pH, concentration of inorganic compounds, clay composition, and prevailing temperature (Bakal and Stoskopf, 2001; Jacobsen et al., 2005). Tolls (2001) suggests that K_{oc} may not be a valid concept for polar compounds that show stronger sorption to dissolved organic matter than that expected/predicted based on K_{ow}. According to that work, sorption is not attributable to hydrophobic partitioning but rather on ionic interactions and hydrogen bonds. More recently, Drillia et al. (2005) used two soil types (i.e., a low and a high organic matter soil) to study the sorption and mobility of six pharmaceutical compounds that are frequently detected in municipal sewage, groundwater, and surface water. Adsorption was also examined on sludge that had been digested aerobically versus that which had been digested under anaerobic conditions. Their findings are summarized in Table 3.2. In general, biodegradation of the six compounds they tested occurs more quickly under aerobic than anaerobic conditions. This has implications in soils but also in aquatic environments in the sense that deeper layers of the water profile contain lower oxygen levels compared to surface waters. Biodegradation was also more rapid in the soil that had lower OM. Work by Gavalchin and Katz (1994), Martens et al. (1996), and Thiele-Bruhn (2003) showed that sorption was accompanied by an expansion of the montimorillonite clay, possibly leading to the compounds penetrating the micropores that effectively protect them from both biotic and abiotic degradation.

3.1.2 Compound Half-life and Clearance

Where the aging phenomenon occurs, it is usually preceded by a duration under which disappearance of the compound is rapid followed by a slow disappearance of the remaining compound (Fig. 3.5a). This clearly shows a biphasic pattern, depicting typical first-order kinetics. First-order kinetics is the simplest kinetic situation in drug metabolism as a typical pharmaceutical compound concentration over time relationship shows this type of kinetics, decreasing at a rate that is proportional to the amount of compound remaining, but never appearing to reach zero. The shape of this relationship is also synonymous to that of the concentration of drug at the absorption site presented in Figure 3.1. When concentration of the drug is plotted on a logarithmic scale against time, a straight line (i.e., linear relationship) is obtained (Fig. 3.5b). Figure 3.5 can be used to introduce the concept of half-life ($t_{1/2}$) as it

TABLE 3.2 Sorption Parameters for Six Commonly Encountered Pharmaceuticals in Two Soils and with Sludge Digested Under Distinctly Different Conditions[a]

Compound	MW	Solubility (mg/mL)	pK_a	log K_{ow}	k_F for Soil High OM	k_F for Soil Low OM	k_F for Sludge Aerobic	k_F for Sludge Anaerobic	K_d for Soil High OM	K_d for Soil Low OM	K_{oc} for Soil High OM	K_{oc} for Soil Low OM
					Adsorption							
Oxfloxacin	361.4	?	5.97–8.28	0.35	3,224	832	1,125	169	3,554	1,192	50,056	322,162
Propranolol	259.4	33	9.53	3.48	207	7.15	576	758	199	16.3	2,803	4,405
Diclofenac	296.2	?	4	1.12	172	0.44	41	745	164.5	0.45	2,310	121
Carbamazepine	236.3	0.01	7	2.45	57	0.51	49	396	37	0.49	521	132
Sulfamethaxazole	253.3	0.01	1.7, 5.6	0.89	36.2	0.28	19.1	BD	37.6	0.23	530	62.2
Clofibric acid	214.6	0.58	2.5	2.57	7.52	BD	BD	BD	5.38	BD	75.8	BD
					Desorption							
Oxfloxacin					5,650	6,172	ND	ND	4,525	4,087	63,732	1,104,595
Propranolol					251	6.89	ND	ND	250	14.8	3,521	4,000
Diclofenac					245	0.34	ND	ND	247	0.64	3,479	173
Carbamazepine					73.4	0.58	ND	ND	63.6	0.44	896	119
Sulfamethaxazole					42	0.57	ND	ND	43.1	0.40	607	108
Clofibric acid					10.5	BD	ND	ND	9.59	BD	135	BD

[a]k_F = Fruendlich distribution coefficient. BD = Below detection as most of the compound remained in the aqueous phase. ND = Not determined. The high OM soil was a sandy loam (15.84% clay; 26.64% silt; 57.52% sand) with 7.1% OC, pH 4.3, and a specific area of 0.415 m^2 g^{-1} whereas the low OM soil was a clay soil (43.28% clay; 27.2% silt; 29.52% sand) with 0.37% OC, pH 6.8, and specific area of 22.98 m^2 g^{-1} (*Source*: Drillia et al., 2005). Solubility values were compiled from Kasim et al. (2004) and for clofibric acid from Emblidge and DeLorenzo (2006).

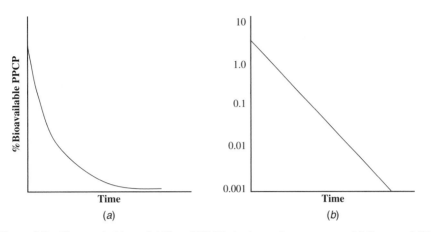

Figure 3.5 Changes in bioavailability of PPCPs in the environment on a (*a*) linear and (*b*) semilogarithmic scale.

relates to pharmacokinetics derived from Eq. (3.21):

$$t_{1/2} = \frac{0.693}{k_e} \tag{3.21}$$

where k_e is the elimination or degradation rate constant represented by the slope of the line on the log concentration versus time curve. From the relationship in Eq. (3.21), the half-life of the drug in the body or environmental matrix ($t_{1/2}$) can be determined from any two points on the plot that describe a 50% decrease in drug concentration.

In clinical settings, k_e is influenced by drug clearance (CL) and the volume of distribution (V) based on the relationship in Eq. (3.22):

$$k_e = \frac{CL}{V} \tag{3.22}$$

Clearance in this instance relates the concentration of the drug in blood to the rate at which it is eliminated from that system. It represents the volume of blood that is irreversibly cleared of the compound per unit time. The volume of distribution (V) is the amount of drug in the body per unit concentration in the blood, that is,

$$V = \frac{\text{Amount of drug in the body}}{\text{Drug concentration in the blood}} \tag{3.23}$$

Thus, volume distribution is the apparent space into which an amount of compound is distributed in the body to result in a given plasma concentration. It can best be described with a hypothetical situation in which the concentration of a particular drug in the blood is 0.2 mg mL^{-1}. If the initial dose that was administered (i.e., amount of drug in the body) is known (e.g., 100 mg), then based on Eq. (3.23), V is 500 mL. This means that 500 mL of blood will be needed to account for all of the drug in the body. From a clinical perspective, chemicals with a high affinity for tissues will have a large volume of distribution, whereas those with a low affinity for tissues will have a volume of distribution (Medinsky and Valentine, 2001). From an environmental perspective, the body and blood are synonymous to the solid and aqueous compartments of an environmental matrix, respectively, and the above relationship may have ramifications in the sense that high volume of distribution drugs under clinical settings may also be more sorbent once they get into the environment. The volume of distribution is dependent on the flow of blood to different tissues, affinity of the drug to different tissues, pH, and the solubility of that drug in lipids. It is important to note that only the unbound drug is available for distribution. By relating Eqs. (3.21) and (3.22), a new relationship [Eq. (3.24)] is generated. Thus,

$$\frac{CL}{V} = \frac{0.693}{t_{1/2}} \implies t_{1/2} = \frac{0.693V}{CL} \tag{3.24}$$

From the relationship in Eq. (3.24), it is apparent that the half-life ($t_{1/2}$) is directly proportional to the volume of distribution and inversely proportional to clearance. In other words, the half-life can be doubled by either doubling the volume of distribution or reducing the clearance by a half. Thus, under environmental settings, the half-life of a PPCP can double by doubling the concentration of compounds in the solid phase or reducing the rate at which it is degraded from the system by a half.

From a clinical perspective, it is desirable that the half-life of the pharmaceutical compound be sufficiently long so as to stay in contact with the intended target. Thus, some efforts have been devoted to controlled release of pharmaceuticals in the body so as to increase the half-life by altering the molecular structure or using more stable carriers (Goole et al., 2007). Furthermore, most of the present-day PPCPs, including antibiotics, are synthetic or semisynthetic and are designed to be quite stable, making them less biodegradable. Clinical studies have shown that it generally takes five to seven half-life cycles to completely eliminate the drug from the body (Medinsky and Valentine, 2001; Jeffrey, 2002). However, if the elimination is based on first-order kinetics, the concentration of chemical remaining will theoretically never reach zero, and these many half-life cycles would have eliminated about 99.2% of the compound, which for practical purposes is considered as complete elimination. Analogous determinations have not been established for PPCPs in the environment. In any case, PPCP half-lives in the environment are likely to vary greatly from one compound to another and possibly from one environment to another as factors such as pH, organic matter content, temperature, biotic status (i.e., existing microbial community), and the like have a significant role in the elimination process. For example, Halden and Paull (2005) used structure and activity relationships to estimate the half-life of triclosan to range between 1 day in air and 540 days in sediments

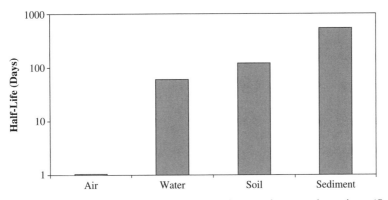

Figure 3.6 Estimated half-lives of triclosan in various environmental matrices. (*Source*: Halden and Paull, 2005.)

(Fig. 3.6). In the environment, a half-life of 15 days is considered to be readily degradable (Schowanek and Webb, 2005).

3.2 DEGRADATION OF PPCPs IN THE ENVIRONMENT

Persistence of any PPCP in the environment increases the potential for long-term varied effects on the ecosystem. Chapter 1 gave us a snapshot of some of compounds that are widely used. However, it should be emphasized that the number of therapeutic agents currently available are quite numerous, and Chapter 1 only highlighted a subset of many PPCPs. It can be quite overwhelming and possibly uneconomical to aim at determining the degradability of each compound or its impact on the environment. Rather, it is more economical to assess representative compounds from substances that have somewhat similar chemical structures, functions, or modes of action. We will thus continue the format that was developed in the previous two chapters by focusing on representative compounds based on their common usage in therapy. Simple experimental setup systems such as the closed bottle test (CBT, i.e., OECD 1992 Method 301D), Zahn-Weltens test (OECD 1992 Method 302B), Erlenmeyer flasks, and vacuum desiccator assemblies have been used to assess the degradability of these compounds with the caveat that the relevance of some of these approaches to translate into their degradability in the environment have been questioned (Jjemba, 2002a; Zweiner and Frimmel, 2004). For one, some of these experiments have used high concentrations of the compound, concentrations that are unlikely to occur in the environment. Most pharmaceuticals and their metabolites that occur in wastewater and biosolids do so in small quantities (nanograms per liter or per kilogram range). On the other hand, other studies have directly looked at the concentration of PPCPs in full-scale reactors. Irrespective of the approach, it has been determined that a variety of compounds of therapeutic origin remain stable and survive treatment processes in sludge and manure. It should be noted that the excreted pharmaceuticals may also directly affect the aerobic and anaerobic (methanogenic) degradation in sewage and farm manure anaerobic decomposition treatments by directly impacting the microbial degraders (Loftin et al., 2005). Comparisons between studies for biodegradation are difficult as the compounds and the study conditions greatly differ. In both aquatic and terrestrial settings, elimination of PPCPs can be due to biotic or abiotic processes. These include absorption (and complexation), hydrolysis, mineralization (or degradation), thermolysis (i.e., temperature effects), volatilization, photolysis, and redox reactions. Intricacies within each of these processes as they relate to the dynamics and fate of PPCPs in the environment are discussed below.

3.2.1 Degradation of Antibiotics in the Environment

Antibiotics can impact microbial communities in sewage systems, negatively affecting the degradation processes (Stanislavska, 1979; Gomez et al., 1996) and other essential processes such as nitrification (Tomlinson et al., 1966; Gomez et al., 1996; Klaver and Matthews, 1994) and sulfate reduction (Hansen et al., 1992). As outlined in Chapter 2, antibiotics emanating from therapeutic and subtherapeutic use occur in various matrices in the environment, with the highest levels detected in hospital and long-term health care home effluents. The biodegradation of the other individual nonantimicrobial PPCPs can also be slowed down by antibiotics by deactivating the microbial degraders. This reality has ecological implications in sewage and wastewater treatment systems where compounds typically occur in mixtures, with antibiotics certainly being some of the components of the mixture. Unfortunately, the degradation of PPCPs in mixtures has not been widely studied.

The effective killing and elimination of the pathogens during the treatment of an infection is often successful in clinical settings as it is mediated by the host's immune defense mechanisms. However, in the environment, such augmentation is nonexistent, leading to the proliferation of organisms in the environment that may otherwise have been negatively impacted if the compound was in a host. Furthermore, concentration of the PPCP in the environment tends to be much lower (i.e., in subtherapeutic concentrations) than those encountered under therapeutic conditions. This, in the case of antimicrobial agents can be a driver for the development of resistance. Antibiotic resistance will be discussed extensively in Chapter 4. The fate of several important categories of antibiotics in the environment is discussed below.

3.2.1.1 Quinolone Compounds   More recent generation quinolones have deliberately been designed to have a higher retention and bioavailability in the body (Jjemba and Robertson, 2006) but substantial amounts are still excreted as the parent compound. For example, more than 90% of trovafloxacin is excreted as parent compound (Stass and Kubitza, 1999). Furthermore, because of their more complex structure, these newer 4-quinolone antimicrobial agents are likely to be more persistent in the environment. The increased complexity of this group of antibiotics comes at an additional risk, that is, increased toxicity. Thus, some of them, after a brief period of use, have already been suspended, put under even more restricted use, or completely withdrawn from the market due to serious hepatotoxicity or cardiac arrest (Ball, 2000; Chen et al., 2000; Emmerson and Jones, 2003). Their unfavorable eco-toxicity profile to a variety of organisms, including bacteria, algae, crustacean, and fish has been documented (Halling-Sørensen et al., 2000; Golet et al., 2002a). Golet et al. (2002b) reported the presence of fluoroquinolones at concentrations of 0.27–0.3 mg kg^{-1} in fields at 20 months after sewage sludge had been applied to the site as a fertilizer, indicating that they can persist for long durations in the environ-ment. In studies by Samuelsen et al. (1994), flumequine and oxalinic acid were per-sistent in sediments for more than 180 days. They are mostly introduced in the environment in animal wastes (livestock manure) from intensive farming and from sewage treatment plants in biosolids used to fertilize fields (Jjemba, 2002a; Hamscher et al., 2002; Díaz-Cruz et al., 2003). The use of these compounds in aqua-culture also introduces them directly into the surface water and surrounding sediments.

In general, 4-quinolones have a very high propensity to sorb to soil, sediments, sewage, and livestock dung (Marengo et al., 1997; Tolls, 2001; Ingerslev et al., 2001; Halling-Sørensen et al., 2003), an attribute that can delay their degradation in the environment. Based on evidence with enrofloxacin accumulated using infrared spectrometry, adsorption onto clay surfaces is attributed to the carboxylic acid (i.e., —COOH) moiety binding the positively charged clay surface, that is, the Stern layer (Nowara et al., 1997). As evidenced in Table 1.9, this moiety is present in all quinolone pharmaceutical compounds. It is interesting to note that the same moiety is essential for gyrase binding together with the ketone of the C4 position in vivo (Appelbaum and Hunter, 2000; Andersson and MacGowan, 2003). Once adsorbed onto environmental matrices, their removal (i.e., desorption) is very slow and not completely reversible as is displayed in Table 3.3 where desorption was studied using a CaCl$_2$ solution at concentrations that are typical to those encountered

TABLE 3.3 Sorption of Enrofloxacin on Five Divergent Soils

Soil Origin	%Clay	Dominant Clay (%)	%OC	pH	%Sorption Adsorption	%Sorption Desorption[a]	CEC[b] (mEq/ 100 g)
Brazil	41.7	Kaolinite (22%)	1.63	4.9	99.7	0.2	13.6
Philippines	17.2	Montimorillonite (21%)	0.73	5.3	99.7	2.6	16.9
Sweden	7.2	Montimorillonite (25%)	1.23	6.0	98.6	0.2	7.0
France	23.4	Kaolinite (8%)	1.58	7.5	95.2	0.9	19.1
Germany	2.5	Montimorillonite (8%)	0.70	5.3	95.5	1.3	12.0

[a]Desorption was conducted using 0.01M CaCl$_2$.
[b]CEC = cation exchange capacity.
Reprinted from Nowara et al. (1997).

in environments such as soils and sediments. Thus, enrofloxacin adsorption rates of 95.2–99.7% were registered compared to meager desorptions of 0.2–1.3% of the compound, irrespective of the clay type and content as well as organic matter content.

The recalcitrance of other 4-quinolones such as ciprofloxacin, sarafloxacin, and ofloxacin, has been established (Marengo et al., 1997). When 4-quinolone compounds are biodegraded, the microbial-mediated reactions include demethylation, oxidation of the side chains (such as cyclopropyl), reduction of the carboxylic acid, and for the 7-(N_4'-alkypiperazinyl)-6-fluoroquinolones, the occasional transformation of the piperazine ring. Their demethylation is catalyzed by peroxidases and cytochrome P450 enzymes that are commonly found in microorganisms (Chen et al., 1997). As indicated earlier, CYP450 enzymes are also involved in phase I metabolism of drugs in our bodies. If the piperazine ring remains intact, these antimicrobial agents preserve their antimicrobial activity in the environment (soils, sediments, water). In other instances, the degradation of fluoroquinolones by the white rot fungus *Phanerochaete chrysosporium* and the brown rot fungus *Gloephyllum striatum* has been demonstrated in vitro (Wetzstein et al., 1997). These fungal species are known to degrade a wide range of polyaromatic compounds (Bumpus et al., 1985; Alexander, 1994; Field, 2003). In the environment, however, degradation of these compounds occurs at a much lower rate, possibly due to sorption, which limits access to the fungal degraders (Martens et al., 1996).

3.2.1.2   Fate of β-lactams and Cephalosporins

As noted in Chapter 1 (see Table 1.8), both β-lactams and cephalosporins have a similar mode of action, inhibiting the synthesis of cell walls by impacting the enzymes involved in the assembly of the peptidoglycan layer and its cross-linkages. In general, β-lactams such as penicillins and tetracyclines precipitate with cations readily, accumulating in sewage sludge and sediments (Stuer-Lauridsen et al., 2000; Christian et al., 2003). However, β-lactam antibiotics have only been occasionally detected in the environment (Kolpin et al., 2002; Lindsey et al., 2001), which suggests that they may not persistent despite the fact that bacteria that are resistant to these antibiotics have been encountered in environments such as sewage. On the other hand, the infrequent detection of the compounds may be a result of the possibility that the methods are not fully adequate in detecting them. The characteristic β-lactam ring can be ruptured by the bacterial enzyme β-lactamase or by chemical hydrolysis. Gilbertson et al. (1990) studied the degradation and mobility of the cephalosporin antibiotic, ceftiofur sodium, in cattle dung and three soils of varying pH and found this antibiotic to degrade to microbiologically inactive metabolites. Its half-life was 22, 41, and >49 days in the soils of pH 6.96, 7.37, and 8.02, respectively, showing some level of association of degradation to soil pH. When the dung was sterilized, the degradation rates severely declined, demonstrating the importance of microorganisms in the degradation of antibiotics in feces and urine.

3.2.1.3   Degradation of Tetracyclines

Tetracyclines are relatively stable under acidic conditions but not under alkaline conditions. They form complexes with chelating agents such as divalent metals, proteins, and silanol groups (Oka et al., 2000;

Hirsch et al., 1999). They are generally sparingly soluble in water and, just like penicillins, tetracyclines also precipitate with cations readily, accumulating in sewage sludge and sediments. Tetracycline has high sorption coefficient values (Tolls, 2001) and this may explain why it is not so mobile in environments such as soils and sediments. Despite the high sorption coefficient, however, the presence of high levels of dissolved organic matter in liquid manure was associated with increased mobility of this antibiotic in soil column studies (Aga et al., 2003). The unexpected mobility of this compound, during those studies may have been influenced by the preferential flow through cracks and worm channels. As a matter of fact, fieldwork by Kay et al. (2004) demonstrated that weak acids such as OTC and sulfonamides are potentially mobile in the environment, mostly through cracks and channels. To that effect, after repeated application of OTC in soil as a drench, residues were not found at any depth below 20 cm in a sandy soil profile in Florida (Gonsalves and Tucker, 1977). Its adsorption was dependent on the soil type with the sandy soil (99.1% sand) showing the least OTC adsorption and the organic muck registering the highest adsorption of the compound. Gonsalves and Tucker (1977) also found 25 μg OTC g^{-1} soil 40 days after application, but this compound ultimately declined steadily and persisted for 18 months at concentrations of less than 1 μg g^{-1} soil. Its immobilization in the soils was attributed to the existing clay and organic matter, which strongly bind the antibiotic to soil particles. OTC was also quite persistent in sediments under laboratory conditions, with no degradation registered after 6 months of incubation (Samuelson, 1994). An earlier laboratory study by this same team had reported a half-life of 30–64 days for OTC in sediments from a fish farm (Samuelson, 1989). These seemingly contradictory observations from Samuelson's work can be due to the differences in sediments, the most likely one being an absence of OTC degrading microorganisms in the sediments used in the 1994 study compared to the 1989 study for which sediments were obtained from a fish farm and possibly already contained OTC degraders.

Tetracyclines generate the metabolite 4-epi-tetracycline, which is known to occur quite frequently in significant amounts during storage of manure (and possibly sludge). It has negligible antimicrobial activity but is reversible to the parent compound. Studies by Kühne et al. (2000) that simulated the stability of tetracycline in manure storage tanks showed an exponential decline in the parent compound in Ringer's solution (a solution that is closely related to water as it is only composed of 8.6 g NaCl, 0.3 g KCl, and 0.33 g CaCl per liter). The decline was especially more prevalent when the solution was not ventilated (Fig. 3.7). An even more rapid decline in the parent compound was registered in liquid manure that was ventilated. In the solution and liquid manure, the tetracycline degradation showed a biphasic trend, with the rates being rapid in the first few days and slowly declining at a steady rate. The apparent differences in the degradation of this antibiotic in liquid manure versus the solution was attributed by those workers to inherent differences in the pH. The pH in the manure increased from 7.6 at the onset of the experiment to 8.3 in the unventilated and 8.7 in the ventilated manure. By comparison, the pH in the Ringer's solution remained stable at 6.2–6.4 with and without ventilation. These pH changes may have impacted the prevailing microbial degraders differently.

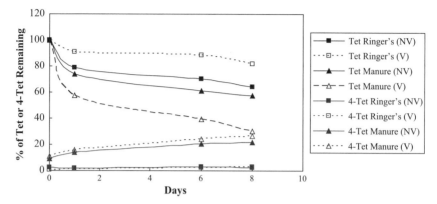

Figure 3.7 Decrease in concentration of tetracycline (Tet) in a Ringer's nutrient solution and liquid manure that were ventilated (V) or not ventilated (NV). Ringer's solution was made by dissolving two Ringer's tablets (Merck, Darmstadt, Germany) in 1000 mL deionized water. (Based on data from Kühne et al., 2000.)

All samples contained the metabolite 4-epi-tetracycline, and this was even more pronounced in the liquid manure. Under strong acidic conditions (pH < 2), tetracycline is metabolized to anhydro-tetracyclines, which are bioactive, whereas alkaline pH conditions (pH > 7.5) can favor the formation of isotetracyclines, which show almost no in vitro activity. Tetracycline can also undergo extensive photodecomposition, forming a variety of products (Oka et al., 1989; Peterson et al., 1993).

3.2.1.4 Degradation of Macrolides   Among macrolides, most of the degradation studies have been done with tylosin, a compound that is widely used in animal production as a growth promoter. Once introduced into soil, tylosin is less firmly bound to soil colloids compared to other antibiotics (Bewick, 1979; Westergaard et al., 2001), and it disappeared from soils after manure was applied (De Liguoro et al., 2003). It is a weak base that is unstable under alkaline and acidic conditions but is relatively stable under neutral conditions (i.e., pH 7). These differences are important as media such as manure and sludge can greatly affect the hydrolysis of this compound during storage or treatment of these materials. Tylosin and a variety of macrolides have very low solubility, which increases with increases in solvent polarity (Salvatore and Katz, 1993). Ingerslev and Halling-Sørensen (2001) studied the degradation of three antibiotics in soil–manure slurries under aerobic laboratory conditions. In those studies, the half-lives for tylosin, olaquindox, and metronidazole were 4.1–8.1 days, 5.9–8.8 days and 9.7–26.8 days, respectively. The half-lives were not concentration dependent or dependent on the type of soil. However, the expectation under field conditions is likely to differ from these laboratory-based results as prevailing conditions such as wetting and drying, temperature, soil pH, redox potential, and abundance of microorganisms in the soil are known to play a major role in the degradation of PPCPs. As a matter of fact, more recent studies by that same group has found the half-life of tylosin to range between 49 and 67 days under field conditions (Halling-Sørensen et al., 2005), a far longer duration than that of 4–8 days in their earlier laboratory-controlled studies. In that later field study, the half-life of chlortetracycline was also found to be 25–34 days. Thus, half-lives under field conditions seem to be substantially higher compared to those obtained under controlled laboratory conditions. By comparison, Donoho (1984) reported a faster degradation rate of monensin under field conditions compared to those detected in laboratory studies indicating the complexities of predicting degradation of PPCPs in the environment.

Erythromycin, another macrolide, is rapidly transformed to anhydro-erythromycin in the environment (Hirsch et al., 1999). A recent study by Chander et al. (2005) compared the sorption of tetracycline and tylosin in two soils that differ in clay and organic matter content. The adsorption of each of the antibiotics was higher in the soil that has a higher clay and organic matter content as these soil traits also provide greater exchange capacity (Fig. 3.8). Those workers speculated that the larger tylosin molecules (MW = 916.1) are adsorbed on the surface of the clay

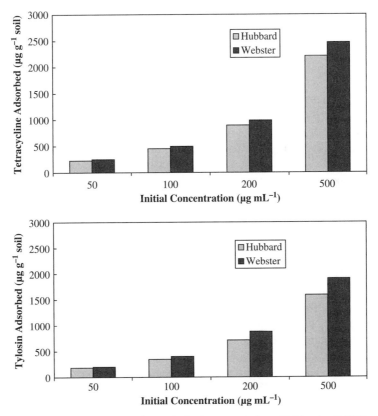

Figure 3.8 Adsorption of tetracycline and tylosin on sandy loam Hubbard soil (10% clay and 2.2% organic matter) and a clay loam Webster soil (34% clay and 4.4% organic matter). (Based on data from Chander et al., 2005.)

particles, whereas the much smaller tetracycline molecules (MW = 444.4) are able to penetrate into the interclay spaces.

3.2.1.5 Fate of Other Important Groups of Antibiotics

As indicated in Chapter 1, there has been a lot of interest in the potential fate of the livestock antibiotic virginiamycin due to its similarity with other streptogramins such as quinupristin-pristinamycin (marketed as Synercid), which are used in clinical settings. Surprisingly, few studies have looked at the degradability of virginiamycin. Interestingly not many field studies to determine its fate in the environment have been conducted, but Weerasinghe and Towner (1997) found virginiamycin persistent in six different soils under laboratry conditions.

Sulfonamides are weak organic acids (pK_a 4.5–7.5) and are quite hydrophilic. Their hydrophilic nature enables them to be easily transferred to aquatic environments. Work by Boxall et al. (2005) shows that the sorption coefficient (K_d) of sulfa-chloripyridazine (a sulfonamide antibiotic) in both a clay loam and a sandy loam soil decreased as pH increased. Those studies have a relevant bearing to the environment as sulfonamides typically used to treat livestock are excreted in the urine and feces, and the animal manure containing the excreted antibiotics is routinely applied to land. The livestock manure, because of inherently high levels of ammonia, increases the pH of the soil. Thus, as the livestock manure is applied to the soil, the increase in pH reduces the sorption of the compound, which in turn increases the nonsorbed fraction that can be released into the soil solution and ultimately into the surface and ground-water. Sulfonamide degradation was suspected in aquifers even under anaerobic (i.e., sulfate reducing and methanogenic) conditions (Holm et al., 1995).

3.2.2 Degradation of Analgesics and Anti-inflammatory Drugs

The degradation of ibuprofen (Motrin) has been one of the most well-studied among drugs that belong to this use-defined category of compounds. In the body, ibuprofen is metabolized to carboxy-ibuprofen, hydroxyl-ibuprofen, and carboxyl-hydrotrophic acid. The degradation pathway for ibuprofen is shown in Figure 3.9 with carboxy-ibuprofen reportedly occurring under anoxic conditions whereas hydroxyl-ibuprofen occurs under oxic conditions. These metabolites have rarely been tracked in environmental studies, but Stumpf et al. (1998) report that they can occur at high concentrations. Thus, the first two metabolites have been reported in the environment (Buser et al., 1999) at concentrations that were approximately 1.5 times higher than those of the parent compound. These metabolites have also been reported in laboratory batch studies by Zwiener et al. (2002) albeit at much lower concentrations. However, some studies that may not necessarily have looked at its metabolites show that ibuprofen typically occurs in water at concentrations that are lower than those encountered for diclofenac (Tixier et al., 2003; Rodriguez et al., 2003). Ternes (1998) assayed the removal of 32 different pharmaceuticals from sewage, some of which were analgesics and anti-inflammatory agents. From those studies, he found that 90, 81, 69, and 66% of ibuprofen, acetylsalicylic acid, diclofenac, and naproxen daily loads, respectively, were removed in the effluent that was

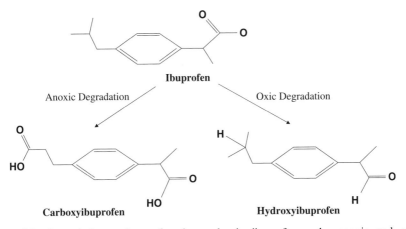

Ibuprofen

Anoxic Degradation

Oxic Degradation

Carboxyibuprofen

Hydroxyibuprofen

Figure 3.9 Degradation pathway for the analgesic ibuprofen under anoxic and oxic environments.

disposed of in German rivers compared to the effluent from municipal waste treatment plants. However, Ternes's (1998) approach did not establish whether some disappearance of the respective drugs from the effluent was due to sorption onto the sludge or to biodegradation. If just sorbed onto the colloids, the compound can still potentially become available to nontarget organisms in the environment.

In the same regard, Zwiener and Frimmel (2003) compared the degradability of ibuprofen with that of diclofenac and clofribic acid, the metabolite for several lipid-lowering drugs (i.e., fibrates). Those comparisons were prompted by the fact that all three pharmaceuticals are high volume drugs in Europe, have frequently been detected in water, and reportedly differ in persistence. From these studies, the elimination of ibuprofen in a pilot sewage plant under a steady state was 40–43% of its initial concentration compared to both clofibric acid and diclofenac for which 94–99% was not degraded (Fig. 3.10). Degradation in biofilm reactors was also comparable to that attained in the pilot sewage plant as ibuprofen was reduced to 30–36%,

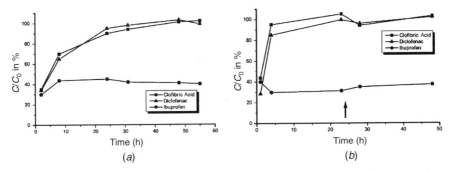

Figure 3.10 Elimination of three pharmaceuticals from wastewater in (*a*) a pilot sewage plant and (*b*) a biofilm reactor. Figures adapted from Zwiener and Frimmel (2003).

whereas 96–99% clofibric acid and diclofenac still persisted in the same duration. This short-term experiment confirmed what those workers had also observed in long-term experiments of Zwiener et al. (2000). Their pilot sewage plant study lasted 23 days, whereas the biofilter reactors were monitored for 120 days. The degradation of ibuprofen in soil can also be substantial (Heberer and Adam, 2005) with the degradation products differing under anoxic and oxic conditions.

Another common analgesic, acetaminophen, is also readily degraded and removed during normal treatment of sewage. It is also excreted as a conjugate but is easily deconjugated in the sewer back to its parent form (Henschel et al., 1997). A number of reports indicate that aspirin is readily biodegradable in the environment. As a matter of fact, its removal or degradation rates in wastewater can be more than 90%, but despite such high removal rates in a recent study Nakada et al. (2006) found acetylsalicylic acid (aspirin) as the most abundant PPCP in the wastewaters of Tokyo (Japan), and the compound was still detected in the treated effluents, suggesting that it can also persist. Where degradation occurs, it is of a microbial of chemical nature, generating salicylic acid. As mentioned in Chapter 2 diclofenac has been encountered in some rivers in relatively high concentrations, suggesting that it is also fairly persistent. It is also quite mobile in soil. Other analgesics such as ketoprofen, naproxen, and fenoprofen are also fairly resistant to microbial attack (Boyd et al., 2005; Nakada et al., 2006).

3.2.3   Degradation of Estrogens and Other Reproductive Hormones in the Environment

Estrogens are used as birth control pills primarily as 17β-ethinylestradiol (EE2) and in drugs (hormones) that relieve menopausal symptoms. The 17α-ethinylestradial is eliminated from the body as conjugates with sulfates and glucuronic acid (Ranny, 1977). The conjugated form does not have any estrogenicity. For example, Alder et al. (2001) found that on average, 58% of the total E1, 50% of E2, and 26% of EE2 in various raw sewage in Germany STPs were conjugated. Once in the environment, these are cleaved quite readily by existing microorganisms, releasing the EE2, becoming hydrolyzed into free, and more potent estrogens and glucuronide together with sulfuric acid in the presence of fecal bacteria, notably *Escherichia coli* (Khanal et al., 2006). This conversion requires glucuronidase and sulfatase enzymes. The clearance occurs following pseudo-first-order kinetics (Henschel et al., 1997; Panter et al., 1999; Ternes et al., 1999a, 1999b). Thus, EE2 was not significantly reduced in aerobic batch experiments with activated sludge in the studies conducted by Ternes et al. (1999a, 1999b). By comparison, Andersen et al. (2003) registered a 50 and 70% reduction in estrone (E1) and EE2 concentrations, respectively, in a denitrification–nitrification STP with a long (i.e., 11–13 days) retention time (Table 3.4). The longer retention time enables the growth of microorganisms that are capable of degrading EE2. Those results show that most of the removal occurred during the denitrification process. It is notable that the concentrations of E1 in the digested sludge increased compared to what was detected in the secondary effluent, possibly as a

TABLE 3.4 Fluxes of Estrone (E1), 17β-estradiol (E2), and 17α-ethinylestradiol (EE2) in the Wiesbaden STP, Germany

Mass Flux of	Influent	Primary Effluent	Denitrification Tank 1	Denitrification Tank 2	Nitrification	Secondary Effluent	Digested Sludge
E1 to E2 (g/d)							
Dissolved			11.4	<0.9	<0.7	<0.13	0.04
Adsorbed			8.6	6.2	5.2	0.003	0.55
Total	5.4	5.8	20.0	<7.1	<5.9	<0.133	0.59
EE2 (g/d)							
Dissolved			0.4	0.3	<0.24	<0.07	<0.0005
Adsorbed			1.3	1.5	<1.0	0.001	<0.03
Total	0.54	0.35	1.7	1.8	<1.2	<0.071	<0.03

Based on data from Andersen et al. (2003).

result of the more nonconjugated E1 and E2 being discharged into the digester from the primary clarifier.

Some bacteria such as *Rhodococcus zopfii*, *R. equi*, *Bacillus thuringiesis*, *Pseudomonas fluorescens*, *Corynebacterium* spp., and *E. coli* have the ability to degrade estrogenic compounds present in wastewater into harmless products (Khanal et al., 2006). According to that review, the degradation of these compounds was reportedly faster in municipal WWTPs (i.e., 84% degraded within 24 h) than in industrial WWTPs (only 4% within 24 h), clearly suggesting a dramatic difference in the microbial diversity between these two wastewater systems. A degradation pathway for E2 has been elucidated by Lee and Liu (2001) and is shown in Figure 3.11. E2 is initially oxidized on the cyclo-pentane ring D at C17 into E1. This is further degraded into a metabolite (X1) and, through the TCA cycle, ultimately to CO_2. The degradation of EE2 in the environment is possibly similar to that of most aromatic compounds, including hydroxylation of the ring with oxygen as the reagent. As a parallel, its metabolism in humans is initiated by hydroxylation at position 2, forming 2-hydroxy-ethinylestradiol (Colucci and Topp, 2001). Thus, ultimate degradation of estrogens seems to depend on the destruction of the phenolic ring. The fungus *Paecilomyces lilacinus* has also been implicated in the degradation of estrogens and other aromatic compounds as it has the ability to cleave the phenolic ring of biphenyl into di- and trihydroxylated metabolites (Gesell et al., 2001).

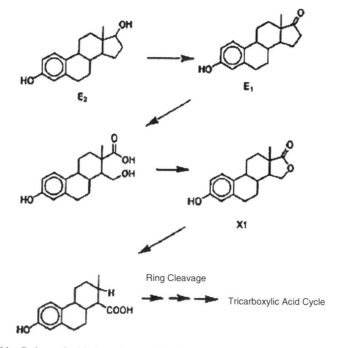

Figure 3.11   Pathway for biodegradation of E2 by bacteria in sewage. (Adapted from Lee and Liu, 2002.)

Natural estrogens, notably estrone (E1), estradiol (E2), and estriol (E3) have log K_{ow} values that range between 2.81 and 3.94 (see Table 1.15); log K_{ow} values for synthetic estrogens are even higher. These higher log K_{ow} values render the estrogens highly hydrophobic and readily adsorbable onto organic solids (Holthaus et al., 2002; Jürgens et al., 2002), a trait that directly affects the concentration of these compounds in environmental matrices such as livestock manure and biosolids as well as organic matter in general. However, estrogens have been detected in groundwater and drinking water (see Chapter 2). Freundelich isotherms have been used to model sorption behavior of steroidal estrogens (Casey et al., 2003). Typical estrone (E1) and 17β-estradiol (E2) adsorption coefficient (K_f) values in various soils are summarized in Table 3.5. From those data, it is apparent that K_f values are generally lower in sandy and/or low organic matter soils compared to clayey and silt loam soils. As is seen from those data, increases in organic carbon also contribute to increasing adsorption coefficient values for the respective estrogens. Thus, bentonite, kaolinite, and sand, which have no organic carbon (OC) content, also show noticeably low sorption as is signified by the low K_f values. Specific surface area also plays an important role in the adsorption of PPCPs in the environment. To that effect, despite the absence of organic carbon in bentonite, its extraordinarily high specific surface area of 654 $m^2 g^{-1}$ enabled it to register a K_f value of 110, which is comparable to those registered by the two soils that had 7.5%OC content (Table 3.5). The concentration of E2 in Tokyo River declined from 4 to 10 nM to below detection in 5 days during summer and 7 days during winter (Matsuoka et al., 2005). Somewhat similar declines were reported in three rivers (i.e., Aire, Calder, and Thames) in the United Kingdom by Jürgens et al. (2002) following first-order kinetics, with a half-life of at least 10 days. However, degradation of E2 to E1 still poses a problem as the latter still retains estrogenicity.

3.2.4 Degradation of Other Important Pharmaceuticals

Other compounds, such as the anesthetic propofol, which as we saw in Chapter 2 is excreted in substantial amounts, can undergo a series of repeated oxidation and reduction processes ultimately degraded to CO_2. It is also interesting to note that the unipolar lipid regulators clofibrate, etofibrate, and fenofibrate are not detectable in sewage treatment plant (STP) effluents, whereas their polar metabolites clofibric acid and fenofibric acid occur in effluents (Ternes, 1998; Steur-Lauridsen et al., 2000). Fibrates and statins are some of the most large volume drugs in developed countries (see Chapter 1). Work by Ternes (1998) in German rivers showed removal of a variety of antihypertensive drugs and their metabolites, notably clofric acid (51%), fenofibric acid (64%), bezafibrate (83%), propranolol (96%), metoprolol (83%), and gemfibrozil (69%) from effluents compared to the loads in municipal water treatment plant influents. In that same study, the degradation removal of the antipsychotic drug carbamazepine was only 7% of approximately 125 g day^{-1}. It generally had the lowest removal rates compared to all of the compounds Ternes (1998) evaluated. Fortunately, very low (1–2%) of the drug is excreted unmetabolized (see Table 3.6). The degradation of other sedatives and

TABLE 3.5 Adsorption Coefficient Values for E1 and E2 in Various Soils

Matrix	%OC	Specific Area (m^2 g^{-1})	pH	Constituents			E1	E2	Reference
				Sand	Silt	Clay			
Soil	1	ND[a]	ND	ND	ND	ND	ND	6.2	Mansell et al. (2004)
Soil (silty clay loam)	2.9	ND	7.2	13	65.8	21.2	44.8	43.4	Das et al. (2004)
Soil (sandy clay loam)	3.3	123	ND	38.3	49.4	12.3		86	Casey et al. (2003)
Soil (Sioux loam)	7.5	106	7.8	35.5	51.6	12.9		135	Casey et al. (2003)
Soil (silt clay loam)	7.5	175	6.4	18.9	37.1	44		151	Casey et al. (2003)
Soil (clay loam)	5.3	154	6.4	30	50	20		332	Casey et al. (2003)
Soil (silt loam)	9.2	151	7.9	18.7	55.2	26.1		667	Casey et al. (2003)
Bentonite (pure clay)	0	654	ND	0	0	100	ND	110	Casey et al. (2003)
Kaolinite	0	43	ND	0	100	0	ND	10	Casey et al. (2003)
Sand	0	ND	ND	100	0	0	ND	4	Casey et al. (2003)
Sand	0.22	ND	7.5	93.9	0	6.1	3.4	3.6	Das et al. (2004)
Sediments (sandy clay loam)	1.1	ND	ND	0	70	30	53.7	36.3	Lai et al. (2000)

[a]ND = not determined.

TABLE 3.6   Dissipation of 17α-ethinylestradiol in Three Distinct Soils at Various Moisture Contents

Soil	%Moisture Content	%17α-Ethinylestradiol Remaining	%Estrogenicity Remaining[a]
Loam	5	27	57
	12	13	22
	20	BDL[b]	BDL
	30	BDL	BDL
	40	BDL	BDL
Silt loam	5	21	26
	15	7	6
	55	BDL	BDL
Sandy loam	5	30	36
	10	BDL	BDL
	24	BDL	BDL

[a]Estrogenicity was determined using a yeast estrogenicity screening assay after 22 days of incubation.
[b]Below detection limit.
Data adapted from Colucci and Topp (2001).

antipsychotic drugs (e.g., diazepam, fluoxentine, and setraline) in the environment has not yet been widely determined.

3.2.5   Degradation of Surfactants

Surfactants are major ingredients in PPCPs and their degradation is also impacted by the same processes that apply to other organic pollutants, including advection, dispersion, and capillary forces (Marchesi et al., 1991; Allred and Brown, 1996; Liu and Roy, 1995). Soil parameters exert a great deal of influence on the mobility of surfactants and their presence in soils reduces the soil's hydraulic conductivity. The extent to which conductivity is affected depends on the concentration of the surfactant and the clay content of the soil (Jafvert, 1991; Jafvert and Heath, 1991). The degradability and general behavior of surfactants is also markedly influenced by temperature. For example, the half-life of a linear alkylbenzene sulfonate (LAS) in soil greatly depended on the season, that is, 68–117 days in winter compared to only 5–25 days in summer (Litz et al., 1987).

3.3   ROLE OF PHYSICOCHEMICAL FACTORS IN THE FATE OF PPCPs IN THE ENVIRONMENT

Gilbertson et al. (1990) generally concluded that the fate of PPCPs in the environment is dependent on the physicochemical properties of the compounds. Such properties include molecular structure, shape, size, speciation, solubility, degradability, and hydrophobicity as well as toxicity. An appreciation of these factors and their effects on the fate of PPCPs in the environment compels us to have a clear understanding of the basic chemistry of PPCPs. Fortunately, there is a good amount of information about these properties that is generated at the time when these compounds are being developed for clinical purposes. We just have to examine, integrate, and synthesize that information from an environmental perspective. Thus, there is a need to link what information has been accumulated through the years of preclinical research ranging from the basic chemistry of the compound to its mode of action with their likely fate in the environment. The physicochemical characteristics directly and indirectly affect other attributes such as sorption and complexation as well as mobility of the compounds in the environment.

3.3.1 Molecular Size as an Attribute to Absorption and Persistence

Molecular weight is a convenient way of expressing the size of a compound. It is closely related to other properties such as surface area, molar refractivity, ovality, van der Waals volume, molar volume, parachor, and McGowan's characteristic volume. Furthermore, compounds with a large molecular size tend to be less easily permeable across the membrane. As with almost every rule in biology, however, there are some exceptions to this generalization. For example, some large molecules such as cyclosporine (MW = 1203) have surprisingly good permeability across membranes though the actual mechanisms are not clearly understood. According to the rule of 5, pharmaceutical compounds of molecular weight >500 have more than 5 H-bond donors, more than 10 acceptors or a calculated partition coefficient (C LogP) of more than 5, and are most likely to have poor oral absorption (Van de Waterbeemd, 2002). A recent study examined 80 PPCPs and found that all of them contained C and H, with a majority of them also containing N and/or O (Jjemba, 2006). Their molecular weight ranged between 120.2 and 1154.4, the two extremes being for the fragrance compound acetophenone and the anticholesterol drug artorvastatin, respectively. However, molecular weight did not correlate with any of the other traits deemed relevant for occurrence in the environment, notably bioavailability, proportion of PPCP that is excreted under normal usage, as well as the concentration of the compound in the aquatic environment. Other electronic properties such as molecular orbital calculation display different atomic and molecular properties that are related to the distribution of electrons within the molecule. Thus, many surface properties such as molecular electrostatic potentials (MEP), molecular lipophilic potentials (MLP), polar surface area (PSA) (Palm et al., 1998; Clark, 1999a, 1999b) have been used to model interactions between ligand and target in pharmaceutical chemistry, but their usefulness in the fate of PPCPs in the environment has not yet been critically examined.

3.3.2 Solubility and Hydrolysis

The solubility of a solute in a solvent is the maximum amount of that compound that dissolves in a given amount of solvent at a specific temperature. Solubility is a physical constant, and each compound has a different solubility in every solvent. For environmental systems, the default solvent is water, a substance that covers approximately 70% of Earth's surface. As a global solvent, water transports reactants and products from one locality to another. Water also happens to be the most polar compound and serves as a reactant or product in its own right in various biochemical reactions. Its polarity and hydrogen bonding give it some of these extraordinary properties as a solvent. In general, polar compounds dissolve in polar compounds and vice versa. Reactions with water are sometimes referred to as hydrolysis. Thus, both inorganic and organic compounds can undergo hydrolysis. For organic compounds such as PPCPs, reactions with water occur at functional groups that may or may not be readily reactive. For example, anhydrides react readily with water to from carboxylic acids, whereas esters are hydrolyzed only very slowly.

A clear understanding of the solubility properties of various organic compounds, including PPCPs, directly relates to the partitioning of these compounds as defined by several physical properties such as the distribution coefficient normalized to the percent organic carbon (K_{oc}), octanol–water partition coefficient (K_{ow}), base association constant (K_b), and Henry's law constants. Its relationship with K_{ow} and the melting point are summarized by Eq. (3.25):

$$\log S = -\log K_{ow} - 0.01 \text{ mp} + 1.2 \qquad (3.25)$$

where S is the solubility in milligrams/milliter or grams/liter; $\log K_{ow}$ is the partition coefficient in the compound's neutral state, and mp is its melting point in degrees celsius (°C).

Aqueous solubility is typically inversely proportional to lypophilicity. Lypophilicity represents the affinity of a molecule (or moiety) for a lypophilic environment. Drug solubility is also an important attribute to its oral absorption and subsequent permeability through membranes. In medicinal chemistry, solubility for a 1-mg kg^{-1} dose of 5, 50, and 500 μg mL^{-1} is estimated to be low, medium, and high, respectively (Van de Waterbeemd, 2002). Solubility characteristics have long been recognized as important in pharmaceutical chemistry and are part and parcel of the drug design process. For example, the solubility can be used to determine the maximal absorbable dose (MAD), that is,

$$\text{MAD} = S \times K_a \times \text{SIWV} \times \text{SITT} \qquad (3.26)$$

whereby S is the solubility (mg L^{-1}) at pH 6.5, K_a is the transintestinal absorption rate constant (min^{-1}), SIWV is the small intestinal water volume, and SITT is the small intestinal transit time. For humans, both SIWV and SITT are approximately 250 mL and 270 min, respectively. The maximum absorbable dose can be used in the drug design process to determine the minimum acceptable solubility (MAS) from

$$\text{MAS} = \frac{\text{MAD}}{K_a \times \text{SIWV} \times \text{SITT}} \qquad (3.27)$$

Solubility is driven by forces that determine the extent of dissolution of a compound in a liquid solvent (in this instance water). Dissolution increases the degree of entropy (i.e., randomness) or disorder in the system. It also increases the compatibility of intermolecular forces of attraction. Thus, increased dissolution is accompanied by the breaking up of solute–solute and solvent–solvent intermolecular bonds as well as the subsequent formation of solute–solvent intermolecular bonds. In reality, solubility is affected by the prevailing buffering capacity and ionic strength (salting out effect), pH, supersaturation, purity, as well as thermodynamic (i.e., equilibrium) effects. Thus, most pharmaceutical compounds are less soluble in physiologic saline than in pure water due to the salting-out effect. From a practical standpoint, the aquatic environment, including groundwater, surface water, and marine waters, are closer in terms of chemical composition to physiologic saline than to pure water. Abraham and Lee (1999) developed a model that uses molecular structure to predict solubility in an aqueous environment:

$$\log S_w = 0.52 - 1.00R_2 + 0.77\pi_2^H + 2.17\sum\alpha_2^H$$
$$+ 4.24\sum\beta_2^H - 3.36\sum\alpha_2^H\sum\beta_2^H - 3.99V_x \qquad (3.28)$$

where R_2 is the excess molecular refraction, π_2^H is the solute depolarity/depolarizability, $\sum\alpha_2^H$ is the summation hydrogen bonding acidity, $\sum\beta_2^H$ is the summation of hydrogen basicity, $\sum\alpha_2^H\sum\beta_2^H$ is a mixed term hydrogen bond interactions between acid and basic sites in the solute, and V_x is McGowan's characteristic volume. It is not clear whether that model is only applicable for clinical settings or can also be applied to predict solubility of compounds in the open environment.

As indicated earlier (Section 3.1.1) the sorption of organic compounds onto dissolved matter can significantly increase the solubility of the compound. This can in turn affect the fate of these chemicals in the environment. We can use physicochemical parameters such as distribution coefficients (log D), aqueous acid dissociation constants (pK_a), and octanol–water partition coefficients (pK_{ow}). These attributes are also linked to the acidity and alkalinity of the environment as well as lipohilicity of the compound. The mathematical relationships between these attributes are outlined below to explore how each of these impacts the fate of PPCPs in the environment.

3.3.3 Effects of Chemical Dissociation, Partitioning, and Lipophilicity on PPCP Degradability

The dissociation constant (pK_a) describes the degree of ionization of the compound at a known pH. The pH of relevance in the environment ranges between 4 and 8, with activated sludge typically presenting a pH in the range of 7–8 (Christofi et al., 2003). By comparison, the pH in the digestive system ranges between 6.5 in the mouth cavity (i.e., salivary pH) to 2 in the stomach (i.e., gastric juice; Marieb, 2001). However, as the gastric juices transit through the small and large intestines, their pH is elevated to slightly alkaline (pH 7.4–7.8) and is isotonic with blood plasma.

The intestinal juice is mostly water but also contains some mucus that is secreted by the duodenal glands and the goblet cells of the mucosa. Blood pH is usually in the range of 7.35–7.45 (Ignatavicius and Workman, 2002), whereas urine has a typical pH of 4.2 (Hansch and Leo, 1979). The body has many mechanisms that ensure minimal changes in pH. Many microbial processes are inhibited by acidic conditions. For a standard ionizing compound:

$$HA \leftrightarrow H^+ + A^- \tag{3.29}$$

$$K_a = \frac{[H^+][A^-]}{[HA]} \tag{3.30}$$

When K_a equals $[H^+]$, half of the compound will be dissociated and its pK_a (i.e., $\log K_a$) is the acid dissociation constant. The corresponding base association constant (pK_b) is 14 pK_a. Thus, when pK_a is equal to pH (i.e., $K_a = [H^+]$), the mixture will contain 50% ionized and the other 50% will be associated (or nonionized) species. The pK_a values that are less than 7 indicate that the compound is negatively charged under acidic conditions and vice versa. Most pharmaceutical compounds are acids or bases with pK_a values of 2–12. pK_a is the state at which 50% of the species is protonated. Electron-donating groups such as alkyl have a pK_a-increasing effect, whereas electron-accepting groups such as nitro have a pK_a-lowering effect. Compounds with both acidic and basic functional groups can exist as an anion, cation, neutral species, or zwitterions. Several programs such as ACD/pKa, PALLAS, and ZPARC are available to calculate pK_a. However, note that for the same compound, different programs give a different value (e.g., Testa et al., 1997). Extensive ionization increases the solubility of the compound in water and reduces its ability to partition with lipidlike (i.e., lipophilic solvents) making the knowledge of pK_a an important parameter in determining or predicting the sorption and toxicity of the compound. Adsorption increases with a decrease in pH (i.e., increasing acidity). By contrast, adsorption of basic PPCPs is high as is exemplified by flouroquinolones (see Golet et al., 2002b). Thus, weakly acidic pharmaceuticals such as the NSAIDs (i.e., naproxen, ibuprofen, and acetysalicylic acid) with pK_a values of 4.2, 5.2, and 3.5 as well as clofibric acid ($pK_a = 2.95$) have a low tendency to adsorb onto sludge.

Lipophilicity is an important physicochemical descriptor used in correlating chemical structure with biological activity. It is traditionally expressed as the logarithm of the partition coefficient (i.e., $\log K_{ow}$) of the solute between n-octanol and the aqueous phase, thus representing the rate of concentrations of a compound (X) in n-octanol and water, that is,

$$P = \frac{[X]octanol}{[X]aqeous} \tag{3.31}$$

The n-octanol is used as a surrogate for lipids. It was chosen by Hansch et al. (1962) as the reference organic phase as those workers felt that it simulates the lipid components of biological membranes. This relationship also considers the concentration in water because that medium represents the aqueous phase of biological systems. When both phases are in equilibrium, the ratio of the concentration of a diluted chemical in the two phases, that is, n-octanol and water, represents the octanol–water distribution coefficient (D_{ow}). Thus, D_{ow} exemplifies the tendency of the organic chemical to distribute among various environmental components and to bioconcentrate in organisms. In a sense, it also represents the ability of the organic chemical to partition into lipids (or fats) and sorb to particulates such as soil, sediments, sludge, or biomass. It is important to note that it is the nonionized fraction of the organic species that predominantly partitions into octanol (or lipids since n-octanol is a lipid surrogate), which in turn largely depends on the prevailing pH. Thus, for environmental risk assessment, the octanol–water distribution coefficient (D_{ow}) is typically corrected for the ionization of the compound to represent only the nonionized fraction, giving what is typically referred to as the octanol–water partition coefficient (K_{ow}). The relationship between these two coefficients is presented in the Henderson-Hasselbach equations. For acids

$$\log D = \log K_{ow} - \log (1 + 10^{pH - pK_a}) \qquad (3.32)$$

For bases

$$\log D = \log K_{ow} - \log (1 + 10^{pK_a - pH}) \qquad (3.33)$$

Thus, the octanol–water partition coefficient (K_{ow}) is derived from the octanol–water distribution coefficient (D_{ow}) after correcting for ionization. It needs to be carefully evaluated with regard to multiple ionization sites. Most pharmaceuticals are unique in that they have several functional groups that ionize to different extents under various conditions. These give the respective pharmaceuticals several pK_a values, which in turn affects the bioavailability and chemical activity of the compound, ultimately affecting the fate in the environment. In simpler terms, the log K_{ow} of a compound applies in the neutral unchanged state, whereas log D is the distribution coefficient at a selected pH (often pH 7.4). Note that log K_{ow} is also referred to as log P in some instances. For biological processes, the distribution coefficient (log D) values at pH 7.4 (intracellular uptake) or pH 6.5 (gastrointestinal absorption) are more relevant. The difference in these two coefficients has been shown to be useful in absorption studies (i.e., brain absorption and gastrointestinal absorption) (Young et al., 1988). For environmental studies, D_{ow} at pH 7 is used, although D_{ow} values at pH 5, 7, and 9 are all usually determined.

Thus, the log K_{ow} values describe the hydrophobicity of chemical compounds and, coupled with other physicochemical characteristics such as dissociation constants, vapor pressure, Henry's law constants, pK_a values, K_{ow}, and the like may help to determine whether a particular pharmaceutical compound is likely to concentrate and persist in the aquatic, terrestrial, or atmospheric environment (Jjemba, 2002a; Díaz-Cruz et al., 2003). For example, the fragrance ethylhexyl methoxycinnamate (EHMC) is also quite sorbable to sediments and sludge due to its high log K_{ow} of 6–6.3 as well as its large K_{oc} of approximately 1230 (Straub, 2005). The octanol–water relationship can be effectively used to predict the behavior of compounds toward biological systems as high log K_{ow} values of 5 and higher signify considerable lipophilicity of the compound, which in turn corresponds with a high potential of that compound to bioconcentrate and bioaccumulate in tissues of living organisms (Dimitrov et al., 2003).

3.3.4 Effects of Moisture and Oxygen in the Fate of PPCPs in the Environment

Colucci and Topp (2001) examined the dissipation of 17α-ethinylestradiol at different moisture contents in three distinct soils. Their findings are summarized in Table 3.6 and show a trend in all three soils of enhanced degradation of this compound coupled with reduced estrogenicity as the moisture content increases. The effects of moisture on PPCP degradation widely investigated for a variety of other compounds but low moisture content is known to limit biochemical processes (Jjemba, 2004). However, to be beneficial, the moisture content has to be sufficient so as not to cause anoxic conditions. The biodegradation of chemicals in surface water is facilitated by the free dissolved oxygen (DO). Thus, DO is an important water quality parameter. However, in practice the amount of dissolved oxygen in the water is influenced from time to time by the sinks and sources. A more informative and easily measurable parameter to reflect oxygen consumption dynamics is the biochemical oxygen demand (BOD). BOD represents the oxygen required by microorganisms for respiration as they degrade the pollutant. It is simply measured by taking two water samples from the same source, determining the DO in one of the samples and incubating the other sample in the dark for 5 days. The DO in the incubated samples is thereafter determined after 5 day. The difference between the DO at the time of sampling and after the 5-day incubation period is the 5-day BOD (i.e., BOD_5).

3.3.5 Effects of Temperature in PPCP Dynamics and Degradation in the Environment

Temperature affects the rates of reactions. In the body of homiothermic organisms, the effects of temperature are possibly not as important as these organisms maintain a somewhat constant temperature. By contrast, metabolism of pharmaceuticals in poikilothermic organisms is dependent on ambient temperature, trends that may also affect the toxicity of the compound (Kendell et al., 2001). Similarly, the environment can undergo a series of temperature extremes. To that effect, temperatures of relevance in the environment range from subzero (in temperate regions) to the lower thermophilic range (55°C). The upper temperature range becomes relevant in thermophilic systems under which some waste such as sewage sludge can be incubated during treatment. It is also applicable in landfills and compost piles (Jjemba, 2004). The rate of degradation of PPCPs can depend on the prevailing temperature. For example, the half-life for ivermectin in the environment was 6 times greater in winter than in summer (Bull et al., 1984; Halley et al., 1993). Ternes et al. (1999a) conducted degradation experiments at 20°C with initial concentrations of 1 μg estrogen L^{-1} and total suspended solids (TSS) concentrations of 0.52 g of TSS/L in which they assumed a pseudo-first-order reaction [Eq. (3.34)]:

$$\frac{dC_{E2}}{dt} = -k_{E2}(TSS)C_{E2} \qquad (3.34)$$

Based on this equation, when the pseudo-first-order kinetic constant (k_{E2}) was estimated at 150 L g^{-1} of (TSS)d, the half-life of E2 was established to be 0.2 h, with nearly all of the E2 being converted to E1. E1 was removed more slowly at a half-life of 1.5 h and a kinetic constant of approximately 20 L g^{-1} of (TSS)d, and EE2 was not significantly degraded under those same conditions. By comparison, in similar experiments conducted by Layton et al. (2000) at higher temperatures (30°C), at least 40% of the EE2 was mineralized in activated sludge within 24 h.

In another study, the removal of 17α-ethinylestradiol from a loam soil was accelerated with increasing temperatures ranging between 4 and 30°C (Colucci and Topp, 2001). The time to dissipate 50% of the initial compound (DT_{50}) ranged between 7.7 days at 4°C and 3 days at 30°C. The decreases in percent 17α-ethinylestradiol remaining in the soil were also accompanied by a corresponding decrease in estrogenicity. Dissipation rates were much lower in oxygen-deprived environments. Gavalchin and Katz (1994) have also shown the influence of temperature on the degradation of various antibiotics, notably bactricin, bambermycins, chlorotetracycline, erythromycin, streptomycin, and tylosin at 4, 20, and 30°C in a sandy soil mixed with chicken manure. In general, higher concentrations of the respective antibiotics were recovered from the soil at the low temperature. More recently, Jacobsen et al. (2005) showed a greater persistence of ^{14}C-testosterone in soils at 4 and 12°C compared to 30°C. Thus, cooler temperatures generally enable the PPCPs to persist for longer in the environment. This has notable implications in the temperate regions during subzero ambient temperatures. The degrading testosterone forms intermediates such as 4-androstene-3,17-dione, 5α-androstan-3,17-dione, and 1,4-androstadiene-3,17-dione, all of which are a result of transformation of the phenolic moiety at position 17 on the pentose ring, forming a ketone. Subsequent studies by, Jacobsen's group (2005) as well as Lorenzen et al. (2005) showed that, where it occurred, degradation was a biological process so as no mineralization of ^{14}C-testosterone occurred in treatments that had been sterilized. The latter group's studies in soil microcosms at 30°C also showed that 50% of the testosterone was dissipated in 8.5 h (loam soil) to 21 h (silt loam soil), with androgenicity declining faster than extractable radioactivity. Those observations suggest that the extractable ^{14}C may be partitioned between the parent compound and its intermediates of less androgenic activity. The production trends of [14]CO_2, the rate at which extractable radioactivity declined, and the loss in androgenic activity collectively suggested a slightly faster dissipation of testosterone in the loam soil (pH 7.4; 3.2%OM) compared to the sandy loam (pH 5.8; 0.8%OM) and silt loam (pH 6.7; 2.9%OM) (Colucci et al., 2001). The observed trends are shown in Figure 3.12. Note that the soils widely differed in texture and chemical properties. In general, adsorption increases with increasing organic matter content of the soil. This observation has important ramifications in instances where manure and biosolids

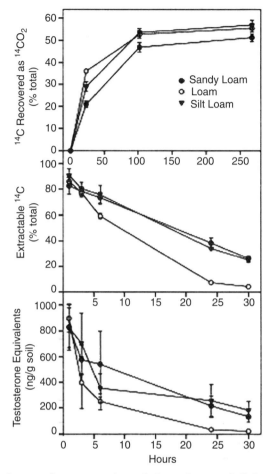

Figure 3.12   Persistence of testosterone in sandy loam, loam, and silt loam soil spiked with 1 mg [^{14}C]-testostrone/kg. The middle panel shows ^{14}C residues in soil extracts, whereas the bottom panel represents a loss in androgenicity in the soil extract. Androgenicity was measured using a human androgenicity receptor recombinant yeast strain. (Adapted from Lorenzen et al., 2005.)

are applied in late fall and winter months when temperatures are quite low. That provides an opportunity for the PPCPs in these materials to persist for longer in the soil.

3.3.6   Other Factors That Determine PPCP Fate and Persistence in the Environment

3.3.6.1   Effect of the Presence of Other Compounds   The response (pharmacokinetics and pharmacodynamics) of an individual pharmaceutical compound may be predictable. By comparison, the response of the same compound in the presence of other drugs is often less predictable. The presence of other drugs in the system can stimulate the metabolism of a specific compound and hasten its loss or interfere with its absorption. The fate of PPCP mixtures has not been widely researched.

3.3.6.2 *Photolysis of PPCPs*

Ultraviolet (UV) radiation has a wavelength of $<200-400$ nm, a range that is commonly divided into UV-A (400–320 nm), UV-B (320–280 nm), and UV-C (280– <200 nm) wavebands. High-energy UV radiation damages organic compounds, which in turn interferes with their function. The photodegradation process occurs when a compound absorbs light. Most of the light absorbed is usually transformed into thermal energy or emitted as light through fluorescence or phosphorescence. However, some of the absorbed photons can induce the transformation of the compound into different products. The efficiency with which these energy changes occur is expressed as the reaction quantum yield. Thus, quantum yield refers to the number of molecules of the parent compound that are transformed as a fraction of the total number of photons absorbed. It is related to the photochemical rate constant ($k_{p,direct}$). In general, photosensitive drugs have substituents of chlorine atoms that, during photodegradation, are substituted or reduced (Glass et al., 2001; Konstantinou et al., 2001). However, photodegradation can also result from other processes such as sulfoxidation and dealkylation. Photolysis is the direct absorption of light by a compound followed by a reaction that transforms the parent compound into one or more products. It is important in determining the residence time and fate of compounds in the environment. Photolysis rates and phototransformation products are dependent on the distribution of the wavelength used and its intensity. Under laboratory conditions, most of the work that evaluates photolysis has used artificial sources of irradiation, including xenon arc or mercury lamps of approximately 254 nm. Where UV light has been used, it has often been combined with catalyst particles such as Fe_2O_3 or TiO_2. Photodegradation reactions are attributed to UV-A (315–400 nm) and UV-B (280–315 nm), with visible sunlight (400–750 nm) not contributing to this process (Doll and Frimmel, 2003).

In environmental science, photolysis has been more frequently studied under aquatic compared to terrestrial conditions. The approach of using arc or mercury lamps causes problems when extrapolating to environmental conditions as, in the environment, wavelengths of ≥ 290 nm are relevant compared to the 254 nm that is provided by the lamps used. Thus, the full impact of photolysis under natural conditions is still not fully cataloged. A more realistic approach would involve using natural sunlight by, for example, taking aliquots of water from transparent containers run in parallel with control (e.g., similar containers wrapped with aluminum foil) containers. Incident solar radiation has to be monitored and matrix aliquots periodically drawn to determine the status of the compound(s) of interest. Using this approach for various pesticides (i.e., atrazine, propanil, molinate, propachlor, propazine, and prometryne), Konstantinou et al. (2001) found that photodegradation rates of all six herbicides in different natural waters followed a first-order degradation curve:

$$C_t = C_0^{-kt} \tag{3.35}$$

where C_t is the concentration of the chemical at time t, C_0 is the initial concentration of the chemical, and k is the photolysis rate constant. The photolysis rate constant is important as it describes the decrease in concentration of the compound with time. To exclude the effects of other processes such as volatilization, hydrolysis, as well as sorption, the photodegradation constants were calculated by subtracting the exponents of the different degradation curves from the control (i.e., blank experimental containers that were wrapped in aluminum foil). From these results, the duration it takes the concentration of the compound to be reduced by a half (i.e., half-life), can be determined from Eq. (3.12). Using this procedure, Konstantinou et al. (2001) found the photodegradation rates of herbicides to decrease in the order marine water $>$ lake water $>$ river water $>$ distilled water $>$ groundwater. Thus, half-lives were consistently longer in marine water than groundwater (Fig. 3.13). This trend of lower photodecomposition rates in natural compared to distilled waters is possibly attributed to increasing optical filter effect (quenching) of organic matter. The quenching organic matter, in this instance, includes sediments and microorganisms that act as an important sunlight-absorbing component in natural aquatic systems. The organic matter in surface waters can also act as a precursor of reactive species, generating singular oxygen, superoxides, hyroxyl radicals, and solvated electrons (Frimmel, 1994, 1998; Frimmel and Hessler, 1994). However, just like biodegradation, photodegradation can, in some instances, generate products with higher potency (Burhenne et al., 1997). Sorption also protects the compounds from photolysis as the sorbed compounds migrate into regions of the sorbent that is not adequately penetrated by light.

Photodegradation occurs by direct or indirect absorption of light. The latter involves instances whereby the photons are absorbed by other constituents of the media (e.g., dissolved organic matter, soil, etc.), whereas during the direct transformation processes, the substance absorbs light energy and undergoes transformation. Irrespective of the type of absorption, the excited species can thereafter transfer the energy to the substance, undergoing an electron transfer or forming a reactive species such as hydroxyl radicals or singlet oxygen. The rates of photochemical processes in an aquatic system are affected by solar spectral irradiance at the water

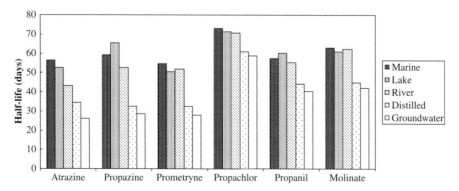

Figure 3.13 Trend of half-life for six pesticides after exposing to photodegradation in marine, lake, river, distilled, and groundwater. (Based on data from Konstantinou et al., 2001.)

surface, radiative transfer from air to water, and the transmission of radiance (e.g., sunlight) in the water body. A wide range of organic compounds and some heavy metals have been shown to succumb to phototransformation by UV in surface water (Tixier et al., 2002). Tixier et al. (2003) described the phototransformation of PPCPs using the relationship presented in Eq. (3.36):

$$K_{p.\text{direct}} = \ln(10) \int_\lambda \varepsilon(\lambda)\Phi(\lambda)\left[\frac{dI(\lambda)}{d\lambda}\right] d\lambda \qquad (3.36)$$

where $K_{p.\text{direct}}$ is the rate constant for the direct photodecomposition of the compound in a given volume element, λ is the wavelength of the light, $\varepsilon(\lambda)$ is the decadic molar extinction coefficient of the compound in meters squared/mole, $\Phi(\lambda)$ is the reaction quantum yield (mol einstein^{-1}), and $dI(\Phi)/d\lambda$ is the irradiation spectrum (einstein m^{-2} s^{-1}). The quantum yield is the fraction of absorbed light resulting in a photo-reaction at a fixed wavelength (λ).

Photolysis processes in soils and other porous media have also been recently studied by Ciani (2003) who examined the light intensity in such media, the molar absorption coefficient of the absorbed photodegradable compound, and the reaction quantum yield. Photodegradation in these types of media can be unpredictable due to the very shallow and highly nonlinear light penetration. Equation (3.37) is applicable to aqueous and solid (or porous) surfaces, including soil with the notion that for porous surfaces, there is a much stronger decrease in the intensity of light with depth as compared to aqueous systems. It is important to note that for porous surfaces, the overall rate of disappearance of a compound on irradiation does not only depend on the phototransformation rate but also on the diffusion kinetics of the compound. Ciani (2003) showed that soils with much smaller particle size (e.g., clays and silt) absorb more photons, possibly because the photons will hit surfaces more often (and thus have a higher probability of being absorbed) when passing through layers of small particles than when passing through a layer of larger particles. The transmittance of light in soil is, however, more complex as the presence of moisture in the soil can have a tremendous impact. For very sandy soils, moisture increases the transmittance of light, whereas moisture in a clayey soil has the opposite effect. These differences may result from the possibility that the water causes a reorientation of the soil particles at the surface, leading to a change in the optical properties of the soil.

Zepp and Cline (1977) developed a computer program, GCSOLAR, that can be used to determine the rate constant for the direct photodecomposition of compounds using Eq. (3.37) under terrestrial sunlight as a function of season, time of day, changes in latitude, changes in the depth of the water body (in case of aquatic systems), as well as the thickness of the ozone layer. Ozone layer data are readily available from NASA (see http://jwocky.gsfc.nasa.gov/teacher/ozone_overhead_v8.html) for very specific locations if the latitude and longitude information is submitted to the searchable database. To that effect, it is important to note that phototransformation can greatly vary between locations. Furthermore, where it occurs, it is often compli-cated by the fact that other processes such as sorption as well as competing biotic and abiotic processes are involved and their effect is hard to separate out.

What is often not realized is the fact that sunlight is polychromatic, with a wide spectrum of wavelengths. Furthermore, light from the sun toward a particular location can be diffuse and therefore coming from all directions, as is the case on a cloudy day, or collimated. Thus, sunlight can be a mix of diffuse and collimated character, a situation that can also vary at different times of the day (e.g., noon versus sunset or sunrise) and/or weather conditions (cloudy versus sunny). Photodegradation is more relevant in surface water as groundwater, soils beyond a certain depth, and sediments that lack light. UV can only effectively penetrate the water column to a limited depth. However, this limitation can be offset by the repeated mixing and turbulence in flowing natural waters.

Photolysis has been shown to significantly contribute to the removal of diclofenac in surface water (Buser et al., 1998b). Similarly, ofloxacin, propranolol, and sulfamethoxazole were removed by direct and indirect photolysis under laboratory conditions, whereas carbamazepine and clofibric acid were reported to be photodegraded slowly (estimated $t_{1/2} = 100$ days) in winter at latitude 50°N (Andreozzi et al., 2003). By comparison, clofibrate is estimated to persist in water for more than two decades (Buser et al., 1998a). In general, the efficiency of photolysis largely depends on the properties of the compound, intensity of the solar radiation, and on the constituents in the environmental matrix, which may act as photosensitizers (such as humic acids and nitrates) producing singlet oxygen and hydroxyl radicals. Under adequate light, quinolones are also readily photodegraded (Burhenne et al., 1999). However, the rate at which these antibiotics photodegrade in the presence of humic acids decreased (Volmer et al., 1997). Reduced photodegradation is also reported in some pesticides in the presence of liquid suspended natural organic matter due to competitive light absorption by colored organic matter (Frimmel and Hessler, 1994). Photodegradation can play a role in the photic aquatic systems although its effect on individual pharmaceuticals can be greatly hampered if these compounds exist in the environment as mixtures due to competitive inhibition compared to their photodegradation as sole contaminants (Doll and Frimmel, 2003).

Photooxidation of the chemical can be somewhat enhanced in the presence of humic substances due to the hydroxyl radicals generated by the humates. Photooxidation was estimated to eliminate 80% of triclosan in a lake in Switzerland compared to <5% of the compound eliminated by degradation (Tixier et al., 2002). However, photooxidation may also generate some undesirable products in surface water. Biological contaminants, including viruses, are also susceptible to degradation/inactivation by UV, making UV one of the most important means by which surface water is naturally self-purified. However, the overall effects of UV on the self-purification of natural waters depends on the season, latitude, day length, thickness of the ozone layer, and the turbidity of the water.

3.4 CONCLUSION

In conclusion, it is apparent that there is some removal of PPCPs from effluents once they get into the environment. However, such removal may only be temporary as compounds get sorbed onto surfaces and environmental matrices. It is most desirable that our efforts be devoted to developing techniques that enhance the total mineralization (i.e., complete conversion into CO_2) of these compounds in the environment. Achieving such an end requires investing into research to develop more efficient waste treatment techniques, some of which will be discussed in Chapter 5. However, before we dive into such investments, it is imperative to determine the potential ecotoxicity of these compounds in the environment, a task to which Chapter 4 is devoted.

4

ECOTOXICITY OF PPCPs

Today, we are faced with an ever-increasing assault on the environment by PPCPs together with a variety of other pollutants that emerged with the rapid advances in technology and medicinal chemistry (manipulation of chemicals for medicinal purposes). Advances in sample extraction and analytical instrumentation described in Chapter 2, as well as increased surveillance for the presence of PPCP, have improved the detection of these compounds in the environment. These improvements have to be accompanied by understanding not only the fate and transport of PPCPs but also the ecological effects and implications of their presence. Such scrutiny involves considerations of PPCP toxicity in the environment. Toxicology entails investigating deleterious effects of chemicals. For most PPCPs of toxicological concern in the environment, exposure to organisms occurs by extravascular routes such as absorption (dermal entry), ingestion (oral entry), and, in the case of volatile compounds, inhalation. However, absorption of these compounds into the circulation system through each of these three routes of exposure is incomplete. That observation has important ramifications for the bioavailability of individual chemicals as toxicity is not as critically influenced by the dose but rather the concentration of the compound at the site of action.

Ecotoxicology is a natural extension of toxicology that studies the fate and effects of toxic substances on an ecosystem. It is based on scientific research that employs both laboratory and field methods. Ecotoxicology requires an understanding of ecologic principles and theories pertaining to how chemicals can affect individuals, populations, communities, and ecosystems. Individuals are single entities, whereas a

Pharma-Ecology: The Occurrence and Fate of Pharmaceuticals and Personal Care Products in the Environment. By P.K. Jjemba
Copyright © 2008 John Wiley & Sons, Inc.

group of individuals of the same species living together in a specific location at the same time comprises a population (Jjemba, 2004). An assemblage of populations of all the organisms in an area with individuals of different species that live in a designated location at the same time and exploiting the same resources comprise a community. By this token, therefore, communities are collections of interacting populations. At the top of the ecologic organizational levels is the ecosystem, and this primarily refers to a group of self-sustaining communities that interact with each other and with their physical environment (i.e., both living and nonliving components) as energy flows and cycles within the ecosystem (Odum, 1983). The cycling and flow of energy within the ecosystem maintains varying levels of connectivity to the extent that disturbance to one component may be realized at another seemingly distant component. To that effect, ecosystems are generally in a state of communication, and this can potentially facilitate large-scale effects when pollution occurs. Thus, ecotoxicology illuminates the effects of exposing contaminants on terrestrial and aquatic biota. By virtue of its complexity, it heavily relies on interdisciplinary scientific exploration that encompasses a variety of topics that relate to testing the toxicity and resulting responses of various ecological levels to chemicals at lethal and sublethal concentrations. However, unlike standard toxicological tests that aim at defining the cause–effect relationships with certain concentrations of toxicant exposure at a sensitive receptor site of an individual, ecotoxicology does not lay as much emphasis on the individual but rather on the higher ecological levels (i.e., populations, communities, and ecosystems). In essence, therefore, ecotoxicological testing tries to evaluate the causes and effects at these higher ecological levels. These evaluations are best addressed from field studies, but integrating laboratory assays in ecotoxicology is still necessary as they define the impact on individual organisms, their biochemistry as well as physiology. Laboratory studies are more attractive than field studies as they are less expensive to conduct and less variable. They can be interpreted more meaningfully when the parameters are limited. It is important to understand the complex parameters that an organism can deal with so as to survive and/or reproduce when it is under ecological pressure, as would be the case in the presence of a chemical toxicant. To that effect, integrating both laboratory and field research ensures that ecotoxicologic testing methods generate more relevant data. After all, for most chemicals, the toxicity is a result of nonspecific changes in cellular functions that lead to organ-specific effects.

During the early stages of drug development, toxicity studies are typically done in vitro using established cell lines. However, it is important to point out that cell lines during in vitro studies do not express their tissue-specific functions and may therefore not entirely reflect the toxicity at the tissue or, certainly, the organ level. On the other hand, in vivo toxicity studies are, owing to resources and time constraints, typically limited to a single sex (e.g., females), single species, and a single route of administration. Furthermore, under regulatory toxicology studies, the toxicity is typically determined at high doses (i.e., lethal or acute toxicity). The numerous tests done during drug development under these conditions have generated a lot of acute toxicity data for various PPCPs, and a good number of them are available in the public domain. However, such data may not bear much relevance in addressing the question of effects of these compounds in the environment and the associated risks since, as we have seen in the previous chapters, PPCPs mostly occur at low (i.e., sublethal) concentrations. Clearly, assessing and analyzing the risk at such low concentrations demands nonconventional strategies that include probabilistic approaches and have an increased emphasis on ecological health. Risk is the probability of a situation or substance causing an adverse (i.e., hazardous) effect under specific conditions. It emanates from exposure to a hazard. Exposure refers to the contact between the substance or situation. Thus, there is no risk if exposure does not occur and/or if the material one is exposed to is not hazardous. Hazard can be in the form of degradability, toxicity, or physicochemical characteristics that are intrinsically stressful, causing changes in the homeostasis of an environment. The stress can cause direct (i.e., primary) or indirect (i.e., secondary) effects to the receptor. The receptor is simply an ecological component, and it can be a tissue, organ, organism, population, community, or an entire ecosystem.

Because of the inherently low PPCP concentrations in the environment, strategies such as those that identify induction of enzyme systems, monitoring immune function, genotoxicity, and reproductive end points seem to bear more relevance in ecotoxicology. PPCPs in the environment may also trigger adverse immunological reactions (Witte, 1998). These effects may not result in immediate mortality but rather affect fecundity and reproductive success, which in turn affect population structure and function as well as growth rates and body size. Determining the effects of sublethal concentrations is an important component of assessing risk as this may provide information that is not attainable from measuring the concentrations of the chemical in the organism since such concentrations might be too low to detect. Furthermore, since chemicals in the environment almost always occur in mixtures, the toxic effects of some chemicals may not be readily predictable amid other chemicals that are present. The effects of these compounds on organisms in the environment may also occur across generations in life-cycles (i.e., multigenerational), at various stages, or even throughout their entire life. Thus, we have a responsibility to focus on the long-term exposure and on the specific mode of action of these PPCPs in the environment. Risk assessment involves the characterization of the harmful response. It identifies potential hazardous consequences with their likelihood and severity. In principle, risk assessment is composed of three phase, that is:

1. Problem formulation
2. Risk analysis
3. Risk characterization

The problem formulation process involves generating and evaluating hypotheses about ecological effects as well as evaluating management goals. It is important to realize that complex problems often do not have simple solutions and thus need extraordinary insight and foresight by placing the problem in perspective. The risk analysis phase involves collecting and evaluating data about exposure and how it relates to ecological effects. In essence this involves characterizing the exposure and its effects.

The risk characterization phase involves interpreting the data so as to quantitatively or qualitatively estimate the risk that is posed to the individual, population, or community.

Risk assessment is quantitative and differs from risk management, which involves weighing options to reduce the risk. The risk assessment process begins with identifying the potential hazards and their occurrence in a specific environment (i.e., exposure assessment), their toxicity (i.e., dose–response), and a characterization of the risk (NRC, 1994). Risk assessment determines the probability of realizing harm as a result of exposure to a given hazard.

Of environmental concern are not necessarily the quantities of the respective compounds per se but their biological activity (i.e., potency and toxicity) as well as their persistence. Of the 80 PPCPs recently examined, 31% contained a halogen in the form of chloride, fluoride, or bromide (Jjemba, 2006). As a parallel, the effects of some chlorinated hydrocarbon insecticides and other industrial chemicals such as DDT, PCBs, and PAHs on wildlife are well known (Carson, 1962; Ames, 1966), but the risks associated with PPCPs that get into the environment still remain unclear. Risk analysis is the process by which science-based public health decisions are made. It generally includes identifying the health effects and their adverse consequences, assessing the likelihood of such adverse effects. In populations that have been exposed, managing the health risks also includes communicating such risks to the interested parties and stakeholders, as well as controlling or eliminating the risks. Once in the environment, only a portion of the total quantity of the PPCP present is potentially available for uptake by organisms. This is characterized as the bioavailable fraction and is the fraction that ultimately dictates toxicity. A fundamental challenge in ecotoxicology is relating the presence of a chemical in the environment with a valid prediction of its hazardous effect on biological receptors. The resultant hazard can be in the form of alteration in organelle, tissue, or cellular function. Therapeutic agents are designed to be biologically very active chemicals and their activity, when they get into the environment, could impact various organisms. The fact that therapeutic agents have been used for decades is no reason for complacency in evaluating their impact on the ecosystem.

Dosing with an adequate concentration of the chemical generates a continuum of responses that culminate into delivery of the chemical to a critical site. The effects of PPCPs to organisms in the environment will depend on the dose that reaches the target organ and the physiologic response to that dose. Several widely prescribed pharmaceuticals bind to nuclear receptors, affecting important physiological processes such as development, reproduction, differentiation, and cellular homeostasis. Receptors are of two main types, specifically nuclear hormone receptors and 7-trans-membrane receptors. Nuclear receptors act as transcriptional switches, turning on and off the complex circuitry of gene expression within cells. Wellknown lipophilic hormones such as estradiol, testosterone, progesterone, cortisol, aldosterones, thyroxine, and retinoic acid control these transcriptional switches. Pharmaceutical compounds act principally by binding to cellular constituents and proteins that are often their molecular target. The four classes of regulatory proteins that drugs typically interact with are (i) enzymes, (ii) ion channels, (iii) carriers (or transporters), and (iv) receptors. For example, aspirin acts by inhibiting the enzyme cyclooxygenase, local anesthetics act by inhibiting Na^+ channels, Prozac acts by inhibiting serotonin transporter, whereas the antiallergic medicine Zyrtec acts by blocking the histamine H1 receptor. Enzymes are proteins that contain chiral recognition sites for specific substrates, and they affect a variety of enzyme-catalyzed reaction-impacting biochemical processes that lead to growth, viability, and metabolism. Once in association with recognition sites, the enzyme enhances the rate of chemical reaction severalfolds by reducing the activation energy of the reaction, increasing the rate of reactions as depicted in the following reaction:

$$E + S \underset{}{\overset{K_s}{\rightleftharpoons}} ES \overset{k_{cat}}{\longrightarrow} ES^* \longrightarrow EP \underset{}{\overset{K_p}{\rightleftharpoons}} E + P$$

where ES^* is the transformed chemical that forms as enzyme–product complex. Enzyme inhibition is an important strategy in drug design. When enzymes are the target of a particular drug, more questions than answers often persist. For example: (a) Does the target enzyme have only one substrate or will its inhibition affect other biological processes? (b) Is the enzyme unique to the infecting organism or is it present in the host as well? (c) Does the substrate of the enzyme have biological activity and if so, will inhibiting the enzyme cause the substrate concentration to increase, resulting into undesirable side effects? (d) Are there related enzymes that may be inhibited as well? Ion channel modulators have a wide range of therapeutic areas and accomplish their functions by forming an ion-specific membrane-spanning aqueous ion conduction pathway with the ion movement down an electrochemical gradient generating a net charge and creating a voltage (or potential) that ultimately controls cellular responses. They are ubiquitous proteins that are involved in various physiological processes such as degradation of immune cells, salivary secretions, muscle contractions (including the heartbeat), electrolytes balance within the kidneys, blood pressure regulation, sensory perception of smell, touch, and light, as well as the complex integration of electrical signals in the brain and nervous system. Ion channels utilize only four ions as charge carriers, that is, Na^+, K^+, Ca^{2+}, and Cl^-.

The potency of a pharmaceutical compound is typically expressed by its IC_{50} value. IC_{50} is the concentration that inhibits 50% of a particular condition. Ideally, this is determined by testing a range of concentrations ranging from one that is high enough to give 100% inhibition of the condition and concentrations that are low enough to have a minimal effect on the condition that is being inhibited. Drug potency can also be predicted from the equilibrium dissociation constant (K_i) (also referred to as the distribution constant) of the compound–ligand complex based on the Cheng–Prusoff equation, that is,

$$IC_{50} = K_i \left(1 + \frac{F}{K_d} \right) \tag{4.1}$$

where F is the concentration of the free ligand used in the binding study, K_d is the dissolution constant, and K_i is independent of the receptor and ligand concentration used in the assay (Van de Waterbeemd, 2002). The free energy of binding of the ligand to its target (i.e., enzyme, nucleic acid, ion channel, or receptor is given by

$$\Delta G = 2.303RT \ \log K_i = \Delta H - T\Delta S \qquad (4.2)$$

If K_i is known, the difference in binding energy (ΔG) can be easily estimated as 2.303RT is approximately 1.4 kcal mol^{-1}.

4.1 CONVENTIONAL ASSESSMENT OF THE RISK

Risk assessment and management require knowledge of the hazards encountered and assessment of the exposure. In Chapter 2, we focused on the quantities of various PPCPs that are typically detected in the environment. Those detected (albeit at nanogram to microgram concentrations) contribute to the potential exposure in the respective environments, but the actual hazards attributed to these compounds in the environment are less certain. The International Committee on Harmonization (ICH) of pharmaceuticals for human use has, in the face of globalization, made a tremendous effort to harmonize drug testing guidelines (see www.ich.org). However, because of differences in existing local and national regulations, environmental risk assessment of pharmaceuticals has not been part of that harmonization process.

Toxicity has conventionally been determined by looking at end-point toxicity using parameters such as the lethal dose or lethal concentration that eliminates 50% of the population (i.e., LD_{50} and LC_{50}), as well as effective dose or effective concentration against 50% of the population (i.e., ED_{50} and EC_{50}). LD and ED values apply mostly under terrestrial systems where the chemical is introduced via ingestion or inhalation, whereas LC and EC values mostly apply to aquatic systems where the subjects are immersed in water or a solution that contains the chemical. However, considering the low concentrations at which PPCPs typically occur in the environment, assessing their risks has to accommodate gaps in knowledge and aim at determining relevant end points. Thus, in acute studies, survival is often the main end point determined, whereas in chronic toxicity studies the effects on growth and reproduction are also determined. Ecotoxicity testing under this set of circumstances should aim at assessing pharmacological functions and activity rather than adapting uniform (standardized) testing such as ED_{50} or EC_{50}, which are designed for high dose exposure situations. Thus, other parameters such as egg hatchability, sex ratios, neonate survival, biochemical responses by tracking certain biomarkers, and the like appear to be more relevant in PPCP ecotoxicity investigations where chronic toxicity prevails. Likewise, ecotoxicity studies that involve plants should consider parameters such as germination percentage, root elongation, height, grain yield, and leaf area index. Other generic parameters such as maximum allowable toxicant concentration (MATC), lowest observed effect concentration (LOEC), and no observed effect concentration (NOEC) can also provide useful information in this context. The choice of parameters is generally guided by professional judgment, nature of anticipated exposure and their effects, as well as available resources. The nature of the study's focus (i.e., acute or chronic toxicity) also informs the decision as to the kinds of end points to be determined.

To have meaningful context, chronic toxicity studies should be designed to span the entire life cycle of the organism, that is, from egg or zygote stage to the age of first reproduction. Chronic toxicity studies take a longer time, which possibly makes them less popular unless organisms such as *Daphnia magna*, the small planktonic crustaceans *Ceriodaphnia dubia*, or the green algae *Selenastrum capricornutum*, which have inherently short life spans, are used. More recently, the predicted environmental concentration (PEC) has been adapted to assess the risks associated with PPCPs in the environment (EU, 1994; U.S. FDA, 1998). This relationship is developed from the quantities of individual compounds used and how efficiently they are removed during treatment. More specifically for pharmaceuticals used in the treatment of humans, PEC is based the removal of these compounds during sewage treatment by using the model outlined in Eq. (4.3):

$$\text{PEC (g/L)} = \frac{A(100 - R)}{365PVD \times 100} \tag{4.3}$$

where A is the averaged annual consumption of the compound, R is the percentage removal during waste treatment, P is the population that consumes the compound, V is the per capita volume of wastewater per day, and D is the dilution factor of the final effluents to receiving water (EU, 1994). A similar approach has been developed for determining the predicted environmental concentration of pharmaceuticals that are used in the livestock industry with the assumption that the compound is excreted in urine and feces, the livestock manure is applied on a particular location once a year, and that the quantities applied are primarily limited by the maximum allowable annual nitrogen application rates [Eq. (4.4)]:

$$\text{PEC} = \frac{(\text{Nr} \times Q)/(\text{Pn} \times \text{EO})}{\text{CM}} \tag{4.4}$$

where Nr is the nitrogen requirement (kg ha^{-1} y^{-1}), Q is the total quantity of active ingredient given to the animals (mg y^{-1} location^{-1}), Pn is the maximum annual manuring rates, EO is the excreta output (kg place^{-1} y^{-1}), and CM is the mass of the contaminated environment (Spaepen et al., 1997; Jjemba, 2002a). The total quantity of active ingredient that is given to the animals (Q) is computed as a product of the individual dose (mg kg^{-1} body weight), the body weight of the animal (kg), the number of individual treatments per animal, and the number of animals raised per year at each location where the animals are housed (Spaepen et al., 1997). The predicted environmental concentration has been used to assess the risk for various PPCPs and is a widely adapted approach of conducting risk assessment for new drugs prior to approval. However, PEC assumes the application rate of the compound in question will be uniform during it normal usage, an assumption that is not very defendable as the compounds can be concentrated at specific locations within the ecosystem. Furthermore, its trigger values of 0.001 g L^{-1} in aquatic systems and

$10 \, \text{g} \, \text{kg}^{-1}$ soil in terrestrial systems seem to have been set arbitrarily. Biological activity of the respective PPCPs will greatly differ, a reality that these preset triggers do not take into account (Jjemba and Robertson, 2005). PEC predictions have also been coupled with the predicted no effect concentration (PNEC) with more information being generated by considering the ratio PEC/PNEC, referred to a hazard quotient (HQ) ratio (Lange and Dietrich, 2002). In that instance, $HQ > 1$ ratios signify a low risk. PEC/PNEC ratios that are less than 1 are assumed to be environmentally safe (Schowanek and Webb, 2005). Using the GREAT-ER model, Schowanek and Webb (2005) obtained ratios for paracetamol (0.74), aspirin (<0.01), clofibrate (0.3), and dextropro-poxylene (<0.01) as compared to ratios of 7 and 1.3 for oxytetracycline and ethynl estradiol, respectively.

Pharmaceuticals are inherently biologically active and very potent. They are also, by design, quite exquisitely resistant to biodegradation as their metabolic stability is paramount for their pharmacological action. Pollution of the environment with PPCPs may not only affect human health but also the health and well-being of other organisms. Several effects of exposure of PPCPs to biota have been identified or speculated in recent works, and some of them are discussed in this chapter. The previous lack of interest in recognizing their ecotoxicity potential following normal therapeutic use is fairly puzzling but excusable as the detection of most of the negative effects is subtle and, therefore, difficult to recognize. Until recently, they were not subject to environmental risk assessment by regulators. However, even under the current regulations, the regulatory measures that have been set may not be deemed satisfactory, elevating the need to pursue scientifically sound approaches in assessing the risks associated with these compounds in the environment. It so happens that some of the receptors in the so-called lower organisms are similar to those in "higher" organisms and can be affected in a similar way. When PPCPs are introduced in the environment, they may affect the same pathways in organisms that have identical or similar target biomolecules, cells, tissues, or organs. On the other hand, some receptors that are in the lower organism are different, and exposing them to PPCPs can translate into totally unexpected modes of action compared to those elicited in higher organisms. Similarly, targets may govern different processes in different species or similar targets may govern different processes in different species. It is also important to note that the same mode of action across species may lead to differing end points in different organisms or species. An example of this is summarized in Table 4.1 for the serotonin neurotransmitter 5-hydro-tryptamine (5-HT). Serotonin is present in all mammals and a range of other phyla. As noted in Chapter 1, 5-HT analogs include Prozac, Paxil, and Luvox (fluvoxamine).

When interest about these compounds in the environment has existed for a comparatively longer time, it has focused on the potential transfer of resistance to opportunistic pathogens and other bacteria (Pursell et al., 1996; Herwig and Gray, 1997; Threlfall et al., 1997; Chee-Sanford et al., 2001; Petersen et al., 2002; Davies and Amábile-Cuevas, 2003; Emmerson and Jones, 2003; Hawkey, 2003; Ohlsen et al., 2003). However, other negative effects that are linked to microorganisms but have not received as much attention include perturbations in biogeochemical processes such as nitrification in sewage treatment systems (Halling-Sørensen, 2001)

TABLE 4.1 Effects of Serotonin Neurotransmitter 5-HT

Organism	Reported Effect	References
Humans	Regulates appetite, controls depression, disorder, sleep, sexual arousal, etc.	Gutierrez and Queener (2003)
Fish	Stimulate the release of gonadotropin, which in turn stimulates the synthesis of sex steroids, controlling oogenesis and vitellogenesis.	Brooks et al. (2005)
Crustaceans	Regulates ovarian growth.	Fong et al. (1998)
Protozoa	Ciliary reaction and regeneration.	Czaker (2006)

and soil, suppression of nitrogen fixation, changes in microbial respiration and microbial diversity (Westergaard et al., 2001; Halling-Sørensen et al., 2002; Jjemba and Robertson, 2003), creating vacant niches as they affect these sensitive commensals. Reports of negative effects extend beyond microorganisms to include undesirable changes in the growth of plants (Migliore et al., 1996, 1998, 2000; Jjemba, 2002a, 2002b), earthworm casting (Gunn and Sadd, 1994), algae (Lanzky and Halling-Sørensen, 1997), soil dwelling arthropods (Jensen et al., 2003), and nematodes. These issues will be explored in this chapter with specific examples. It is logical to address the potential risks from PPCPs in the environment by starting from the seemingly simpler organisms and work our way to the most evolved ones. Thus, the effects on microorganisms, invertebrates, aquatic organisms (mostly fish and amphibians), wildlife and livestock, as well as humans will be the logical order followed in presenting risk-associated findings. This approach is followed only for convenience and not as a strict categorization since some evidence overlaps across this classification. For example, some of the wildlife is of aquatic origin whereas livestock species have some evolutionarily close relatives among wildlife. To complete the picture, the ecotoxicity of PPCPs on vegetation is also considered.

4.2 ECOLOGICAL IMPACT OF PPCPs ON MICROORGANISMS AND MICROBIAL PROCESSES

4.2.1 Antibiotic Resistance

Antimicrobial agents have played an important role in the control of infectious diseases since the 1940s, more than a decade after Alexander Fleming's accidental discovery of penicillin. A large number of synthetic and semisynthetic antimicrobial agents have been developed since then. The increase in antibiotic-resistant bacteria worldwide has led to the social and scientific concerns about the widespread use of antibiotics in human medicine and agriculture. Various sources estimate that in most developing countries, 70–80% of the antibiotics are used for therapeutic and/or subtherapeutic treatment of animals (FEDESA, 2003; UCS, 2003). The concerns for widespread usage of antibiotics are not unfounded, considering the fact that antibiotics are in most instances prescribed improperly or, in the livestock industry, used at subtherapeutic levels to prevent rather than cure diseases and as growth promoters. From a practical perspective, antibiotic concentrations of ≤ 0.2 g kg^{-1} in the livestock industry are regarded as subtherapeutic. Most antibiotics are poorly metabolized after administration by humans and livestock. Thus, relatively high fractions of the drug are excreted (Table 2.6). Furthermore, some excreted metabolites can also be transformed back into the parent compound. For example, on administration, sulfamethazine is conjugated with sugars in the liver. On excretion, the sugars are degraded by microorganisms quite rapidly allowing the now nonconjugated to reassume its pharmaceutical activity (Renner, 2002). Deconjugation may be a very common phenomenon for pharmaceuticals once they are excreted. For most of these, the mechanisms through which such conversion occurs have not been fully investigated, but a pathway hypothesized by Ingerslev and Halling-Sørensen (2003) for conjugated 17β-estradiol deconjugated back to the biologically active E2 is shown below (Fig. 4.1). In any case, those metabolites that are not transformed back can also still have some bioactivity. Fortunately, the nonconjugated PPCPs are generally less hydrophilic and bind more strongly to solids compared to those that are conjugated.

As is expected, the susceptibility of microorganisms to antimicrobial agents greatly varies between microbial groups, and such differences always stimulate questions such as what the difference is between resistance and tolerance. Resistance in this instance refers to the relative nonsusceptibility of an organism to a specific treatment under a defined set of conditions. Resistance is either intrinsic or extrinsic.

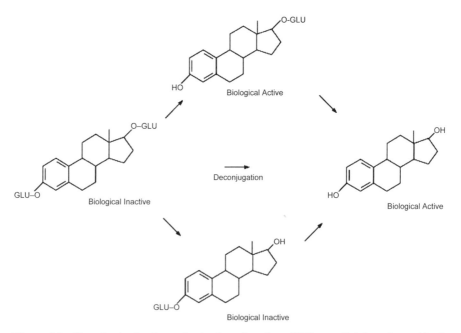

Figure 4.1 Hypothesized pathway for the deconjugation of 17β-estradiol deconjugated back to the biologically active E2. (From Ingerslev and Halling-Sørensen, 2003.)

There are various approaches to determining the extent of resistance and also in deter-ring the breakpoint between susceptible and resistant populations. These approaches often become a center of controversy about the prevalence of resistance in bacteria. Those controversies naturally affect the interpretation of resistance data, ultimately contributing to some uncertainty. It is typically quantified by determining the minimum inhibitory concentration (MIC) of the antibiotic that is required to effect a particular outcome. The outcome can, in this instance, be inhibition of growth of a bacterial population, for example. MIC is, in practice, obtained by introducing a disk that is already impregnated with a known concentration of the antibiotic or mixture of antibiotics mounted on an agar plate with a test bacterial inoculum. The agar plate is incubated at a set temperature (typically 35–37°C so as to represent mam-malian body temperature). After several hours of incubation (typically 24–48 h) the zone of clearing, which demarcates the extent of failure of the organism to grow due to the influence of the antibiotic, is measured. This approach has worked well for decades to study resistance under clinical conditions, and its use has, almost wholesale, been exported to environmental microbiology and microbial ecology. However, it is riddled with inherent limitations if used to answer questions about the transfer and persistence of resistance in the environment. Some of these include:

1. Whatever medium is used as part of the agar, some bacteria that are present in the respective environment will not grow as they will be nonculturable. It has been shown repeatedly that the number of bacteria in any environmental sample that grow on medium are several orders of magnitude less that those enumerated using direct counting.

2. The 35–37°C temperature is applicable to microorganisms that are associated with poikolotherms and is not applicable for most other microorganism in the environment.

3. Minimum inhibitory concentration determinations are usually done using pure cultures. The environment almost always contains a milieu of microbial genera that interact and influence each other. Some organisms in the environment will have intrinsic resistance to the antibiotic by nature of their physiology and bio-chemistry. These can neutralize the antibiotics or even exchange genetic infor-mation that may enable susceptible organisms to eventually acquire resistance and survive too.

Based on the above analysis, resistance to antibiotics is a relative term depending on the growth conditions in laboratory settings under which resistance is determined (Herwig and Gray, 1997). However, the presence of resistance can be established on a molecular basis by using DNA probes to examine the presence of specific resistance genes. For example, tylosin acts against gram-positive bacteria in vitro, including mycobacterium, and some gram-negative bacteria by inhibiting the synthesis of pro-teins as it binds on the ribosomes (McGuire et al., 1961; Weisblum, 1995). The expressed proteins can be determined using molecular approaches. Using this approach, two morphotypes were found to dominate the soil at 33 days after tylosin treatment, constituting 64% of the total colony forming units (CFI) compared to only 27% in soils that were not treated with this compound (Westergaard et al., 2001). Treatment with tylosin led to the evolution of a different set of morphotypes compared to the non-tylosin-treated soil. Similar shifts in microbial diversity have been documented in soils within 24 h with varying concentrations of ciprofloxacin and piromidic acid (Jjemba et al., 2006). The transfer of antibiotic-resistant genes happens more frequently in compartments such as biofilms that have high densities of microorganisms. Biofilms are present in soils, sewage, water distribution pipes, sediments, and other innocuous surfaces including hospital equipment. A prerequisite for transfer to occur is that the bacteria (or their genetic material) are stable and able to cope in a new environment. The environments can be as divergent as the human gut and wastewater, biosolids and soil, or a plant and drinking water.

The assessment of risk from antibiotic resistance is somewhat distinct from regular microbial risk assessment (MRA). As in the former instance, the focus is on the determinants of antimicrobial resistance carried by bacteria of interest, whereas MRA focuses on pathogenic microorganisms. Thus, in assessing the risk of antibiotic resistance in microorganisms, some of the microorganisms that have to be considered are not pathogenic and indeed can be commensal benign carriers of the determinants for resistance. Furthermore, the existence of such resistance may not surface until treatment with an antibiotic of interest is attempted. It is important to note that the determinants for resistance can be transmitted within and even across species. Available evidence indicates that the transfer of antibiotic-resistant genes is more likely to occur through horizontal transfer rather than single or multiple mutations (de la Cruz and Davies, 2000).

Farm animals and farming practices in general serve as a reservoir for antibiotic-resistant pathogens, causing rising numbers of infections in humans (Simonsen et al., 1998; Borgen et al., 2000; Wegener, 2003). Heifers are infected by *Staphylococcus aureus* early in life and the infection appears to originate from reservoirs in the environment rather than from the diary herd (Zadoks et al., 2000; Robertson et al., 1994; Saperstein et al., 1988). Although often opposed by industry, it is logical to suspect that, in light of the increasing occurrence of community-related MRSA infections, some transfer of these resistant microorganisms from livestock to humans does occur through the environment (manure, water), animal to human contact, and in contaminated food such as meat and milk (Spika et al., 1978; Tacket et al., 1985; Robertson et al., 1994; Fey et al., 2000). A study of 93 *S. aureus* isolates from cattle and humans originating from the Netherlands and the United States, 22.6% shared more than 90% similarity irrespective of their origin (Zadoks et al., 2000). Similarly, resistance, particularly to the first-generation quinolones such as nalidixic acid and flumequine develops readily following several passages in the laboratory, although it is slower to attain in the environment (Emmerson and Jones, 2003). It is linked to chromosomal mutations. Such slow development of resistance to these antibiotics that have relatively simpler chemical structures may be due to the possibility that they are more easily degraded by nonsusceptible microorganisms in the environment. However, resistance to floroquinolone antibiotics has increased over the years. For example, resistance to ciprofloxacin in an integrated livestock–fishing pond system in Thailand was found in 80% of the opportunistic pathogen, *Acinetobacter* spp., in water and sediment samples (Petersen et al., 2002). There has also been a strong temporal association between increased resistance to ciprofloxacin in *Salmonella* spp. and the use of enrofloxacin at subtherapeutic doses in animal feed (Threlfall et al., 1997). As indicated in Chapter 1, ciprofloxacin has also been implicated as a mutagen in hospital-derived wastewaters in Switzerland (Hartmann et al., 1998). Because of their dual activity against DNA gyrase and topo-isomerase IV, the third- and fourth-generation quinolones do not select for resistance as rapidly as the earlier (i.e., first- and second-generation) quinolones (Hawkey, 2003; Emmerson and Jones, 2003). Because of their more complex chemical structure, these third- and fourth-generation antibiotics are likely to be more persistent than other microorganisms in the environment, transferring resistance to other microor-ganisms in the environment. Considering the trend and past experience from their predecessors, it is predicted that resistance to these will be slower than that of the first- and second-generation quinolone antibiotics. A recent study of *Streptococcus pneumoniae* collected from 13 countries revealed that 0.8% of 5015 isolates were already resistant to the third-generation quinolone, levofloxacin (MIC 8 mg/L). The highest rates of resistance to this antibiotic was found in the isolates from Hong Kong (8%), China (3.3%), and Spain (1.6%) (Critchley et al., 2002). Past experience from the use of subtherapeutic levels in the livestock industry should serve as a lasting lesson to avoid mistakes that were done in the past. Because of the high molecular weight of these compounds, they are not volatile.

The low concentrations of antibiotics in the environment, which are typical of instances where antibiotics are used at subtherapeutic concentration in the livestock feeding industry, is believed to more effectively select for resistance among the indi-genous microbial population. This premise was clearly demonstrated in the work pub-lished by Onan and LaPara (2003) whereby they obtained environmental samples from three farms in Minnesota and Wisconsin where antimicrobial agents are used routinely at subtherapeutic levels in livestock feed and compared the presence of anti-biotic bacteria to samples obtained from three other diary farms where antibiotic use is restricted only to instances when they are therapeutically necessary. A summary of the occurrence of resistance to tylosin, a commonly used growth-promoting macro-lide antimicrobial in the areas from where they obtained the samples is shown in Table 4.2. It is interesting to note that of the culturable bacteria they isolated, only *Streptomyces* sp. were detected on selective media from the diary farms that use anti-biotics only for therapeutic purposes (rather than growth promotion). By comparison, the nonselective culture medium used isolated *Streptomyces* sp. together with other bacterial strain (e.g., *Haloanella gallinarum*, *Moraxella* sp., *Arthrobacter sulfoni-vorans*, *Microbacterium*, and *Planococcus psychrophilus*). From the sites that use tylosin as a growth promoter (the subtherapeutic users), resistant strains isolated were composed of a whole range of bacterial strains including *Variovorax* sp., *Rhizobium* sp., *Sinorhizobium xinjiangense*, and *Kaistia adipata* but no *Streptomyces* sp. These results are quite revealing as tylosin was originally produced by *Streptomyces* spp. (McGuire et al., 1961), and, therefore, it is not surprising to find *Streptomyces* sp., which are presumably inherently resistant to tylosin as the only

TABLE 4.2   Occurrence of Tylosin-Resistant Bacteria from Several Sites that Differ in Antimicrobial Usage in Minnesota and Wisconsin

Site	Characteristics of Site from Where Sample Obtained	Tylosin Resistance Prevalence (%)
A	Small cattle feedlot where animals were supplied with feed containing 240 g tylosin per ton of feed.	2.1% in field soils but as high as 25.8% in cattle manure
B	Swine farm where animal feed contains various antibiotics at subtherapeutic levels over time.	7.5–16.5% in soil depending on location on the farm; 69% in piglet manure
C	Soil from cornfields that have historically received swine and poultry manure. Both poultry and swine were fed antibiotics at subtherapeutic levels.	5.8–7.6% in cornfield soils
D	Diary farm where antibiotic use is restricted to instances of disease treatment as to minimize exposing humans to antibiotics through milk consumption.	Soils collected from pens showed 0.9% resistant bacteria
E	Diary farm where antibiotic use is restricted to instances of disease treatment so as to minimize exposing humans to antibiotics through milk consumption.	Soils collected from pens showed 2.5% resistant bacteria
F	Diary farm where antibiotic use is restricted to instances of disease treatment so as to minimize exposing humans to antibiotics through milk consumption.	Soils collected from pens showed 0.7% resistant bacteria

Source: Based on information from Onan and LaPara (2003).

tylosin-resistant bacteria at the farms where this macrolide (and other antibiotics) have only been used to a minimum out of necessity. On the farms where macrolides have been overused for purposes of promoting growth, it is apparent that resistance has spread to the other indigenous bacteria (i.e., lateral gene transfer), possibly through mobile genetic elements. Such elements include plasmids, conjugative transposons, phages, IS elements, or gene cassettes. As a mechanism, the resistant bacteria contain efflux pumps that expel the tylosin from the cells or a 23S recombinant RNA (rRNA) dimethyltransferase, which prevents macrolides from binding by methylating-specific adenine residues in the ribosomal RNA (Retsema and Fu, 2001). Other evidence of resistance transfer in the environment has been documented, with the interspecies transfer of *S. aureus* in diary cattle (Sischo et al., 1993) where oxacillin is widely used as treatment. Oxacillin is very closely related to methacillin and bacteria that are resistant to it are possibly selected for with the continuous use of oxacillin (Chambers et al., 1997).

The development of drug resistance has more widely been looked at from both an in vitro and a clinical perspective, with minimal connections to its reservoirs in the environment. It has been reported that the bulk of antibiotic use in humans is mainly aimed at respiratory diseases (Andersson and MacGowan, 2003; Jones and Biedenbach, 2003; Wise, 2003). However, it is also increasingly becoming evident that most of these infections are viral in nature for which most antimicrobial agents are ineffective. A variety of studies suggest the linkage between the use of antibiotics and the development of antibiotic-resistant bacterial strains that can be transferred to humans (Borgen et al., 2000; Smith et al., 1999a; Rhodes et al., 2000; Sengeløv et al., 2003). In an effort to look for reservoirs in the environment, Selvaratnam and Kunberger (2004) investigated the presence of human pathogens and drug-resistant bacteria in treated sewage sludge. The goal of that study was to determine whether the frequency of drug-resistant and indicator bacteria in Sugar Creek, which is used for recreational purposes, was influenced by proximity to a farmland routinely amended with treated sludge (site E). Surface water from three sites along Sugar Creek [site E, one upstream site (site C) and one downstream site (site K)] were tested for the presence of ampicillin-resistant (Amp^R) bacteria and fecal and total coliforms over a period of 40 days. Site E consistently had higher frequencies of Amp^R bacteria and fecal coliforms compared with the other two sites. All of the tested Amp^R isolates were resistant to at least one other antibiotic. However, no isolate was resistant to more than four classes of antimicrobials. These results suggest that surface runoff from the farmland strongly correlated with higher incidence of Amp^R and fecal coliforms at site E. Since one of the main methods of treated sewage disposal is by application to agricultural land, the presence of these organisms is of concern to human health.

At subtherapeutic concentrations, antibiotics can also impact cell functions and cause changes in the expression of virulence factors or even facilitate the transfer of antibiotic resistance. Thus, the process of feeding antibiotics at subtherapeutic levels can actually accelerate the development of resistance to antimicrobial agents because bacteria are able to survive the challenges of prolonged and/or repeated exposure to these compounds (McAllister et al., 2001). Such long-term subtherapeutic doses of antimicrobial agents are more likely to produce resistant bacteria because these minute concentrations are low enough as to enhance the growth of (resistant) bacteria and sufficiently high as to exert selective pressure that favors the resistant strains to predominate. As a matter of fact, the resistance that is established in this manner is highly stable and enables the resistant bacteria to compete effectively against their susceptible counterpart (Schrag et al., 1997).

4.2.1.1 *Acquisition of Antibiotic Resistance*

There is constant acquisition of resistant genetic sequences among bacteria, and through such acquisitions bacteria transfer resistance to others (Ochman et al., 2000; Seveno et al., 2002; Ohlsen et al., 2003). Thus, tetracycline-resistant bacteria have been found in waste lagoons and nearby groundwater in the vicinity of all pig farms in Illinois that were tested (Chee-Sanford et al., 2001). Antibiotic-resistant bacteria have also been detected in sewage treatment effluents (Ohlsen et al., 2003; Halling-Sørensen, 2001). Sediments can also act as a reservoir for antibiotic-resistant microorganisms after completion of the chemotherapy programs (Herwig and Gray, 1997). Some level of antibiotic spread also occurs in the vicinity of animal houses in dust

(Thiele-Bruhn, 2003). The direct transfer of genetic material from one cell to another through conjugation involves transposons and plasmids. It has the potential to convey resistance to classes of antibiotics and can lead to a relatively broad dissemination of resistance in bacteria that are susceptible. Through such mechanisms, antimicrobial agents can cause resistance to pathogens directly or indirectly through the transfer of resistance plasmids from nonpathogenic to pathogenic bacteria. Resistance to the compound can occur shortly after the compound has been introduced for general clinical use as was the case for penicillin, which was introduced in 1945 and its resistance noticed a few months later. It can alternatively occur after several decades of introduction as was the case with vancomycin resistance, which emerged after 30 years of use (Walsh, 2003).

4.2.1.2 *Mechanisms of Antibiotic Resistance*   Resistance to antibiotics within microorganisms occurs primarily through three different mechanisms, that is:

1. Inactivation of the compound
2. Modification of the antibiotic-susceptible site(s)
3. Explusion of the compound from the cell usually through active efflux pumps

The first mechanism is particularly common with natural product antibiotic classes and is rare among synthetic classes such as fluoroquinolones and sulfamethaxazole. Modification of the antibiotic target is an important mechanism of resistance to antibiotics. Such modifications can lead to noncompatibility between the antibiotic and the target and/or a reduction in the affinity between the two. This mechanism is very commons in the development of methicillin-resistant organisms whereby the methylation of one adenine in the 23S of the 50S ribosomal subunit reduces the affinity of streptogramin B and in the development of VRO phenotypes. Resistance by the expulsion (i.e., active export) of antibiotics by efflux pumps leaves concentrations within the cells that are below those needed for therapeutic efficiency. The efflux is mediated by the transmembrane proteins often against a concentration gradient. This mechanism of resistance has been reported for a wide range of β-lactams, tetracyclines, fluoroquinolones, and macrolides.

Vancomycin resistance in enterococci has been determined to be transferable by conjugation and is, most likely, mediated by plasmids (Noble et al., 1992). There is also subtle evidence that such resistance is on the rise in clinical settings. Transfer of resistance through plasmids is inducible and requires prior exposure to the antibiotic in question to be fully expressed. Furthermore, the induction of resistance to vancomycin occurs concomitantly with the appearance of novel membrane proteins as compared to those bacteria that are intrinsically resistant where those novel proteins have not been detected. Vancomycin resistance has also been reported in other organisms including *Lactobacillus* sp. and *Leuconostoc* sp., some of which were isolated from sites that should normally be sterile, notably blood and intraabdominal abscesses (Ruoff et al., 1988). Even though these vancomycin-resistant lactobacilli and *Leuconostoc* sp. are clinically insignificant, they can comprise some of the nosocomial pathogens or vectors of the resistance genes, passing them on to other pathogens, such as enterococci. To date, five other resistance genes, that is, *vanB*, *vanC*, *vanD*, *vanE*, and *vanG*, have been reported (Levine, 2006). Irrespective of the type of vancomycin-resistant gene in play, the phenotypes have target peptidoglycan chains that possess a reduced affinity for vancomycin, thus requiring much higher MICs.

4.2.2 Biogeochemical Perturbations

Biogeochemistry is the study of the exchange of materials between the living and nonliving components of the biosphere. It involves the physical transportation of nutrients and their transformation through chemical and biochemical mechanisms. These transformations are cyclic in nature. Microorganisms are quite central in the cycling of such nutrients (Jjemba, 2004). Both livestock manure and biosolids are often collected and applied to arable land. The environmental risks associated with applying manures or sludge to arable land has, in the past focused on the potential spread of pathogens and heavy metals in soil, groundwater, and on the existing vegetation (U.S. EPA, 1995; Walker et al., 1997). The health concerns in the EU member countries also put a lot of emphasis on nitrogen, phosphorus, and heavy metals (Smith, 1996). However, the presence of therapeutic agents in these materials is becoming an increasingly recognized environmental risk as well (Halling-Sørensen et al., 1998; Hirsch et al., 1999; Jørgensen and Halling-Sørensen, 2000). Some of the therapeutic agents can maintain residual activity in biosolids and farmyard manure (Migliore et al., 1995, 1997; Hirsch et al., 1999) for a long time. PPCPs (and other organic compounds) in sewage may severely affect microbial biomass. For example, in response to sewage sludge amendments, decreases in microbial biomass have been reported (Boyle and Paul, 1989; Chander and Brookes, 1993). Such changes are normally reflected by changes in vital biological processes such as nitrification and nitrogen fixation. Besides introducing antibiotic resistance, antimicrobial agents may also adversely affect the microbial activity in wastewater treatment processes, soils, and sediments by directly impacting important processes such as nitrification and biogeochemical cycles. Sengeløv et al. (2003) reported changes in microbial populations in the soil when manure that contained antibiotics was applied. The microbial population rebounded several weeks thereafter, but it is not clear from that study whether the changes were exclusively due to antibiotics or some other constituents of the manure. These processes are performed by a specialized number of microbial species and because of their low diversity can be used as indicators of pollution. However, such decreases are routinely almost automatically associated with heavy metals, irrespective of whether sufficient evidence is accumulated to implicate heavy metals. A distinct example of biogeochemical perturbation is nitrification. Nitrification involves the conversion of ammonium to nitrate, through a couple of intermediates, that is, nitrite and hydroxylamine. The process is conducted by slow growing beta and gamma proteobacteria commonly referred to as *Nitrosomonas* and *Nitrosospira*:

$$2NH_4^+ + 3O_2 \rightarrow 2NO_2 + 4H^+ + 2H_2O \tag{4.5}$$

$$2NO_2^- + O_2 \rightarrow 2NO_2^- \tag{4.6}$$

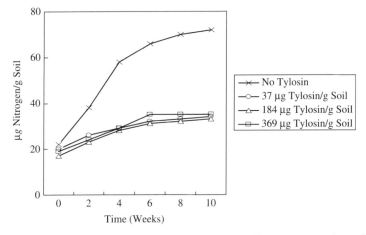

Figure 4.2   Concentration of nitrogen in soils subjected to different concentrations of tylosin. (Based on data from Bewick, 1978.)

Several organisms are able to utilize nitrification as an energy source. Nitrification is also important in releasing ammonium-based nitrogen (NH_4^+-N) for uptake by plants as nitrate nitrogen (NO_3^--N). Bewick (1978) found that concentrations of 37 μg tylosin g^{-1} soil or higher reduced nitrification over a 10-week period (Fig. 4.2). This decrease in N mineralization also corresponded to a depressed oxygen uptake as well as carbon dioxide release from the soil for about 7 weeks, signifying an inhibition of microbial activity. The degree of microbial inhibition did not appear to be proportional to the amount of tylosin.

4.3 EFFECTS OF PPCPs ON INVERTEBRATES

Wilson (1999) estimates that invertebrates comprise 95% of the terrestrial and aquatic animal species. Most widely used among invertebrates in toxicology studies are daphnids, mollusks (gastropods and biovalves), as well as rotifers. Oysters, arthropods (i.e., insects and crustaceans), and annelids (i.e., earthworms and polychaetes) and nematodes have also been used to some extent. Some invertebrates—for example, mollusks—have a highly integrated neuroendocrine system just like all other animals. Besides microorganisms, invertebrates are deemed to get the full blunt of the occurrence of these compounds in manure and biosolids. For example, antihelminthic compounds shed in livestock feces have also been documented to affect ecologically important invertebrates (McKellar, 1997). These dwellers play a major role in the decomposition process. Broadly included among invertebrates of interest in this category are earthworms, snails, and a whole range of soil dwellers. These organisms mainly become exposed to PPCPs and other terrestrial contaminants through ingestion of the contaminated media. For example, Prosobrach snails (*Marisa cornuarietis*) and freshwater mud snails (*Potamopyrgus antipodarum*) have been reported to be affected by estrogenic compounds (Duft et al., 2003). *Potamopyrgus antipodarum* feeds on plants and detritus and has been used in aquatic ecotoxicology to study endocrine disruptions by counting the number of embryos it carries in the brood pouches (Duft et al., 2003).

The presence of such highly integrated neuroendocrine systems in some invertebrates has made invertebrates ideal candidates for studying effects of endocrine disrupting compounds (EDCs) and antidepressants in the environment. For example, Nice et al. (2003) showed changes in the sex ratios in oysters (*Crassostrea gigas*) in favor of females and hermaphrodites on exposure to 1–100 μg nonylphenol L^{-1} a week after fertilization. Similarly, Fong et al. (2001) highlighted the effects of antidepressants on reproduction in mollusks. Similarly, exposure of the amphipod *Gammarus pulex* and the freshwater chiromonid *Chironomus riparius* to 17α-ethinylestradiol (EE2) led to a significant increase in the proportion of females in the former and delayed emergence of males compared to females in the latter (Watts et al., 2002). Experiments by Schulte-Oehlmann et al. (2004) showed increased frequency of imposex (development of male parts) in female *M. cornuarietis* when exposed to a 0.1–1 μg EE2 L^{-1} or 0.1–1 μg 17α-methyltesterone L^{-1} over 6-month period. The onset of these features was much earlier with 17α-methyltesterone (MT) than with EE2. By 6 month, the intensity of imposex was concentration dependent at concentrations of 0.1, 0.25, 0.5, and 1 μg EE2 L^{-1}, whereas the treatments that had received similar concentrations of MT were all equally affected irrespective of the concentration of MT. There was also an increased incidence of degenerative ovaries and york lysis in porticlellogenic oocytes in the female snails exposed to EE2 and MT, with the two highest concentrations of both hormones significantly reducing fecundity. In male snails, most of the EE2 and MT concentrations tested also significantly affected spermatogenesis over an 8-week duration compared to the control. In this regard, these invertebrates differed from vertebrates, which have been reported to, in some instances, develop "supermales" on exposure to MT. Virilization of female Nigorobuna (*Carassius carassius*) following exposure to 0.01–1 μg MT L^{-1} have also been reported (Fujioka, 2002) and in the snail *Lymnaea stagnalis* L (Czech et al., 2001). Similarly, exposure of 0.1–0.32 g EE2 to amphipod *Hyalella azteca* generated significantly smaller second-generation gnathopods of males (Vandenberg et al., 2003).

The oligocheate *Lumbriculus variegatus* and the nonbiting midge *Chironomus riparius* have also been widely used in toxicity studies. Their use is particularly attractive in sediment toxicity assays (West and Ankley, 1998). The latter is one of the most widespread insect species in freshwater and an important prey for fish in those waters. In chronic toxicity studies conducted by Nentwig et al. (2004), carbamazepine had no significant effect on *L. variegatus* over a 28-day period, whereas levels of 1.25 mg carbamazepine kg^{-1} sediments reduced the emergence of *C. riparius* to 12%. In that same study, higher concentrations of the drug in the sediments reduced the emergence of *C. riparius* even further with 5 mg carbamazepine kg^{-1} sediments registering no emergence of the midge. The NOEC and LOEC were 0.625 and 1.25 mg kg^{-1} for this organism, and the *C. riparius* larvae hatched at 1.25 mg carbamazepine kg^{-1} sediments were convulsive, demonstrating that this drug designed to affect humans can also affect nontarget midges in the environment.

TABLE 4.3 Effects of Various Statins on Ootheca and Larval Development in German Cockroach

Parameter	Control[a]	Simvastatin 50 µg	Atorvastatin 50 µg	Atorvastatin 10 µg	Fluvastatin 50 µg	Fluvastatin 10 µg	Fluvastatin 1 µg
Female cockroaches tested	79	10	8	8	40	43	19
Dead cockroaches	1	1	0	0	14	4	0
Days to first ootheca formation	7.5 (0.1)[b]	7.8 (0.1)	7.4 (0.3)	7.1 (0.1)	7.8 (0.2)	7.5 (0.1)	7.3 (0.1)
Number of first viable ootheca as fraction of total first ootheca	73/75	3/9	1/8	5/7	0/38	0/41	19/19
Duration of embryogenesis (days)	15.6 (0.1)	16.5 (0)	15	15.6 (0.4)	N/A[c]	N/A	15.8 (0.1)
Larvae emerging from the first viable ootheca	37.5 (0.9)	39 (0.6)	25	22.2 (4.8)[*,d]	N/A	N/A	29.6 (2.2)[†]
Average of larvae obtained per survivor	35.1	13	3.1	13.8	0	0	30
Total reduction in fertility as a percent of the larvae obtained per survivor	0%	63%	91%	61%	100%	100%	15%

[a]The control treatment received the vehicle (i.e., acetone) used as a solvent for the statins.
[b]The numbers in parentheses represent standard error mean.
[c]N/A = not applicable.
[d]Significantly different from the control (t test, [*]$P < 0.05$, [†]$P < 0.01$).
Source: Zapata et al. (2003).

In other studies, Liebig et al. (2004) reported the uptake of radiolabeled 17α-ethinylestradiol (^{14}C-EE2) by earthworms (*Lumbricus variegates*) over a 35-day period, excreting 50% of the bioaccumulated EE2 into the environment. Earlier studies also showed various sublethal effects such as interrupted molting, pupation, and failure of the adults to emerge. Presence of the avermectins in the dung of cattle led to a general reduction in the growth rate of dung beetles (Strong and Brown, 1987; Strong, 1993).

Statins affect the synthesis of various hormones in juvenile insects (Zapata et al., 2002, 2003) and lobsters (Li et al., 2003). As indicated in Chapter 1, several statins inhibit HMG-CoA reductase, an enzyme that is important in the cholesterol pathway and is an important component of the fat body and reproductive capabilities of insects and lobsters. Zapata et al. (2003) applied various statins topically on the dorsal part of cockroach (*Blattella germanica*) abdomen to study the effects of these compounds on the ovaries. HMG-CoA reductase is involved in the glycosylation and export of vitellogenin, and this reductase enzyme is present in the eggs and embryos of insects. In cockroachs, HMG-CoA reductase activity was significantly reduced by various concentratios of lovastatin, simvastatin, atorvastatin, and fluvastatin (Zapata et al., 2003). Embryos were also affected as the fucandity of vitellogenic females decreased due to delayed viability. The formation of ootheca and production of larvae are also affected (Table 4.3). The data in Table 4.3 show that atorvastatin and fluvastatin, both of which are totally synthetic, have a more extensive effect on larvae and embryos as they were still active at levels as low as 1–10 µg. However, these were laboratory-based controlled studies, and the effect of these compounds on roaches and similar insects in the environment still remains unexplored.

4.4 ECOTOXICITY OF PPCPs ON AQUATIC ORGANISMS

Reproductive and developmental abnormalities have been documented in aquatic organisms for several decades (Colborn et al., 1993) and recent increases in the reporting of diseases in aquatic organisms has caused concerns among scientists, managers, politicians and the general public about deteriorating aquatic life. For example, the increasing trends in mortality in marine organisms was recently compiled by Gulland and Hall (2007; Fig. 4.3). Threats to the health of aquatic organisms can directly or indirectly affect human health and may account for some of the emerging zoonotic and wildlife diseases. However, these threats have been mostly attributed to conventional contaminants such as PAHs, (heavy) metals, and a whole range of organic contaminants of which PPCPs have not been considered as potential culprits. However, since PPCPs are quite frequently encountered in aquatic systems, albeit in low concentrations, it is just a matter of time before they join the list of significant culprits. The organisms in aquatic ecosystems are frequently restricted in their home range and habitat to the extent that they often cannot avoid contaminated areas. This reality presents a unique challenge and circumstances to aquatic organisms compared to those in terrestrial ecosystems. Unlike terrestrial dwellers that mostly exposed through ingestion, aquatic organisms become

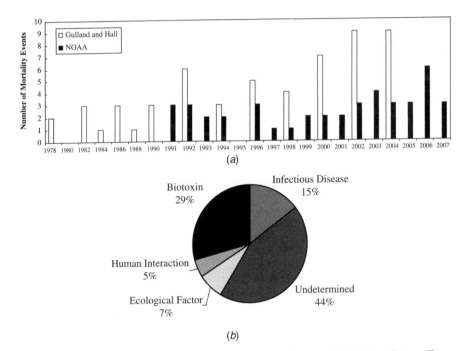

Figure 4.3 Trend of unusual mortality of marine mammal events in the United States. (From Gulland and Hall, 2007.)

exposed to PPCPs and other contaminants through their permeable bodies. Significant aquatic dwellers such as fish and amphibians are unique among vertebrates as their eggs lack a shell and amniotic membrane (i.e., anamniotic egg). This anamniotic egg structure remains quite permeable to assault of chemicals, a situation that is even more compounded by the fact that the embryos develop while the egg is completely immersed in water. Furthermore, their larval stages are unique compared to other vertebrates as they are aquatic. As noted above and in Chapter 2, aquatic ecosystems are where most PPCPs, and other contaminants in general, are deposited, a situation that is compounded with the high sensitivity of the embryo and larval stages to chemical insults. These observations collectively place aquatic organisms in a unique place that adamantly exposes them to PPCPs and other related contaminants.

Henschel et al. (1997) determined a 96-h EC_{50} of 89 mg clofibric acid L^{-1} using the green algae *Scenedesmus subspicatus*, whereas EC_{50} of 106 μg clofibric acid L^{-1} immobilized *Daphnia magna* (Daughton and Ternes, 1999). Algae are the basis of the food chain in aquatic environments. Thus, their decrease affects the balance in that system. However, environmentally relevant concentrations of clofibric acid (i.e., 0.1–1000 μg $^{-1}$) did not alter the growth rate or cell density of another type of algae called *Dunaliella tertiolecta* (Emblidge and DeLorenzo, 2006), and EC_{50} values of approximately 80 mg clofibric acid L^{-1} have been reported by Ferrari et al. (2003). *Daphnia* sp. has been used extensively in ecotoxicology as it is easy to maintain and has a short life cycle. However, for PPCPs in which the mode of action affect the vascular system, using *D. magna* may not give very useful information as these organisms do not have a vascular system. A cumulative dose of 50 mg clofribirc acid kg^{-1} was also shown to decrease the lipids, phospholipids, cholesterol, and triglycerides in the crab *Pachygraspus marmoratus*, the effect being more dramatic against male than female crabs (Lautier et al., 1986). Likewise, the exposure of 10 and 50 mg clofibric acid L^{-1} to the catfish (*Ictalurus punctatus*) increased the hepatic cytochrome 450 expression and liver somatic index after a 24-h exposure (Perkins and Schlenk, 1998). The ecotoxicology of some of the major PPCPs in aquatic systems that have been studied to some extent is discussed below.

4.4.1 Endocrine Disrupters in the Aquatic System

Endocrine disruption has been known to biologists and the medical community since the 1930s. Endocrine disruptors are exogenous compounds that alter the functions of the endocrine system, ultimately affecting the health of that individual organism, its progeny, or subpopulation. Endocrine disrupting compounds are exogenous substances that cause adverse effects in an intact organism or its progeny, causing changes in endocrine function. They function by mimicking and/or antagonizing endogenous hormones. They may also elicit a negative response by changing the degradation or synthesis of endogenous hormones or hormone receptors. There is evidence that they may interact with multiple targets and can act at any level of hormone synthesis, secretion, metabolism, site of action, and transport. The effect of EDCs on hormone synthesis and metabolism can manifest itself by adversely altering the levels of critical endogenous hormones by inhibiting or inducing biosynthetic or metabolic enzyme activities. EDCs came to the limelight in science in the early 1950s when the organochloride pesticide DDT was implicated in the disruption of the endocrine system. However, endocrine disruptive effects of compounds did not cross into the realms of environmental sciences until the early 1980s when Fry and Toone (1981) reported changes in sex ratios and deformities in sex organs of seagulls living in areas contaminated with DDT. That finding was soon thereafter accompanied by similar trends in other areas where DDT and other pesticides had been used (Fry et al., 1987). EDCs have a high ecotoxicological relevance as they can potentially influence the reproduction of organisms adversely, consequently affecting the survival and existence of entire populations. Most of their effects can be quite subtle. It turns out that subtle changes in systems can be fairly complex and intricately interwoven, which sometimes makes it very hard to distinguish from "normal" systems. Several PPCPs, including the natural female hormone 17β-estradiol, estrone, as well as the male hormone testosterone have endocrine disrupting capabilities. For example, the exposure of the natural hormone 17β-estradiol (E2) and the synthetic birth control ingredient 17α-ethinylestradiol (EE2) at concentrations as low as 2 ng^{-1} can induce changes in reproduction in fish (Snyder et al., 2003). However, it is important to note from the onset that not all EDCs are of pharmaceutical origin, and EDC activity has been reported in a number of other compounds of natural or anthropogenic origin, including alkylphenols, some pesticides, polychlorinated biphenyls (PCBs), PAHs, phthalates, and bisphenol-A. Schlumpf et al. (2002) also showed estrogenic effects from various UV screens.

For obvious economic reasons, the toxicity of a variety of organic and inorganic substances in fish has been conducted. To that effect, the effects of EDCs have been more extensively studied in situ with fish in aquaculture and wild marine conditions (Piferrer, 2001; Donaldson and Hunter, 1982). Most widely used as the measure of estrogenic activity is the production of an estrogenic-inducible egg yolk protein precursor called vitellogenin (VTG). VTG is a large glycoprotein of approximately 156 kDa that is synthesized in the liver of female oviparous vertebrates. It is released into the bloodstream by the oocytes where it is cleaved to yield phosvitin and lipovitellin in the yolk. The VTG gene is also present in male fish, but it is not normally expressed due to the low concentration of estrogen in the bloodstream. However, if stimulated with estrogenic compounds, VTG concentrations in the blood of male fish can also increase to levels that are equal to those of mature females. However, it increases substantially in the blood system owing to estrogen during the development of oocytes. It is only marginally expressed in males or juveniles under normal conditions to the extent that its detection in unexpectedly high concentrations in male fish plasma leads to males exhibiting female characteristics (i.e., feminization), making it an excellent marker at the biochemical and molecular level for detecting the presence of EDCs in aquatic systems (Purdom et al., 1994; Jobling et al., 1998; Routaledge et al., 1998; Tyler et al., 1999; Folmer et al., 2000; Korte et al., 2000; Denslow et al., 2001). It is very sensitive in the presence of mixtures, a situation that is more typical of environmental matrices. Ankley et al. (2001) showed the effect of estrogen on fathead minnow (*Pimephales promelas*). Zerulla et al. (2005) used RT-PCR to show an increase in VTG concentration in *P. promelas* after exposure to estrogens and documented premature development of strong male sexual characteristics when the aromatase inhibitor (Fadrozole) was combined with methyltestosterone. Using a quantitative approach, Schmid et al. (2005) developed a fitness factor [Eq. (4.7)] or fish exposed to estrogens:

$$\text{Fitness factor (g/cm}^3) = \frac{100W}{L^3} \tag{4.7}$$

where W is the weight (g) and L is the length (cm). Exposure of male fathead minnows to 50 ng 17α-ethinylestradiol L^{-1} for 35 days significantly ($P < 0.05$) reduced the fitness of the fish within 2 weeks. The significant reduction persisted even after the 17α-ethinylestradiol (EE2) was discontinued (i.e., replaced with EE2-free water) for 3 more weeks (Schmid et al., 2005). Mortality, which occurred only between 20 and 36 days of the 70-day duration was only 3.5 and 2.8% in the control and solvent control (i.e., 0.0001% *N,N*-dimethylformamide used as a vehicle to dissolve EE2) but was as high as 13.3% in the EE2-spiked water. Mortality ceased 7 days after the EE2 was discontinued (Schmid et al., 2005). This high mortality in EE2-exposed fish was possibly a result of the high concentration of the large VTG protein causing blockage of the kidney glomeruli and the liver.

This contention is advanced based on evidence from histological studies that showed changes in the kidney and liver of carp (*Cyprirus carpio*) and toads (*Xenopus laevis*) following exposure to EDCs (Lewis et al., 1976; Schwaiger et al., 2000). Exposure to EE2 and the antiandrogen cyproterone acetate during larva development caused feminization in toads (*X. laevis*) (Bögi et al., 2002). Treatment with these compounds elevated by the estrogen and androgen receptor mRNA (messenger RNA) in the toads indicating stimulatory functions of estrogens for gene expression of both of these receptors.

The VTG was accumulated in a biphasic fashion with the accumulation in the first 2 weeks following an exponential model:

$$C_T = C_0 e^{kT} \tag{4.8}$$

where C_T is the concentration of VTG (mg L^{-1}) in the plasma at time T (in days), C_0 is the initial concentration of VTG at time $T = 0$, and k is the rate constant of the exponential increase (day^{-1}). The second phase at 14–38 days displayed a one-component saturation model, that is,

$$C_T = C_{15} + C_{sat}(1 + e^{-kT}) \tag{4.9}$$

where C_{15} is the concentration of vitellogenin in the plasma (mg mL^{-1}) at day 15 (i.e., at the beginning of the second phase), C_{sat} is the concentration of VTG in the plasma (mg mL^{-1}) at saturation, and k is the rate constant for the saturation phase. The initial rapid (exponential) phase is indicative of rapid induction of gene expression in response to EE2 exposure. On exposure to estrogenic compounds, the half-life of VTG mRNA is increased severalfold, leading to an increase in VTG protein in the plasma. The concentration of 4 pg EE2 mL^{-1} in water caused developmental and reproduction problems in fish (Länge et al., 2001) compared to levels of 25–100 pg mL^{-1} that are considered tolerable in humans (Hümpel et al., 1990), strongly suggesting that PEC risk assessment is based on human concentration safe levels may not necessarily be safe for other organisms. Some effects may not emerge until generations later. For example, studies in fish have shown F1 zebrafish exposed to 5 ng EE2 L^{-1} developed sexual deficiencies that resulted in a 56% reduction in fecundity as the males lacked functional testes (Fig. 4.4).

Vitellogenin production in trout liver cells was decreased by acetaminophen (Miller et al., 1999) showing a classic example of the unpredictable effects of pharmaceuticals to nontarget organisms in the environment. Thus, whereas acetaminophen is reported to exert estrogenic agonistic effects on human breast cells (see Chapter 1), it seems to exert the opposite effects, that is, antagonistic effects, on fish. EDC activity in aquatic systems that are under the influence of sewage and wastewater have been well documented (Atkinson et al., 2003). Estrogenic compounds have also been detected in biosolids (see Chapter 2). This observation may have important implications on the application of biosolids on land, though appropriate risk assessment to consider this aspect has not been conducted. Besides estrogenic compounds, androgenic compounds such as testosterone have not also been encountered, causing musculinization of female fish and causing

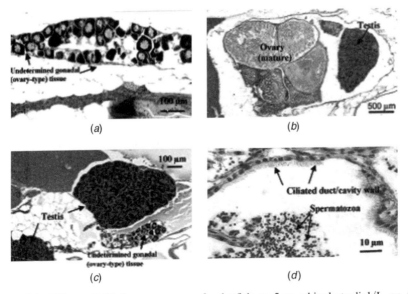

Figure 4.4 Effects of life-long exposure of zebrafish to 5 ng ethinylestradiol/L on the development of gonads. (*a*) Persisting juvenile undifferentiated gonads in a presumptive male zebrafish. (*b*) Shows intersex zebrafish with one ovary and one testis. (*c*) Shows intersex zebrafish with two testes and a smaller juvenile tissue. (*d*) Shows a ciliated sperm duct in testis of a mature male. (Adapted from Nash et al., 2004.)

biases in eelpout embryo sex ratios (Thomas et al., 2002). Adronergic receptors are also present in frogs (Devic et al., 1997) and birds (Yardeny et al., 1986). Recently, Lorenzen et al. (2004) surveyed various municipal biosolids and manure from swine, poultry, as well as diary and beef cattle and found estrogenic and androgenic activity using a hormonal receptor binding assay. They encountered the highest estrogenic activity in manure from finishing pigs (5965 ng estradiol equivalents g^{-1} dry weight) and the highest androgenic activity in manure from pregnant diary cows (1737 ng testosterone equivalents g^{-1} dry weight).

4.4.2 Effects of Antibiotic Resistance to Aquatic Organisms

In Section 4.2.1, focus was on the mechanisms of antibiotic resistance and understanding the microorganisms involved. The current section focuses on the effects of such resistance to aquatic organisms including increased susceptibility to diseases. A recent analysis by Gulland and Hall (2007) has shown an increased ability of more fastidious pathogens such as bacteria, protozoa, and algal toxins as agents of unusual diseases in marine mammals. Where mass mortality has been detected, it has been mostly attributed to biotoxins, viruses, parasites, bacterial infections, human interactions, oil spills, and changes in oceanographic conditions. However, in a number of instances, the actual cause for toxicity remains unidentified and PPCPs— antibiotics in particular—may be indirectly contributing to this mortality by increasing susceptibility to diseases. On the other hand, specific compounds such as antibiotics may not show detrimental effects on algae (except at concentrations that are not likely to be encountered in the environment), and indeed some have been proposed to be included in PPCPs remediation programs in aquatic systems.

4.4.3 Ecotoxicological Effects of Cosmetics on Aquatic Organisms

A variety of fragrances such as HHCB and AHTN are highly lipophilic (log K_{ow} 5.4–6.3) and therefore have a high propensity to accumulate in aquatic organisms such as eels (Balk and Ford, 1999; Fromme et al., 1999), fish and mussels (Gatermann et al., 2002), as well as finless porpoises and sharks (Nakata, 2005). Thus, HHCB was present in the livers of five hammerhead sharks (*Sphrna lewini*) and in all of the finless porpoises (*Neophocaena phocaenoides*) analyzed off the Japanese coastal waters at levels of 13–149 ng g^{-1} wet weight (Nakata, 2005). A fetus of the porpoise also contained notable (i.e., 26 ng g^{-1} wet weight basis) concentrations of HHCB, suggesting transplacental transfer of this compound. HHCB was also detected in various tissues of the porpoises, with the highest concentration detected in the blubber followed by the kidney. These compounds have also been detected in terrestrial wildlife such as birds (Kannan et al., 2005). Veith et al. (1979) developed a model for predicting the bioaccumulation of various organic chemicals using the following relationship:

$$\log \text{ BCF} = 0.85 \log \ K_{ow} - 0.7 \tag{4.10}$$

where BCF is the bioconcentration factor. This relationship has been adapted by the EU for use in instances where the log k_{ow} is less than 6. However, its use by Balk and Ford (1999) to estimate bioaccumulation of the cosmetic products AHTN and HHCB in fish gave values of 13,946 and 20,654 L kg^{-1}, respectively, compared to the actual measurements of 597 and 1584 L kg^{-1}. These overestimated BCF values using this theoretical approach show our inability to extrapolate information about what works for other contaminants to pharmaceutical and personal care products in the environment. It is worth noting that from work by Kannan et al. (2005) these compounds were noticeably undetected in polar bears and sea otters. HHCB and AHTN have also been found in birds. They have been detected in human adipose tissue (Table 4.4) and several other animal species. In the assays conducted by

TABLE 4.4 **Mean ($\pm$SD) Concentrations of HHCB and AHTN (ng/g) in Adipose Tissue for 49 Liposuction Patients in New York City**

		Wet Weight		Lipid Weight	
	Fat (%)	HHCB	AHTN	HHCB	AHTN
Male ($n = 12$)	34 ± 9	72 ± 75	20.5 ± 14	136 ± 143	39 ± 29
%Positive		100	83	100	83
Female ($n = 37$)	31 ± 7.3	105 ± 90	23.5 ± 11	192 ± 170	43 ± 22
%Positive		100	87	100	87
Overall ($n = 49$)	32 ± 7.8	96 ± 88	22.8 ± 12	178 ± 166	42 ± 24
%Positive		100	86	100	86

Kannan et al. (2005) HHCB was found in all human fat tissue examined where 86% contained AHTN, and the former was always detected in greater concentration than AHTN possibly because HHCB is produced/manufactured and used in larger quantities than AHTN. High concentrations of HHCB also have been encountered in Lake Michigan compared to AHTN (Peck and Hornbuckle, 2004), and both HHCB and AHTN have also been detected in mothers' milk (Rimkus and Wolf, 1996). Musks have some antiestrogenic effects (Schreurs et al., 2004).

4.4.4 Ecotoxicity of Other PPCPs in Aquatic Organisms

Various common lipid regulators such as clofibrate, etofibrate, and etoyllinclofibrate act as antiauxins and also have some pesticidal activity. As a matter of fact, their metabolite clofibric acid is an S isomer of the phenoxyalkanoic acid herbicide mecoprop. Pfluger and Dietrich (2001) suspect clofibric to have some endocrine disruption activity in the environment and adversely affecting the synthesis of cholesterol as well as the genesis of steroids. This contention has some basis as fibrates enhance β-oxidation of lipids, increasing the amount of reactive oxidative species in cells, which in turn can lead to cytotoxicity (Laville et al., 2004). Nunes et al. (2006) studied the effects of various concentrations of diazepam, clofibrate, clofibric acid, and the surfactant sodium dodecylsulfate (SDS), which is widely used in detergents, on five important antioxidant enzymes. Some of their findings in that study are summarized in Table 4.5. These studies were done only after 48 h exposure but with exposure concentrations that were way above those typically encountered in the environment. Similarly beta blockers have also been reported to impact fish (Nickerson et al., 2001). Mostly implicated among this group of PPCPs is propranolol because of its high log K_{ow} compared to other beta blockers and its ability to stabilize membranes (Iwai et al., 2002). Further evidence of the potential effect of lipid regulators on other aquatic organisms is generated from the presence of peroxisome proliferator-activated receptors (PPARs) in some of these organisms. PPARs bind fibrates and are known to stimulate the expression of various lipid regulator proteins (Staels et al., 1998) and have, besides humans, also been found in plaice fish (*Pleuronectes platessa*) (Leaver et al., 1998); Atlantic salmon (*Salmo salar*) (Ruyter et al., 1997), and zebrafish (*Danio rerio*) (Ibabe et al., 2002, 2005) where they are induced by bezafibrate and clofibrate.

Analgesics and anti-inflammatory drugs might also have some untold effects on aquatic organisms. For example, a gene that induces COX-2 and COX-1 that is >80 and 77%, respectively, homologous to that found in humans has been located in rainbow trout (*Onchorhynchus mykiss*) and in goldfish (Zou et al., 1999). Similar COX homologs have been reported in sharks as well (Yang and Carlson, 2004). Implicit from all of these findings is that traces of COX inhibitors in water can potentially have an impact on aquatic life. It is not clear at this point whether such an impact is negative (e.g., causing tumors) or positive (e.g., pain relief). Fish have a γ-aminobutyric acid (GABA) system, a system which as discussed in Chapter 1 is inhibited by most of the popular benzodiazepines.

Recently, the effects of some antipsychotic drugs on aquatic communities have also been studied. From some of those studies, sublethal concentrations

TABLE 4.5 Effects of Four PPCPs on Some Key Antioxidants in *Artemia parthenogenetica*[a]

Enzyme	Diazepam	Clofibrate	Clofibric Acid	SDS
Glutathione peroxidase (GPx)	Activity significantly increased by some concentrations $> 6.1\,mg\,L^{-1}$	Activity not affected	Activity not affected	Activity not affected
Glutathione reductase (GRed)	Activity not affected	Activity not affected	Activity not affected	Activity significantly reduced with increasing concentrations ($\geq 4\,mg\,L^{-1}$)
Superoxide dismutase (SOD)	Activity not affected	Activity not affected	Activity not affected	Activity not affected
Glutathione S-transferases (GST)	Activity not affected	Activity not affected	Only $35.1\,mg\,L^{-1}$ significantly reduced activity. Concentrations higher and lower than this had no effect	Activity not affected
Cholinesterases (ChE)	Activity significantly decreased at only $\geq 7\,mg\,L^{-1}$	Not altered	Not altered	Activity decreased at $\geq 8.46\,mg\,L^{-1}$
Selenium-reduced GPx	Activity not affected	Activity significantly reduced at $\geq 3.1\,mg\,L^{-1}$	Activity significantly reduced at $\geq 17.6\,mg\,L^{-1}$	Activity not affected

[a]Diazepam was at 4.07, 4.89, 5.86, 7.04, and $8.44\,mg\,L^{-1}$; clofibric acid was at 29.21, 35.05, 42.06, 50.47, and $60.57\,mg\,L^{-1}$; clofibrate at 12.27, 14.73, 17.67, 21.21, and $25.45\,mg\,L^{-1}$, and SDS was at 4.08, 4.9, 5.88, 7.05, and $8.46\,mg\,L^{-1}$.

of fluoxentine have been implicated in behavioral responses such as appetite, reproduction, and immunity in fish (Brooks et al., 2005; Fong, 1998, 2001; Fong et al., 1998; Hernandez-Rauda et al., 1999). Increasing concentrations of fluoxentine also led to a four to fivefold increased incidence of abnormalities in Japanense medaka (*Oryzias latipes*) including nonresponsiveness, edema, and development of pectoral fins (Brooks et al., 2005). That study also reported a stimulated release of gonadotrophin in fish by serotonins, which in turn stimulates the synthesis of sex steroids, controlling oogenesis and vitellogenesis. Serotonins such as Prozac may induce/stimulate ecadysone, ecdysteroids, and juvenile hormones in invertebrates. These hormones control oogenesis and vitellogenin in insects (Nation, 2002). Brooks et al. (2005) also reported cell deformities and shrinkage in the green alga *Pseudokircheriella subcapitata* with 13.6 µg fluoxentine L^{-1}. These effects were suspected to be due to its antimicrobial properties, which can inhibit cellular efflux pumps (Muñoz-Bellido et al., 1997). The responses to serotonins appear to be very variable across species, clearly indicating the need for further research into these compounds.

4.5 ECOTOXICITY OF PPCPs ON TERRESTRIAL WILDLIFE

Humans and wildlife overlap in their physical environment and can thus be exposed to PPCPs from the environment through somewhat similar routes. Thus, some studies have shown that when highly conserved systems are targeted by contaminants, both wildlife and human health may suffer with processes such as utero, neonatal, pubertal, lactational, and menopausal stages being affected. In the same regard, reproductive and developmental abnormalities have also been documented in wildlife for several decades (Colborn et al., 1993). Furthermore, where human data are of low quality or totally unavailable, wildlife sentinels have provided a useful role in assessing human risk.

Most frequently reported in wildlife are differences in sex ratios or behavior following exposure to chemicals—a reflection of the intense interest in the presence of EDCs in the environment (Brunström et al., 2003). There are fundamental differences between species in how sex in offsprings is determined. In mammals, for which most of the PPCPs are primarily intended at both therapeutic and subtherapeutic concentrations, sex determination is based on the XY/XX system with the male as the heterogametic and vice versa. This system requires the synthesis of testosterone and its derivatives in some target tissues and the presence of functional androgen receptors in the undifferentiated gonads, brain, and other secondary sexual tissue. By contrast, birds have a WZ/WW system whereby the female is heterogametic whose development depends on the ability to synthesize and recognize 17β-estradiol for female sexual development to occur. With just these two contrasting examples, it is not hard to imagine that sex determination in mammals and birds can be affected in totally different ways when they are exposed to EDCs. Overall, wildlife seems to be more sensitive to the effects of EDCs compared to humans. As a matter of fact, more evidence of EDC-related impacts have been reported on wildlife, and indeed endocrine disruption was first linked to environmental contamination based on observations in wildlife. Endocrine disruption and other abnormalities have also been reported in wildlife exposed to wastewater effluents (Folmer et al., 1996, 2000; Jobling et al., 1998). In a more recent example, estradiol and EE2 were found to affect male song behavior in female zebra finches by inhibiting aromatase (Quaglino et al., 2002). Male zebra finches have four highly developed song control nuclei, that is, area X of the paraolfactory lobe, the robust nucleus of the archistriatum (RA), the HVC, and the lateral magnocellular nucleus of the anterior neostriatum (LMAN). It so happens that RA, HVC, and LMAN are greatly reduced in females, and females also do not have a detectable area X. The size of these song control nuclei correlates with singing behavior to the extent that under normal circumstances only the males sing. However, exposure to estradiol early in life enlarges area X, HVC, and RA in females and masculinizes them, enabling them to sing as males do. A summary of the results obtained by that group following oral treatment of the birds with estradiol are presented in Figure 4.5. In those studies, no significant changes in any song nuclear volumes were detected in male finches, although earlier studies by Millam et al. (2001) reported cases of infertility in males exposed to estrogens. Induced singing in females can affect the mating behaviors, which can ultimately impact bird populations. Octylphenol, a personal care constituent that is found in sewage plant treatment effluents (Bennie, 1999; Ferguson et al., 2000; Montgomery-Brown and Reinhard, 2003) also had some effect on area X, HVC, and RA in those female finches.

Significant exposure to PPCPs in wildlife can also occur through the food chain. Thus, species that are not normally in direct contact with the PPCP-contaminated medium can become exposed through ingestion of contaminated prey. A classic example of this is the recently reported diclofenac toxicity that has been linked to a greater than 95% decline in the population of the Oriental white-backed vultures (*Gyps bengalensis*; OWBV) on the Indian subcontinent (Oaks et al., 2004; Fig. 4.6). Other vulture species populations have also been negatively affected (Cuthbert et al., 2006; Swan et al., 2006). The toxicity of diclofenac to this organism is believed to be a result of renal failure and visceral gout. Postmortem of the dead vultures recovered from the wild during 2000–2003 when that study was conducted showed that 85% of the bird carcasses had grossly apparent urate deposits that were characteristic of visceral gout on the surface of their internal organs. In birds, visceral gout is mostly associated with renal failure, which leads to hyperuricemia ultimately resulting in the deposition of uric acid in the internal organs. This chain of events subsequently leads to degenerative, metabolic, and infectious diseases due to the prevailing toxicity. Analysis of the kidney tissue in each of the vulture carcass that had visceral gout showed residual diclofenac concentrations ranging between 0.051 and 0.643 μg g^{-1} of tissue (fresh weight basis) (Oaks et al., 2004). Additional analysis of these kidney tissues did not detect any of the other pharmaceuticals that have similar modes of action (i.e., naproxen, acetaminophen, flunixin, ibuprofen, phenylbutazone, oxyphenylbutazone, indomethacin, ketoprofen, metenamic acid, indomethacin, salicylic acid, and tolmentin) clearly implicating diclofenac in this toxic reaction. The presence of diclofenac in water has been reported in other areas

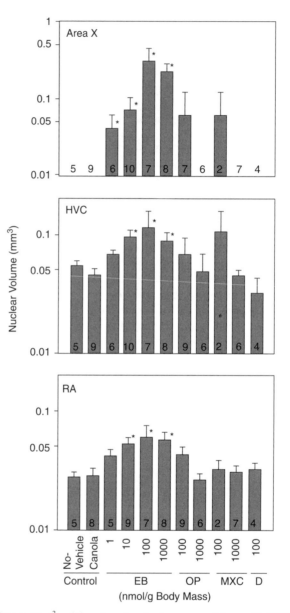

Figure 4.5 Volumes (mm³) of female zebra finch song nuclei area X, HVC, and nucleus robustus of the archistriatum (RA). The finches were orally treated as chicks at the age of 5–11 days with estradiol benzoate (EB), octylphenol (OP), methoxychlor (MXC), or dicofol (D). Canola was used as a control and as a vehicle to EB, OP, MXC, and D. The bars are mean ± SE and the number of replicates per treatment are specified by the number at the base of each bar. Note that the volume (y axis) is expressed on a log scale and the asterisks denote groups that are significantly different from the control (i.e., Canola-treated birds). Instances in area X where there are no bars show that no nuclei were detectable with that specified number of replicates. (Adapted from Quaglino et al., 2002.)

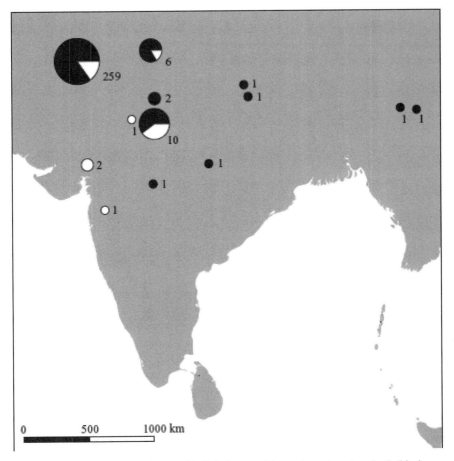

Figure 4.6   Proportion of vultures with diclofenac residues, visceral gout, or both (black area of each pie chart) at 13 locations where dead or dying vultures were found. The numbers next to each circle represent the number of birds assessed.

and not the one that was studied by Oaks et al. (2004). From those studies done elsewhere, its occurrence is at low concentrations that are unlikely to be acutely toxic. Thus, a most likely source of the diclofenac that is decimating these vultures was traced to their consumption of dead livestock that had been treated with this medication under animal husbandry guidelines. This contention was given some weight of truth when several OWBVs that were in captivity were fed goat and buffalo carcasses that had been treated with diclofenac, showing a mortality rate of 65% of the vultures that consumed such tainted carcasses.

As indicated in Chapter 1, prostaglandins are involved in the eggshell synthesis in birds. As a matter of fact, birds that were treated with the COX inhibitor indometacine produced eggs that had thinned shells (Lundholm, 1997; Guillette, 2006). In the same regard, studies done in mice indicate that COX-2 plays a role in pregnancy and is expressed in the uterine epithelium at different times during early pregnancy, where it is supposedly involved in the implantation of the ovum and the establishment of angiogenesis needed to establish the placenta (Hinz and Brune, 2002). Administration of COX-2 to pregnant rats and mice significantly reduced the development of the renal cortex and glomerulogenesis as compared to COX-2 knockout mice.

4.6 LIVESTOCK AND HUMAN HEALTH

Pharmaceutical compounds are deliberately designed to affect biochemical and physiological functions of biological systems in humans and livestock. While the side effects of pharmaceuticals on humans (and livestock in the case of those used in veterinary medicine) are routinely studied during the research and development of the drug, studies during that development stage have not focused on the continued or cumulative exposure to these compounds at low levels that are typically encountered in the environment. The mode of exposure to PPCPs is also quite important. In most clinical research, exposure is tested through injection and diet. By contrast, exposure to PPCPs in environmental settings is mostly through dermal and oral routes, and thus exposure studies based on injection may not be directly transferable to all ecological situations. The health effects of the consumption of PPCPs from the environment, for example, through drinking water, have not been fully elucidated. In most countries, the main risks to human health associated with drinking water are microbiological in nature, although chemical pollutants are also of major concern. From a day-to-day perspective, the threat from waterborne infections of microbiological origin is of major concern, but water-related maladies associated with chemical pollution are also of paramount concern, particularly in industrialized countries. However, the inclusion of PPCPs among typical chemical pollutants is only beginning to be realized, and, currently, no jurisdiction has any specific regulatory guidelines for assessing the risks associated with their presence in drinking water. Similarly, there are no clear drinking water health advisory levels against these compounds. In documents such as the United States Guidance for Industry: Environmental Assessment of Human Drug and Biologics Application (U.S. FDA, 1998), some pharmaceuticals anticipated to have no expressed impact on the environment are categorically excluded. Under that approach, "impact on the environment" is equated with "affecting the quality of the human environment."

Exposure of even low concentrations at critical developmental stages may have profound effects on some organisms. Drinking water is a major component of foodstuff in bottle-fed infants and children. Exposure (E_i) to PPCPs as an integral of time and space can be calculated using the following model:

$$E_i = \sum_{j=1}^{J} c_j t_{ij} \qquad (4.11)$$

where c_j is the exposure concentration in microenvironment j, t_{ij} is the duration organism i spends in microenvironment j, and J is the total number of microenvironments occupied by organism i (Kendall et al., 2001). Christensen (1998), Webb (2001), and

Webb et al. (2005) evaluated the risk of several pharmaceuticals to human health. The latter compiled the estimates based on the concentrations determined in drinking water (or the limit of detection), a lifetime (70 years) (i.e., 25,550 days) consumption of 2 L of water per day, assuming no metabolism on ingestion of the compound and no removal during water treatment or dilution to calculate I_{70} values for various pharmaceuticals. They then compared those values with the typical minimum daily therapeutic doses as provided by Dollery (1991) and Reynolds (1996). In Germany, they found only the asthma drugs (i.e., bronchodilators) clenbuterol, salbutamol, and terbutalin as well as the antimenopausal symptoms hormone 17α-ethinylestradiol (EE2) had I_{70} values that were above the daily therapeutic dose. During analysis of the German data, 64 compounds were included in the computation, 18 of which were antibiotics, 5 were X-ray contrast media, 11 were anti-inflammatory drugs, 10 were antihypertensive drugs, and 4 were bronchiodilators. Both of these workers found no low levels consumed over the lifetime, supposedly with inconsequential effects in humans.

4.6.1 Clinical Antibiotic Resistance Cases

Most of these negative effects of PPCPs in the environment to humans and livestock have been previously discussed in the context of controlling the spread of resistance to antimicrobial agents and endocrine disruptors, and these will form the bulk of this section. However, other subtle effects will also be examined. Some resistance to antibiotics occurs naturally in soil and other environments. It is a natural mechanism by which microorganisms control the populations of their competitors over millions of years and at very low concentrations in very restricted microenvironments. However, also note that an increasing number of antibiotics are synthetic (novel) or semisynthetic compounds without any closely similar cousins in the environment. Thus, microorganisms are increasingly exposed to such novel compounds. The antibiotic-resistant populations that are reduced by the application of antibiotics can persist in the environment for a long duration (Hansen et al., 1992; Pursell et al., 1996; Migliore et al., 2002; Nygaard et al., 1992; Herwig and Gray, 1997). More studies of transfer and spread have focused on its more rapid occurrence under hospital settings (nosocomial) than in the outside community. Such rapid spread is attributed to the more frequent exposure under those hospital settings and the more intensive selective pressure. Indeed the clinical usage of specific antibiotics has been shown to correspond with the emergence of antibiotic resistance in clinical settings (Hiramatsu, 1998; Hiramatsu et al., 2001). One of the most popular examples of emergent antibiotic resistance at a clinical level includes vancomycin-resistant enterococci (VRE), multidrug-resistant pseudomonads (MDRP), and methicillin-resistant *Staphylococcus aureus* (MRSA). However, there is increasing evidence that the development of resistance to antibiotics outside of hospital/clinical settings is also quite significant as was displayed for the VRE complex 17 (Fig. 4.7). Those determinations were obtained using multilocus sequence typing (MLST) to explore the evolutionary origin of epidemic isolates from five categories, notably animal surveillance, community surveillance, hospitalized patients, clinical specimens, and from hospital

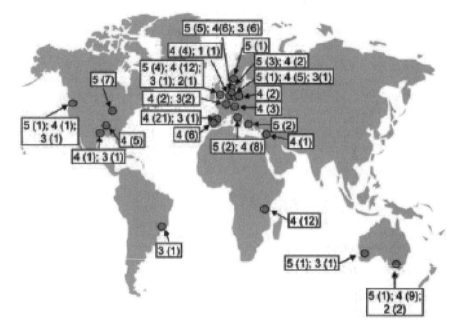

Figure 4.7 Global distribution of complex-17 isolates in various cities (dots). The numbers indicate the type of epidemiologic source of the sample whereby 1 = animal isolates, 2 = human community surveillance isolates, 3 = isolates from surveillance (feces) from hospital patients, 4 = human clinical isolates, 5 = isolates from documented hospital outbreaks. The number of isolates in each instance is indicated in parentheses. (From Willems et al., 2005.)

outbreaks (Willems et al., 2005). Most of the hospital-derived VRE belonged to a single clone, and, besides frequent antibiotic use in hospital settings, the clone may have been favored through selection pressure in the enterococcal subpopulation in hospital effluents. While the development of resistance of these organisms in hospital and clinical settings have been widely researched under the guise of controlling nosocomial infections, the development of such resistance through input of antibiotics in the environment is a fairly new dimension to the discussion. This dimension is the thrust of this section. Antibiotics are some of the most important and widely used pharmaceuticals worldwide with an estimated total usage of more than 100,000 metric tons per annum (Kümmerer, 2004b). A bulk of what is consumed is excreted unchanged (i.e., as the parent compound) into the environment (Hirsch et al., 1999; Jjemba, 2002a, 2006). The excreted moieties end up in sewage and ultimately into biosolids or wastewater. The transfer of resistance is likely to be in environments that have a high density of bacteria such as sewage and biofilms. These residual materials are in some instances applied to arable land. Antibiotic-resistant bacteria have also been reported in drinking water (Kolwzan et al., 1991; Schwartz et al., 2003), where it is possibly transferred to these autochthotounous bacteria in water distribution systems. Pharmaceuticals are similarly used in livestock production (as growth promoters and for therapeutic purposes) and in aquaculture, thus ending up in manure and sediments. Appearance of VRE in animals enables the transfer of such resistance from enterococci to *S. aureus* in the animal reservoir (Wegener, 2003).

The frequent use of antibiotics has aroused concerns about the rise and emergence of new strains of bacteria that are resistant to antibiotics. Such resistance may not only be confined to organisms that are commensal but also expand to a wider range of microorganisms in the environment when the excreted antibiotics increase the range of exposure to the antibiotics. There has been a dramatic increase in the prevalence of vancomycin and ampicillin resistance in *Enterococcus faecium*, and nosocomial enterococcal infections have also increasingly become a major cause of sepsis and wound infections, particularly in hospital settings (Kieke et al., 2006). Vancomycin-resistant enterococci, which emerged in the 1980s and was followed by the emergence of vancomycin-intermediate *Staphylococcus aureus* (VISA) a decade later. To that effect, VRE have emerged as a threat to public health worldwide. Glycopeptides such as vancomycin are the last resort for controlling MRSA infections in humans. Thus, the development of vancomycin-resistant MRSA carries severe consequences, and vancomycin resistance has generally been increasing in the United States (Fig. 4.8) and worldwide. VISA was first detected in Japan and deemed intermediate because their MICs of 4–8 μg mL^{-1} were in the intermediate category of the arbitrarily assigned vancomycin susceptibility breakpoint (Hiramatsu et al., 1997). Despite this arbitrarily set intermediate resistance, treatment of individuals infected with these strains with vancomycin primarily fails. VISA has since been reported in other parts of the world including the United States, United Kingdom, Sweden, France, Poland, China, and Norway (Howe et al., 2004). It is important to note that the occurrence of VISA may actually be currently underreported as there is no standardized method to identify these organisms. In some instance, these are also referred to as *S. aureus* with reduced susceptibility to vancomycin (SA-RVS). Fridkin et al. (2003) used a cutoff MIC $\geq$4 μg mL^{-1} and included MICs of 8–16 μg mL^{-1} among what they referred to as SA-RVS. Treatment of patients infected with SA-RVS with vancomycin (and other glycopeptides) failed in 76% of the patients in a trial, and 40% of those patients died from their infection (Howden et al., 2004). True resistance to vancomycin (i.e., vancomycin-resistant *S. aureus*; VRSA) was first documented in 2002 from community isolates in Michigan and Pennsylvania, followed by isolations in various locations in the world soon thereafter. For the Michigan case, the patient had been receiving intermittent doses of vancomycin for a range of infections (Sievert et al., 2002). By contrast, the case in Pennsylvania had used vancomycin 5 years prior to the isolation of the VRSA (Miller et al., 2002). That latter case is of particular attention as it has been characterized by the medical literature as occurrence of vancomycin-resistant *S. aureus* in the absence of exposure to vancomycin (Whitener et al., 2004; Tenover et al., 2004). Both of these reports do not consider the possibility that the patient could have picked up the resistant organisms from the environment. VRSA is characterized by the acquisition of *vanA* genes, and the first acquisitions were possibly from enterococci (Moellering, 2006). VRE have been isolated in poultry-, cattle-, and swine-generated droppings (Aerestrup et al., 2000; Borgen et al., 2000) and in the ruminal contents of wild animals such as deer (Laukova et al., 2000).

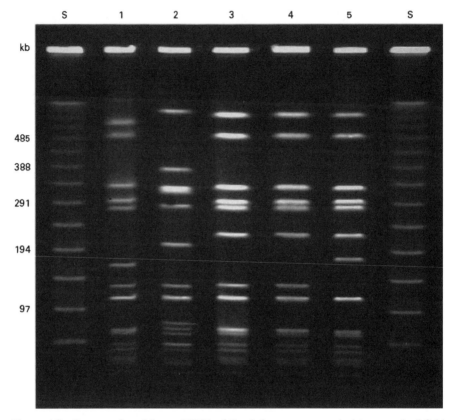

Figure 4.8 Patterns of smaldigested DNA of *Staphylococcus aureus* isolates on pulsed-field gel electrophoresis (PFGE). The isolates were obtained from four patients in diversely distant locations within the United States. Lane 1 = ATCC29213 (positive control), lane 2 = a methicillin-resistant *S. aureus* isolated from a patient in Georgia; lane 3 = a gylcopeptide intermediate *S. aureus* isolated from a patient in Michigan, lane 4 = a methicillin-resistant *S. aureus* from the same patient in Michigan, and lane 5 = a glycopeptidet-intermediate *S. aureus* isolate from a patient in New Jersey. (From Smith et al., 1999b.)

An increase in the resistance to antibiotics associated with feeding fish with medicated food has also been documented (Samuelsen et al., 1992). In another study, Dazzo et al. (1973) reported a buildup of and increased survival of bacteria on arable land that had received repeated applications of manure. Similarly, Esiobu et al. (2002) showed a 70% increase in resistance to penicillin, tetracycline, and streptomycin when manure (presumably from medicated cows) was applied to soil. Campagnolo et al. (2002) detected several antibiotics in the wastewater and waterways near a poultry and swine farm. It is not only important to understand the ecology of the antibiotic-resistant bacteria but also the ecology of the resistance genes themselves. Hayes et al. (2001) recovered high rates of quinupristin-dalfopristin-resistant *E. faecium* from chickens in the United States. This high occurrence was attributed to the rampant use of virginiamycin in the livestock (mostly poultry and swine) industry. The incidence of quinupristin–dalfopristin resistance can, in some instances, be as high as 100% in flocks that have been exposed to virginiamycin (Welton et al., 1998).

Of significance in the livestock industry is virgniamycin. As indicated in Chapter 1, this compound is very similar to another commonly used streptogramin Synercid, which is a mixture of quinupristin and dalfopristin. Various plasmid-borne genetic elements that carry quinupristin–dalfopristin resistance in agriculturally derived *E. faecium* have been shown to cross into *E. faecium* isolates of human origin in vitro (Hammerum et al., 1998). In *E. faecium*, streptogramin A resistance is encoded by the plasmid-based acetyltransferase genes *vatD* and *vatE*, whereas streptogramin B resistance is mediated by erythromycin-resistant methylase (*ermB*) genes (Kieke et al., 2006). The genes for quinupristine–dalfopristin resistance and vancomycin resistance are found on the same plasmid in *Enterococcus* (Bozdogan et al., 1999). Streptogramin-resistant bacterial strains isolated from humans have a much lower tolerance to the antibiotics (i.e., MIC $\leq 4\,\mu g\,L^{-1}$) compared to those isolated from animals (MIC $\approx 32\,\mu g\,L^{-1}$) (Claycamp and Hooberman, 2004). For example, isolates from poultry are often less sensitive to streptogramins compared to those isolated from humans (Aerestrup et al., 2000). Similar findings have been reported in pork and other animal products.

The EU has banned the use of subtherapeutic antibiotics in agriculture as a precautionary measure to minimize the risk of transferring resistant strains from animals to humans. That ban led to a decrease in incidences of antibiotic resistance (Weirup, 2001). It is also possible for the resistant animal-derived microbial strains such as *E. faecium* crossing their resistance to human-derived microbiota. Recently, Kieke et al. (2006) conducted an epidemiologic study of quinudipristin–dalfopristin susceptibility in *E. faecium* that colonizes humans and poultry. During that study, livestock-derived *E. faecium* isolates that were sensitive or had intermediate resistance to quinupristin–dalfopristin (i.e., MIC $< 4\,\mu g\,mL^{-1}$) were detection of inducible resistance by growing them in media containing 0.25 μg virginiamycin mL^{-1}. After growth in the presence of this antibiotic, they were then challenged with high (i.e., 8 $\mu g\,mL^{-1}$) concentrations of quinupristin–dalfoprisitin to determine relative inducement of resistance to this mixture of antibiotics by establishing the ratio of optical density of cells growing with quinupristin–dalfoprisitin to that of the cells growing without exposure to virginiamycin. A similar assessment was also done with isolates derived from humans after preexposure to 0.25 μg quinupristin–dalfoprisitin mL^{-1} instead of virginiamycin. Some of their findings are summarized in Table 4.6. No resistance was found in hospital patients or vegetarians, but susceptible and intermediately resistant *E. faecium* isolates were found in both populations. By comparison, both conventional and antibiotic-free poultry had some resistant *E. faecium* isolates, the highest incidence of resistance occurring in conventional poultry. However, inducible resistance was substantially lower in antibiotic-free retail poultry isolates relative to those in conventional poultry. The only prevalent genetic markers *vatE* and *ermB* encountered in the study were absent in *E. faecium* derived from vegetarians, whereas *vatB* was mostly found in hospital patients and *vat ermB* was present in less than 10% of the patients. *VatE* and *ermB* were found

TABLE 4.6   Quinupristin–Dalfopristin Susceptibility and Genetic Determinants of Resistance in *Enterococcus faecium* from Different Midwestern U.S. Ecologic Sources

E. faecium Source	%Susceptibility to Quinupristin–Dalfopristin			%Resistance Genes	
	Susceptible	Intermediate	Resistant	vatE	ermB
Hospital patients ($n = 105$)	24	76	0	38	9
Vegetarians ($n = 65$)	12	88	0	0	0
Conventional retail poultry ($n = 77$)	14	30	56	47	40
Antibiotic-free retail poultry ($n = 23$)	48	39	13	13	9

in both human- and poultry-derived *E. faecium* and more likely occur in conventional retail poultry than in antibiotic-free retail poultry. The isolates from hospital patients did not grow when challenged with 8 μg quinupristin–dalfopristin mL^{-1} unless they had been preexposed to virginiamycin or quinupristin–dalfopristin. The carriage of *E. faecium* isolates with *vatE* in the group tested was significantly ($P = 0.032$) associated with both touching raw poultry and higher consumption of poultry. The food-borne acquisition of streptogramin resistance genes is plausible as *vatE* is located on plasmids and can transfer by conjugation. Sorensen et al. (2001) showed that ingestion of *E. faecium* led to transient colonization for up to 14 days, a duration that is long enough for conjugative transfer of resistance determinants to endogenous flora to occur.

Available evidence shows that antibiotics were first used in the livestock industry to treat mastitis (McAllister et al., 2001). This disease still accounts for the bulk of therapeutic-related use of antibiotics in diary production worldwide (Zadoks et al., 2000). Fluoroquinolones such as enrofloxacin are frequently used to treat this infection, as well as other bovine respiratory infections (McAllister et al., 2001). Thus, in Europe, *S. typhimurium* DT104, which are resistant to enrofloxacin and other fluoroquinonolanes, have been increasingly isolated. The intramammary infections are greatly associated with both *Staphylococcus* spp. and *Streptococcus* spp. The former microorganism has also been associated with nosocomial (i.e., hospital-related) and community-acquired infections in humans and linked to methicillin resistance (i.e., MRSA). *Staphylococcus aureus* is found as normal flora on humans (Voss, 1975; Noble, 1998). The virulence and epidemiological potential considerably varies between *Staphylococcus* strains. Where *Staphylococcus* sp. causes infection, the infection is expressed as a superficial inflammation of the skin (i.e., atropic dermatitis), although it is not clear whether the symptom is due to *S. aureus* acting as a pathogen or just as an opportunistic colonist (Lever, 1996; Nishijima et al., 1995). 4-quinolone resistance has been associated with MRSA, possibly due to an expansion of the antibiotic resistance of the clone together with an enhanced colonization ability (Hawkey, 2003). The methicillin resistance is associated with homologous mobile genetic element IS257 (also known as IS 431). This insertion sequence is very common in the staphylococcal chromosome (Van Leeuwen et al., 1999). Community-acquired MRSA infection is of major concern in Europe, North America, and Latin America (Acar et al., 2000; Jones, 2002). It is routinely encountered in wastewater and sewage (Asbolt et al., 1993; Casanova et al., 2001; Kümmerer and Henninger, 2004).

The use of disinfectants such as triclosan and quaternary ammonium compounds and triclosan in homes and hospitals also leads to an enhanced selection for antibiotic-resistant bacteria (Russell, 2000; Kücken et al., 2000; McMurray et al., 1998). Besides introducing resistance to bacteria, some antibiotics can also impose other effects. For example, a recent study by Velicer et al. (2004) reported an intriguing association between cumulative exposure to all classes of antibiotics and breast cancer. Floroquinolones have also been implicated in inducing tensions in the cartilage of immature animals (Hunt et al., 2002) and are generally deemed unsuitable for pediatric use. However, their presence in the environment can inadvertently introduce them to such susceptible individuals.

4.6.2 Allergic Reactions Related to PPCPs

As discussed in Chapter 2, several antibiotics can be present in the air around farms (see Table 2.5), posing some effects on the farming population. For example, exposure to chloramphenicol poses serious side effects due to its genotoxicity (Holt et al., 1993). A survey by Takai et al. (1998) estimated the presence of $2.2 \, mg \, m^{-3}$ of inhalable airborne dust in pig-fattening buildings in Europe (Hagenstein et al., 2002). It is possible to develop allergic reactions from these airborne antibiotics. Exposure can also occur through the inhalation of dust particles from livestock facilities and wastewater/sludge treatment facilities (Hamscher et al., 2003). Sulfamethaxazole and tylosin also have the potential to cause allergic reactions in susceptible individuals (Danese et al., 1994; Caraffini et al., 1994; Choquet-Kastylevsky et al., 2002). Hamschler et al. (2003) estimated that farms working 8-h day^{-1} shift in a piggery housing under such conditions inhaled about 6.3 mg of dust that is contaminated with approximately 0.2 μg of various antibiotics totaling to about 3.4 mg of antibiotics kg^{-1} dust (at an average tidal volume of 0.5 L). Estimated intakes would be even higher under more intense working conditions where the individual is taking in far more than the conservative 12 breaths per minute that is typical of individuals under resting conditions.

4.6.3 Endocrine Disruption in Humans and Livestock

As is the case with other organisms, the interest and focus on steroids and other endocrine-disrupting compounds in the environment have also been linked to reproductive health problems in humans, livestock, and wild animals (Colborn et al., 1993). To that effect, various researchers have reported a decrease in sperm quantity and quality in the second half of the twentieth century (Carlsen et al., 1995; Sharpe and Shakkebaek, 1993), and several reports have documented cases of reduced testis and a loss of reproductive capability in men and male rats on exposure to anabolic steroids (Wilson and Wilson, 1943; O'Sullivan et al., 2000). Similar characteristics have been reported in cheetahs under captivity and fed with a diet that was rich in phytoestrogens (Setchell et al., 1987). Paul et al. (2005) also reported a reduced weight in male and female fetuses on gestation day in sheep reared on pastures that were fertilized with biosolids. The male sheep progeny from that study also had significantly reduced testis (32% reduction in testis weight), reduced number of Sertoli cells (34% reduction), Leydig cells (37% reduction), and gonocytes (44% reduction) compared to the control. These reductions could have resulted from the restricted growth of the fetuses or the lowered testosterone action coupled with a reduced immunoexpression of the smooth muscle actin in peritubular cells and of the androgen receptor in the testis. There is substantial evidence to the fact that the sensitivity of individual gonadal steroids depends on the life stage of the individual, with fetal material appearing to be more sensitive to EDC impacts (Kendall et al., 2001).

4.6.4 Association Between PPCPs in the Environment and Some Cancers

Exposure to environmental contaminants in utero has long been suspected to cause cancers such as acute lymphoblastic leukemia and lymphoma in children and young adults (Meinert et al., 2000; Alexander et al., 2001), including cancer increases in offspring (Birnbaum and Fenton, 2003). Cancer incidences in developed countries have increased overall in recent years (Ahlborg et al., 1995), and some of these could be associated with PPCPs in the environment. For example, low levels of mutagenic anticancer drugs such as cyclophosphamide have been detected in sewage (Steger-Hartmann et al., 1997) and presumably end up in drinking water. Cytogenic, mutagenic, and reproductive (CMR) effects can cause mutations in DNA (deoxyribonucleic acid) that can in turn result in somatic effects such as cancers. Genotoxicity relates to the effects of aneuogens, clastogens, mutagenic, and teratogens on the genetic materials of organisms. Such genetic materials primarily include RNA, DNA, and chromosomes, but it may also include the modification of proteins. Genotoxic effects can manifest themselves in the form of DNA and RNA strand breakages, modification in bases, rearrangements or fragmentation of the chromosomes, or aneuploidy. Aneuploidy is the situation where more chromosomes are added or subtracted from the normal, generating progeny that lacks chromosomes than are present in its ancestral genome. Some of the common examples of aneuploidy in humans include Turner syndrome (44 autosomes + 1X), Klinefelter syndrome (genetic make up of XXY or XYY), Edwards syndrome, and Down syndrome. Of interest is the documented observation that some types of Down syndrome occur where there is no family history of aneuploidy and show a maternal age effect whereby older mothers show an elevated risk (Fig. 4.9). This age-related trend has been known for many years, but its cause is not entirely clear. Could it be attributed to cumulative PPCPs and/or other chemical pollutants that are genotoxic and occur at sublethal concentrations in the environment? This possibility is quite plausible as gonads and fetuses are the most sensitive life stages (Kendall et al., 2001).

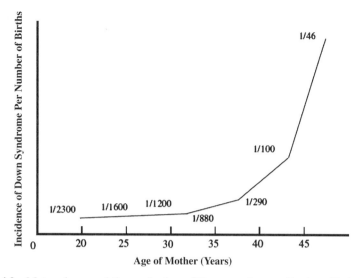

Figure 4.9 Maternal age and the production of Down syndrome offspring. (Redrawn from Penrose and Smith, 1966.)

There is some limited evidence that statins can cause some birth defects that manifest themselves in the form of central nervous system and limb anomalies (Mitchell, 2003; Edison and Muenke, 2004). Beta-blocker antihypertensive drugs also have some effects on glucose and insulin metabolism. Thus, treatment of obese nondiabetic hypertensive individuals with the angiotensin converting enzyme (ACE) inhibitor (i.e., perindopril) or the beta blocker (i.e., atenolol) for up to 6 weeks (Kuperstein and Sasson, 2000). Although both drugs similarly reduced the blood pressure, they had different effects on the left ventricular mass as well as the metabolism of glucose and insulin. More specifically, the beta blocker significantly increased the amount of insulin ($P < 0.006$), glucose ($P < 0.02$), the amount of fasting insulin ($P < 0.03$), amount of fasting glucose ($P < 0.003$), as well as peak glucose ($P < 0.005$) compared to the baseline. The reason for a reduced uptake of glucose in those individuals is not entirely clear, but the beta blocker possibly altered the first phase of insulin secretion and attenuated insulin clearance in susceptible individuals leading to higher insulin levels in the blood, ultimately down-regulating the insulin receptors. Presence of traces of beta blockers could, therefore, potentially become exposed to susceptible individuals (e.g., through drinking water), particularly at a tender age explaining, at least in part, the increasing incidence of diabetes in the young. For example, the incidence of Type 1 diabetes in children 0–4 years doubled between 1990 and 1995 (van Maanen et al., 2000). Similarly, high increases have also been documented in other European countries in some instances at magnitudes higher than 30%, magnitudes and trends that are similar to those in Japan and North America (Bingley and Gale, 1989). These magnitudes cannot be explained by changes in the genetic pool of the population over such a short duration and most likely attributable to environmental factors. Although PPCPs have not specifically been listed among some of the possible environmental factors, considering their occurrence in drinking water, they cannot be entirely ruled out. Among those implicated in childhood onset of diabetes from drinking water include nitrates (van Maanen et al., 2000), acidic water (Stene et al., 2002), as well as dioxins and arsenic (Parker et al., 2002), but no environmental factors have universally been satisfactorily agreed on. More detailed studies are currently ongoing to get a clear idea about the important environmental factors (Anonymous, 2006). Beta blockers are among the most prescribed antihypertension drugs (Table 1.5) and have been detected in effluents (Andreozzi et al., 2003; Miège et al., 2007).

4.6.5   Other PPCPs of Major Concern to Humans and Livestock in the Environment

Medications are rarely tested in children, and most of the medical use for pediatric practice develops through off-label usage and making educated guesses about the dosage and effectiveness (Steinbrook, 2002). Pediatric classifications fall into four categories (Table 4.7). Exposure of children to these PPCPs from the environment is indiscriminate despite the fact that a majority of the compounds are not approved for use in pediatric subjects. For example, between 1991 and 2001, only 20% of the 341 new molecular entities approved by the U.S. FDA were labeled for use in children in that country. A new molecular entity is one that has an active ingredient that has not previously been approved for marketing in any form. Exposing the young to a majority of these compounds poses a great challenge as such exposures at such a tender age can affect their physiology, metabolism, and susceptibility to toxicity changes in unpredictable ways as they go through various developmental stages. Data about the safety of strong medications such as SSRIs in pediatric age groups are quite limited, let alone the effects of low doses that they can be exposed to in environmental matrices such as drinking water over an extended duration. At a clinical level, there is some indication of retarded growth and depressed adrenal function following high doses of asthma-controlling corticosteroids (Agins et al., 2003), but it is unclear at this point whether such effects may extend to levels that are encountered in the environment. While the effects of lifelong exposure to low doses of PPCPs that are present in the environment have not been adequately investigated, some leads can

TABLE 4.7   Standard Pediatric Classifications Based on Age Group

Age Group	Classification
0–1 month	Newborn
1 month–2 years	Infant
2–12 years	Children
12–16 years	Adolescents

Source: Steinbrook (2002).

be derived from those done with pesticides and herbicides. For example, a study by Baldi et al. (2001) showed subtle neuropsychological effects on vineyard workers exposed to pesticides either directly (through mixing and spraying) or indirectly (through harvesting) over several years. The effects could not be easily discerned under normal settings.

4.7   ECOTOXICITY OF PPCPs ON VEGETATION

Relatively few studies have directly investigated the uptake of PPCPs by plants. Studies conducted in vitro (Migliore et al., 1996, 1998) and in vivo (Jjemba, 2002a, 2002b) show that some pharmaceutical compounds are phytotoxic. Recently, Kumar et al. (2005) evaluated to what extent plants take up antibiotics from manure–soil matrices. They used green onions and cabbage as the test plants sowed in a loamy sand soil. The soil was spiked with 20 μg chlortetracycline or tylosin per liter. Plant tissue was harvested at periodic intervals, ground, and assayed for antibiotics. In those studies, tylosin was below detection in the plant tissue, whereas both types of plants accumulated chlortetracycline with time. In another experiment, Kumar et al. (2005) applied manure from pigs that had been fed with antibiotic-laden feedstuff (i.e., 100 g chlortetracycline, 100 g sulfamethazine, and 50 g penicillin per ton of feed). Manure from the fed pig pens and from appropriate (i.e., no antibiotic feed) control pens was collected and applied to soil that had been planted with corn, cabbage, or green onions. The manure was applied at a rate of 200 kgN ha^{-1}, and the plant tissues were harvested after 6 weeks. On analysis, no tylosin was detected in the manure or the plants that had been treated with tylosin-fed pig manure. By contrast, chlortetracycline, which was present in the manure, was taken up by all of the test crops but to different extents. Concentrations were highest in corn and lowest in cabbage. Some PPCPs can impose effects on primary productivity and alter the structure and function of lower organisms such as algal communities (Wilson et al., 2003; Orvos et al., 2002; Kolodzieg et al., 2003). These changes can result in shifts in the processing of nutrients within ecosystems, affecting natural food web structure within specific ecological zones. However, at this point in time, there is no concrete evidence that plant yields are significantly negatively impacted by PPCPs in soils. Shore et al. (1992) reported stimulated growth of alfalfa (*Medicago sativa*). However, those studies were conducted in water and at high concentrations of estrogen (50–500 μg L^{-1}). By comparison, the concentrations of estrogen encountered in the environment range 44–490 ng L^{-1} in sewage influents and are mostly undetectable in surface water (Ingerslev and Halling-Sørensen, 2003).

Where PPCPs are taken up by crops, effects on the consumers (i.e., humans and livestock) cannot be ruled as these kinds of studies have not yet been extensively conducted. Actual uptake of PPCPs by plants has only been studied with a few kinds of compounds, particularly fluoroquinolones and sulfanomides (Migliore et al., 1995, 1996, 1997; Forni et al., 2004). From Migliorie's studies, bioaccumulation of sulfomethoxine by the roots and stems of the plants they tested was, in some instances greater than 2000 mg kg^{-1}. Bioaccumulation was generally higher in the roots than in the stems. However, most of those studies were conducted in artificial media (agar) and at extremely high concentrations that are unlikely to be encountered in the environment (Jjemba, 2002a). More recently, Boxall et al. (2006) also studied the uptake of several classes of pharmaceuticals (i.e., amoxicillin, diazinon, amofloxacin, florfenicol, levamisole, phenylbutazone oxytetracycline, sulfadiazine, trimethoprim, and tylosin) on a root crop (carrot) and a leafy plant (lettuce) from soil.

4.8 GENERAL CONSIDERATIONS IN LONG-TERM PPCP TOXICITY

Overall, polypharmacy has not been widely studied, but it is important to realize that every time one uses a drug one also uses at least one vital organ to process or metabolize that drug. As indicated in Chapter 3, effects on the liver are more often one of the main causes for withdrawing pharmaceuticals off the market. Incidences of liver injury are on the rise possibly due to the overuse of pharmaceuticals and/or the chronic exposure of our bodies to PPCPs and other pollutants (Kaplowitz, 2003; Lee, 2003). Some withdrawn drugs in recent history due to their ability to damage the liver include Rezulin, the antidepressant Serzone, and the painkiller Duract. A liver somatic index [LSI; Eq. (4.12)] has been developed to detect hepatomegaly associated with the proliferation of peroxisomes (Perkins and Schlenk, 1998). This relationship is based on the notion that different stressors can cause the LSI to go above or below the normal range. For example, the liver increases in size to enable the detoxification of some pollutants such as polyaromatic hydrocarbons (PAHs) and polycarbonated biphenyls (PCBs). Other drugs that have been withdrawn in recent history include the COX-2 inhibitors Vioxx and Bextra, the statin Baycol, and the antihypertensive drug Redux.

$$\text{LSI} = \frac{\text{Wet weight of the liver}}{\text{Total wet body weight}} \times 100 \tag{4.12}$$

It is apparent that environmental matrices such as typical sewage treatment effluent and water have probably hundreds of PPCPs, with potentially divergent modes of action for which additive effects to organisms may not be easily deciphered. This phenomenon is somewhat analogous to what is referred to as polypharmacy, the use of several pharmaceutical compounds at the same time by an individual. The effects of most drugs in the presence of others are not always well known. These effects may possibly be better investigated by developing biomarkers and bioassays that focus on the respective modes of action. Biomarkers are xenobiotically induced alternations in cellular components or processes, structures, behaviors, or functions that are measurable in a biological system or sample (Kendall et al., 2001). Effects of drug mixtures in the environment (Richards et al., 2004) mixed fluorentine, ciprofloxacin, and ibuprofen, which reduced biodiversity and led to mortality of fish at the high drug mixture concentrations. Cleuvers (2005) contends that the toxicity of a single compound could be enhanced by the presence of other similarly acting compounds. For example, in his research, the effects of a mixture of diclofenac and ibuprofen (both analgesics) on *Daphnia* sp. was enhanced compared to the effect of each of these compounds used individually. Cleuvers (2003, 2004) evaluated the ecological effects of mixtures of anti-inflammatory drugs (i.e., acetylsacylic acid, diclofenac, ibuprofen, naproxen) on *Daphnia* and algae. Those results showed toxicity of the mixtures to occur at concentrations that were below those of the individual components in the mixture, clearly suggesting an additive effect from the individual components. These findings have important implications as PPCP-contaminated sites tend to have multiple compounds present. Even though each of the compounds may be present at subtherapeutic concentrations, they may generate additive effects on the organisms that are present. Similar findings about the toxicology of mixtures have been reported by others (Pöch, 1993; Altenburger et al., 2000).

In summary, realistic ecotoxicological assessment requires the use of a battery of test species that represent a wide range of organisms. Thus, to test the ecotoxicity of chloroquine, Zurita et al. (2005) used a decomposer (i.e., *Vibrio fischeri*) that has bioluminescence, a primary producer (*Chlorella vulgaris*), and a primary consumer Cladoceran *Daphnia magna* as well as in vitro fish cell line PLHC-1 derived from a secondary consumer, *Poeciliopsis lucida*. In the bioluminescent *V. fischeri*, inhibition of luminescence was used as an indicator of toxicity. Toxicity was highest in *D. magna* > *C. vulgaris* > fish cell line > *V. fischeri*. It is important to compare the ecotoxicity with several species as bioassays from a single species will not provide enough information to transfer across species.

5

TECHNOLOGIES FOR REMOVING PPCPs

Wastewater treatment primarily enhances the adsorption and degradation of PPCPs. Both of these processes were extensively discussed in Chapter 3, but the current chapter focuses on the actual techniques used to rid environmental matrices of PPCPs. Removal of these compounds from wastewater will be used as the model matrix as most of the PPCPs on routine use end up in wastewater, primarily through excretion or in sullage. Wastewater treatment systems gradually first developed in response to the adverse conditions caused by the discharge of raw sewage effluents to water streams. In terms of volumes the per capita amount of wastewater generated from residential dwellings ranges between 100 and 600 L/day with a per capita average of 260 L/day (UNEP, 2002). This range of generated wastewater is, to a good extent, related to the level of economic development, although other factors such as population density and geographic location also play an important role. Thus, average per capita wastewater generated in the US is about 450 L (120 gall), with an average of 240 mg of suspended solids per liter (Xia et al., 2005). Similarly high per capita volumes are typical in most of the other developed countries.

Most reports that document the removal of PPCPs during the treatment of wastewater or sewage are based on measurements of the concentration in the influent and effluent (Ternes, 1998). Based on such studies, it is generally apparent that the average elimination of PPCPs from such environmental matrices can be quite variable not only between different compounds but also within the same compound when present in different matrices. Such tremendous differences in the removal of PPCPs are not entirely surprising as these compounds are quite heterogeneous in

Pharma-Ecology: The Occurrence and Fate of Pharmaceuticals and Personal Care Products in the Environment. By P.K. Jjemba terms of structure, reactivity, and chemical properties. By the same token, removal of the same PPCP from different matrices may also greatly differ due to the range of treatment options that various wastewater treatment facilities pursue. Removal is also impacted by environmental factors such as the prevailing pH, temperature, UV (photolysis), and organic matter content, as well as the type and abundance of microbial degraders. These intricate and often individualized differences can play a role in the outcome of treatment processes. Thus, removal using various treatment technologies will be critically examined, recognizing the fact that PPCPs are indispensable and will continue to enter the environment through the channels, outlined in Chapter 2, ultimately ending up in wastewater. It is, therefore, imperative that minimizing the effects of PPCPs in the environment can be pursued by focusing on better technologies so as to remove these compounds from an environmental matrix that is as important as wastewater. The present chapter will initially focus on systems that use conventional techniques and ultimately examine the efficacy of the more modern techniques developed to remove contaminants from the environment. Adapting this approach should not convey the impression that the wastewater treatment arena is a simple dichotomous field as, with advances in technology, more modern treatment facilities that have documented low or a total absence of PPCPs in their effluent currently tend to use a combination of conventional and modern techniques. Emphasizing the use of technology to address the removal of PPCPs from the environment nullifies the wrongful assumption that expressing concern about the presence and potential effects of PPCPs in the environment could stymie medical and technological development.

From a historical perspective, wastewater treatment processes were developed with the primary objective of removing contaminants other than PPCPs, notably pathogens, nutrients, (heavy) metals, herbicides, and pesticides. That noted, there is ample evidence that the individual processes have varying efficacy in removing these traditional contaminants of concern from environmental matrices. Table 5.1 gives us a flavor of the range of treatment technologies and what they are capable of attaining as far as removing a few important traditional organic and inorganic constituents, which have been studied for a much longer time compared to PPCPs. Aeration (or stripping) involves providing continuous contact of air with the environmental matrix at hand (e.g., water or sewage) that is being treated. Such aeration can, besides enhancing oxidation, sweep the contaminants, particularly the ones that are volatile. As is evident from the table, stripping poorly reduces inorganic constituents and color but can remove some organic constituents to a good or excellent extent. By contrast, sedimentation, coagulation and filtration, ion exchange resins, as well as liming generally poorly remove organic constituents, although they adequately remove some inorganic constituents. Anion exchange resins selectively remove anionic species such as fluoride and nitrate, whereas cation resins replace various cations with protons (H^+), potassium (K^+), or sodium (Na^+) ions, the efficacy of the exchange resins depending on the type of organic and/or its ionic charge. From Table 5.1, it is also noticeable that membrane filtration, particularly reverse osmosis, has a track record of excellent removal efficacy of inorganic constituents and a good to excellent track record for removing organic constituents.

TABLE 5.1 General Effectiveness for Selected Pathogens, Nutrients, Metals, and Various Organics Other Than PPCPs[a]

	Aeration, Stripping	Coagulation, Sedimentation, Granular Media Filtration	Lime Softening	Ion Exchange — Anion	Ion Exchange — Cation	Membrane Filtration — Reverse Osmosis	Membrane Filtration — Ultrafiltration	Chemical Oxidation Disinfection	GAC	PAC	Adsorption — Granular Ferric Hydroxide	Adsorption — Activated Alumina
Inorganic Constituents												
Arsenic	P	G-E	F-E	G-E	P	E	P (for As^{3+}); F (for As^{5+})	P	F-G	P-F	E	F-E
Chromium	P	G-E (for Cr^{3+}); P (for Cr^{6+})	G-E (for Cr^{3+}); P (for Cr^{6+})	P (for Cr^{3+}); E (for Cr^{6+})	E (for Cr^{3+}); P (for Cr^{6+})	E	NA	P (for Cr^{3+}); F (for Cr^{6+})	F-G	F	NA	P
Copper	P	G	G-E	P	F-G	E	NA	P-F	F-G	P	NA	NA
Hardness	P	P	E	P	E	E	NA	P	P	P	NA	P
Nitrate	P	P	P	G-E	P	G	NA	P	P	P	NA	P
Organic Constituents												
Color	P	F-G	F-G	P-G	NA	NA	NA	F-E	E	G-E	NA	P
MTBE	G-E	P	P	P	P	F-E	F-E	P-G	F-E	P-E	NA	NA
TTME	G-E	P	P	P	P	F-G	F-G	P-G	F-E	P-F	NA	P

Source: Crittendon et al. (2005).

[a]P = poor (<20% removal); F = fair (20–60% removal); G = good (60–90% removal); E = excellent (>90% removal); and NA = not applicable or insufficient data.

Ultrafiltration also has a fair to excellent track record of removing organic and inorganic constituents, although data about its performance are not as extensive. These and other related technologies are discussed in detail below, with specific reference to the treatment of wastewater to remove the nontraditional contaminants in the form of PPCPs. However, the accompanying text is not intended to be an exhaustive treatise on wastewater treatment, a subject for which interested parties are referred to more appropriate textbooks.

5.1 CONVENTIONAL TREATMENT SYSTEMS

Conventional processes of wastewater treatment involve primary and secondary treatment. Depending on the extent of debris present, preliminary treatment may be imposed prior to primary treatment. Preliminary treatment basically involves removing large objects such as roots, rags, glass, and rocks by screening the wastewater through a grate or a bar screen. The collected screenings are then disposed of as solid waste, their removal greatly reducing the volume of materials to be treated.

5.1.1 Primary Treatment

During primary treatment, smaller inorganic debris referred to as grit, which typically consists of sandy materials and other particulates, is removed mostly through sedimentation and settlement from the wastewater. Despite this physical removal of approximately 60% of the suspended solid from the wastewater, primary treatment typically removes around 35% of the BOD (Samadpour, 2003). These tremendous reductions in BOD from a purely physical process are primarily due to the removal of microbial particles and attached biofilms that are associated with the settled solids (Jjemba, 2004). If the inorganic grit is not removed prior to more intensive biological treatment, it accumulates in subsequent process units, particularly the sludge digesters, causing excessive wear on the equipment. Primary treatment is typically conducted in settlement tanks, which normally have a deeper section in which the settled particulates are collected and compacted to form sludge. The sludge is then periodically removed from the primary settlement tanks for further treatment and disposal as solid waste. Under these treatment conditions, a small proportion of PPCPs in the wastewater are biodegraded by the existing microorganisms, but an even larger fraction of the compounds remains sorbed onto the settled (i.e., primary sludge) and suspended solids, remaining unavailable for degradation. The fraction that is sorbed in the primary sludge is thus removed when the sludge is disposed of. Some fraction of PPCPs that remain in the aqueous phase is apportioned onto the scum, fats, oil, and grease, mostly through solid–liquid separation by floatation. The scum arises from personal care products such as soaps, whereas the grease, oils, and fats mainly originating from kitchens. No proper studies that have determined these apportioned proportions for various PPCPs are readily available, but a mass balance for an individual compound in the primary treatment process has been proposed:

$$C_0 \times F = (C_1 \times F) + (C_{ps} \times M_{ps}) \tag{5.1}$$

where C_0 is the concentration of the compound in raw wastewater (ng L^{-1}), C_1 is the concentration of the compound in the primary effluent (ng L^{-1}), F is the flow rate ($m^3 day^{-1}$), C_{ps} is the uptake of the compound by primary sludge (ng kg^{-1}), and M_{ps} is the total mass of the primary sludge (kg). The larger materials that settle during primary treatment leave mostly soluble or colloidal material in the wastewater, which is then subjected to subsequent (secondary) treatment.

5.1.2 Secondary Treatment

Secondary treatment is essentially designed to remove dissolved organic matter from the wastewater. Most secondary treatment systems are aerobic although some systems, particularly those designed to treat high-strength industrial wastes (such as wastes from pharmaceutical plants and personal care products industrial plants) are anaerobic. In some instances, systems are designed to alternate between aerobic and anaerobic conditions. During secondary treatment, more organic compounds, including PPCPs, can be degraded by the existing microbes or adsorb onto the floc. There are three main approaches for attaining secondary treatment, notably (i) lagoon (or oxidation pond) systems, (ii) fixed films, and (iii) suspended films.

5.1.2.1 Lagoons Lagoon systems are shallow basins in which the wastewater can be held for several months so as to allow degradation. Under these conditions, the degradation process can be biotic, abiotic, or a combination of both. The biotic phase is facilitated by indigenous microflora and enhanced by aeration. While some lagoons are aerated with the help of agitators, most lagoons depend on the natural flow of the water, which facilitates diffusion and mixing of the contaminant. Studies have shown that the elimination of contaminants in lagoons is enhanced using a raceway design (Fig. 5.1). Such a design provides the back-and-forth flowing of the wastewater through the unit, maximizing aeration, with no dead spots. Where land is not limiting, the use of lagoons can be a cheap and easy technology to maintain. Lagoons are more ideal under tropical and subtropical conditions, their use in temperate regions being limited by subzero temperatures, which can immobilize the rates of degradation. Some lagoons are facultative, with the water layer near the surface being aerobic while the bottom layer (i.e., facultative zone), which includes sludge deposits, is anaerobic.

5.1.2.2 Fixed-Filter Systems Filtration as part of the water treatment process has been practiced for many years. Filtration creates a barrier between the contaminant

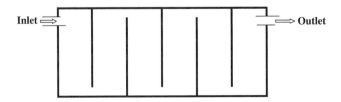

Figure 5.1 Schematic of a typical lagoon raceway design. The number of divided units can vary depending on need.

and the effluent based on size exclusion. Several matrices are used as filtration barriers, and these may range from very simple structures as sand filters to more advanced ones such as activated carbon filters, membrane filters, and ultrafilters. Advanced filters will be discussed later, but it is worth noting that all filters rely on simple sieving to remove particles, including microorganisms and total suspended solids (TSS), reducing turbidity. Filtration primarily aims at removing particles that are larger than a particular size. To get satisfactory filtration of the wastewater, the prefiltration processes have to reduce the suspended solids to several parts per million. While some sorbed PPCPs are removed based on simple sieving as the particulate materials are eliminated, most PPCPs have a large molecular size ranging between 150 and 500 Da. Nearly all water treatment facilities use some form of filtration in conjunction with some other treatment process.

Fixed-filter systems are specifically designed to grow microorganisms on surfaces such as sand, rocks, and plastic, developing a system whereby a form of biological filtration process thrives on attached growth. The flowing wastewater provides nutrients to the attached and thriving microflora, leading to thicker biofilms. Examples of fixed-film systems include sand filters, rotating biological drums, and trickling filters. The air required by this biological purification process is channeled through natural ventilation into the filter. As the biofilm containing the microbial biomass increases, some of the microorganisms will die off, becoming deposited into the flow, leaving the percolating filter. As a matter of fact, maturity of the biofilm in such systems is enhanced as more and more microorganisms die off, leaving a percolating surface. The die-off is continuously replaced by new growth. This filtered flow then passes through a "humus" settlement tank that is similar in design to a primary settlement tank separating the purified effluent from the dead microorganisms and any inert solids that might have come through the system. In essence, the process converts the soluble PPCPs (and other pollutants), providing substrates for microbial growth, thus generating biomass.

Some fixed filtration systems are relatively inexpensive. For example, constructed wetlands and sand filters are generally the most successful methods of polishing the treated wastewater effluent from lagoons. These systems have also been used with more traditional, engineered primary treatment technologies such as septic tanks and primary clarifiers. In such constructed wetlands, the system utilizes the roots of plants to provide substrate for the growth of attached bacteria, which utilize the nutrients present in the effluents and for the transfer of oxygen.

Fixed filtration systems can be a viable option for small water plants in rural areas where lagoons, suspended filters, or advance processes routinely used by large municipalities are not affordable. Rooklidge et al. (2005) investigated the removal of five classes of antimicrobial agents from water using fixed filters and found tremendous differences in the percent removals of various antibiotics (Fig. 5.2). In general, this filtration system was ineffective in removing tripmethoprim, sulfamethoxazole, sulfamethazine, and lincomycin (log K_{ow} 0.73, 0.48, 0.76, and 0.29, respectively) but was relatively effective in removing tylosin (log K_{ow} 1.05). The comparatively higher removal of tylosin was possibly due to its higher log K_{ow}, which enables it to sorb to the existing colloids. This type of filtration was also somewhat effective at

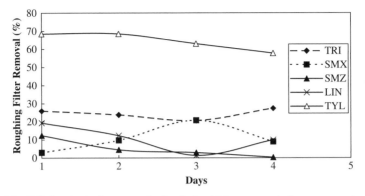

Figure 5.2 Efficiencies in the removal of antimicrobial compounds from water using a fixed filter system. The agents were TRI = trimethroprim, SMX = sulfamethoxazole, SMZ = sulfamethazine, LIN = lincomycin, and TYL = tylosin. (Based on data from Rooklidge et al., 2005.)

removing other pharmaceutical compounds such as diclofenac, carbamazepine, propylphenazone, and primidone (Heberer et al., 2002; Drewes et al., 2002).

5.1.2.3 Suspended-Filter Systems

Suspended-filter systems include systems such as sequential batch reactors, extended aeration systems, and activated sludge systems. The term *activated sludge* refers to sludge that consists of flocs of bacteria that consume the biodegradable organic susbtrates in wastewater. Thus, the activated sludge process occurs when suspended microorganisms break down organic pollutants, reducing the BOD. During this process, nutrients such as ammonia and phosphorus are also reduced in the wastewater. These nutrient removal processes, particularly the nitrification–denitrification reactions, generally seem to enhance the removal of some PPCPs such as steroids (Andersen et al., 2003), particularly if the retention time is long enough so as to enable the growth of microorganisms that have the ability to degrade these compounds. The treated effluent is then ultimately separated from the existing microorganisms, with the latter settling to form sludge. Some of that sludge is pumped back to the inlet of the activated sludge process for re-seeding purposes, whereas the rest of the activated sludge is removed from the system. The activated sludge process is also enhanced by actively pumping air through the system. More than 75% of the wastewater that is treated in the United States undergoes some form of activated sludge treatment (Halden and Paull, 2005). Activated sludge processes are also widely used in other parts of the world (Drillia et al., 2005; Nakada et al., 2006) and have been increasingly studied for efficacy in removing PPCPs. From some of those studies, it has been determined that the removal efficiency of PPCPs tends to be higher in activated sludge than in trickling filters. In a survey of five municipal sewage treatment plants in Tokyo that use primary and secondary treatment with activated sludge, Nakada et al. (2006) registered a 10- to 1000-fold reduction in the concentration of analgesics, thymol, and triclosan as well as some EDCs in the effluent compared to the influent. By contrast, the concentration of various antipsychotic drugs such as propyphenazone and carbama-zepine were not greatly reduced in any of those activated sludge processed effluents.

In another study, Simonich et al. (2002) reported significant removal of HHCB and AHTN from wastewater with 15–50% of these fragrance materials being removed in the primary effluent. The average fragrance materials in the final effluent were significantly different from the primary effluent profile. It was also observed that the final fragrance materials effluent was a function of the design of the wastewater treatment process with the removal of sorptive nonbiodegradable fragrance materials correlating to the removal of TTS in the wastewater. Removal of the nonsorptive, bio-degradable fragrance materials was correlated with the reduction in BOD_5 from the water. Removal of these compounds by activated sludge processes was between 87.9 and 99.9%, removal by lagoons and oxidation ditches was 71.3–99.9%, removal by trickling filters was 71–98.6%, and removal by biological contactors was 81–99.9%. Despite these modest removals, the concentrations remaining in the effluent can still be fairly substantial, collectively showing that these secondary processes cannot adequately reduce most PPCPs in the environment, and thus the need for further treatment.

5.2 ADVANCED TREATMENT PROCESSES

Advanced (or tertiary) treatment is used to improve the physicochemical quality of secondary treatment effluents. Several processes, such as coagulation, flocculation, settling-sand filtration, nitrification and denitrification, adsorption using carbon, ion exchange, and electrolysis, are some examples of advanced treatment processes. From a PPCP removal perspective, tertiary treatment always ends up with an even cleaner product, but the extent of PPCP removal during this stage depends on the efficacy of the respective method used and the type of compound. These methods have evolved as a result of continuous research in water and wastewater treatment to strive for more purity as the list of candidate contaminants and an increased chal-lenge to the industry for meeting tighter effluent discharge requirements have grown. A number of the most promising processes for removing PPCPs in advanced treat-ment systems are discussed below.

5.2.1 Advanced Filtration Systems

Advanced treatment plants employ either granular filters or membrane filters. The former is exemplified by activated carbon, whereas membrane filtration has been developed only in recent memory. Besides the general principle of excluding con-taminants based on size, these advanced filtration systems also have a charge that enables them to exclude particles, with the removal of anionic compounds being higher than that of nonionic ones. Both systems come at a premium.

5.2.1.1 Activated Carbon Activated carbon can be used as granulated activated carbon (GAC) or as powdered activated carbon (PAC). GAC is setup as adsorptive beds, whereas in PAC the activated carbon is added directly to the water. The efficacy of activated carbon depends on the surface area, surface charge, pore size distribution and hydrophobicity of the compound. Thus, smaller activated carbon particle sizes offer more rapid adsorption kinetics, which in turn increases the service life of the activated carbon medium and reduces replacement cost. Because GAC breakthrough curves are generally not symmetrical (U.S. EPA, 1999), a logistic function to model filtration through GAC beds is typically preferred as it incorporates a Freundlich isotherm parameter that enables generating breakthrough curve profiles that are based on adsorption characteristics and the concentration of the compound in the influent. Adapting a logistic function also matches the S-shaped breakthrough curve. Thus, Bond and Digiano (2004) used a logistic function model [Eq. (5.2)] to determine the breakthrough time for various types of GAC.

$$C(t) = \frac{A_f}{1 + Be^{-Dt}} + A_0 \tag{5.2}$$

where $C(t)$ is the concentration of the compound interest at filtration time (t) in days; A_f is the line at which the S-shaped adsorption reaches an asymptomatic point; B and D are constants that depend on the shape of the adsorption curve, and A_0 represents the step that is applied to the logistic function to match moderate to high immediate breakthrough levels. Research with total organic carbon as well as chlorine and its disinfection by-products adsorption measurements showed a better fit for breakthrough curves with this modification, which basically reflects the long initial lag period during which the concentration of compounds (including PPCPs) in the effluent will, prior to breakthrough, be below the detectable limit (U.S. EPA, 1999). Equation (5.2) can be rearranged to generate Eq. (5.3):

$$t = \frac{1}{D}\ln\left(\frac{B(C - A_0)}{A_f + A_0 - C}\right) \tag{5.3}$$

The sorption and removal using activated carbon or some other sorbent is always relying on competition with natural organic matter and other pollutants that may be present. The dissolved organic compounds in water, referred to as natural organic matter (NOM), are composed of several fractions. As a matter of fact, NOM is found in varying concentrations in all natural water sources as a complex mixture of compounds and is formed as a result of breakdown of animal and plant materials in the environment. NOM spectra show three distinct fractions: (i) carbohydrate, (ii) melanin, and (iii) aromatic rings (Newcombe et al., 1997). Most of the compounds that are present in NOM carry a charge that is generally attributed to the carboxylic acid and phenolic groups that are present. These charges cause the larger compounds to behave as polyelectrolytes in the water. Natural organic matter can compete for adsorption sites on the activated carbon, decreasing the capacity of activated carbon to adsorb PPCPs and other pollutants (Newcombe et al., 1997). The adsorption of NOM onto activated carbon is influenced by pH and inorganic materials such as calcium, magnesium, and sodium. It is not easily determined in aquatic environments and thus total organic carbon (TOC) is typically used as a surrogate for NOM. Its absorption on the activated carbon tends to decrease with increasing pH because NOM has a weak acid characteristic that becomes less polar and hydrophilic as the pH decrease, which in turn makes the NOM more prone to adsorb onto the activated carbon surface as conditions become more and more acidic. GAC grain size is also inversely proportional to the rate of adsorption of TOC (Crittenden et al., 1991, 1993). The interference of NOM to adsorption of pollutants by activated carbon has implications in the sense that results that show superb adsorption of compounds in the laboratory test with distilled water, which is, of course, devoid of NOM, may not adequately represent the overall adsorption that can occur under natural environmental settings.

Granulated activated carbon is packed stationary beds with water flowing through continuously. GAC columns can operate quite efficiently for the first few months, but, later as the system continues to run, the incoming and more adsorbable constituents including NOM may displace some previously adsorbed constituents based on sorption capacity. To that effect, the efficiency of NOM removal by GAC seems to be site specific and requires stringent pretreatment calibrations that may vary from one location to another (Bond and Digiano, 2004). The flowing water attains a short contact time of approximately half an hour or even less with the activated carbon. Thus, GAC systems can last long, attaining some level of equilibrium. There are a number of reports that document the removal of PPCPs (Andreozzi et al., 2003; Boyd and Grimm, 2001; Ternes et al., 2002) and a variety of other organic pollutants such as pesticides, herbicides, and PAHs (Kim et al., 2005) from water by GAC. Activated carbon is particularly effective on nonpolar compounds that do not have functional groups or N-heterocyclic structural groups. Thus, GAC effectively removes several pharmaceutical compounds such as the antiphlogistic diclofenac, the antiepileptic carbamezapine, and the lipid regulator bezafibrate (Ternes et al., 2002, 2003; Boyd and Grimm, 2001). By contrast compounds that have carboxyl groups dissociate in water to a good extent, becoming negatively charged. Such dissociation can greatly reduce their propensity to sorb to activated carbon. Examples of such compounds include ibuprofen, diclofenac and clofibric acid. The removal of PPCPs such as iopromide (log $K_{ow} - 2.49$) and sulfamethoxazole (log $K_{ow} = 0.89$), which have low log K_{ow} values, by GAC is quite low. Activated carbon, combined with coagulation or precipitation, also remove substantial amounts of surfactants (Adachi et al., 1990).

The long-term effects of saturation at the adsorption sites on the GAC beds can be minimized if the activated carbon is used as a powder instead (i.e., PAC). Once introduced to the water, the PAC is allowed to stay in contact with the water for a short duration of about 4 h after which it is allowed to settle in the tank and is then disposed of as a sludge. The PAC sludge can alternatively be filtered off. Using this approach, Adams et al. (2002) used various doses of PAC and obtained removal rates of 57–98% for various antibiotics with a Calgon WPH Pulv PAC that is commonly used by drinking water treatment plants in the US (Fig. 5.5 in Adams et al., 2002). The PAC doses tested were up to a maximum of $50\,\mathrm{mg\,L^{-1}}$ with the highest removal of antibiotics in deionized water and Missouri River water at this highest dose. However, PAC doses higher than $20\,\mathrm{mg\,L^{-1}}$ did not significantly increase

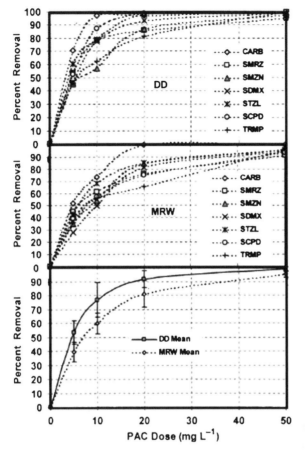

Figure 5.3 Removal of various antibiotics from distilled water (i.e., DD) or Missouri River Water (MRW) by powdered activated carbon (PAC; Calgon WPH Pulv). The bottom panel is a mean for all of the compounds in the two types of water tested. The compounds tested were CARB = carbadox, SMRZ = sulfamerazine, SMZN = sulfamethazine, SDMX = sulfadimethoxine, STZL = sulfathiazole, SCPD = sulfachloropyridazine, and TRMP = trimethoprim. (From Adams et al., 2002.)

the removal of most antibiotics tested (Fig. 5.3). By comparison, better removal of estrogen was reported by Khanal et al. (2006) with PAC compared to GAC, especially when these treatments were combined with an increased retention time. However, PAC can be more expensive than GAC and cannot be used on a sustained basis without substantially increasing the operating costs.

5.2.1.2 Filtration Membranes Membrane bioreactors are a fairly more recent phenomenon in the wastewater treatment industry, and, therefore, a slightly detailed coverage of their architecture and attributes in regard to how these relate to usage to remove PPCPs from water is in order. They are practically a form of activated sludge process that uses membranes to separate the sludge from the effluent instead of a traditional final settlement tank (i.e., clarifier). Membrane filtration has a better ability to remove suspended solids, pathogens (including viruses), nutrients (including N and P), and chemical oxygen demand (COD). This form of filtration involves using a thin semipermeable film (membrane) as a selective barrier to remove contaminants from the water. More studies for the removal of pathogens by MBR in the drinking and wastewater industry to meet environmental standards have been conducted compared to those that have evaluated the removal of PPCPs from such waters (Delanghe et al., 1994; Cicek et al., 1998; Comerton et al., 2005). Numerous pilot and full-scale studies have demonstrated the ability of MBR to produce high-quality effluent water with excellent removal of organics and suspended solids in some instances, providing water with an average turbidity of 0.1 NTU and BOD_5 of zero (Chiemchaisri et al., 1993; Ueda et al. 1996; Cicek et al., 1998). In a study comparing conventional activated sludge treatment and an MBR process, the MBR system removed more nitrogen and phosphorus than the conventional treatment (Bodzek et al., 1996). These excellent outcomes are based on the ability of membranes to remove low-molecular-weight (i.e., $MW \leq 110$) organic compounds (Adams et al., 2003; Kimura et al., 2004), and an increasing number of studies have recently also looked at their efficacy in removing various pharmaceutical compounds from sewage effluents and drinking water. MBRs have been proposed as promising technology to remove PPCPs including endocrine disruptors from wastewater (Khanal et al., 2006) especially if integrated with reverse osmosis (Sedlak and Pinkston, 2001; Xu et al., 2005). The successful use of membranes for PPCP removal is usually enhanced by adapting well-controlled recycle rates and sequencing of aerobic and anaerobic processes (Ahn et al., 2003; Holakoo et al., 2005) or by combining them with reverse osmosis (Comerton et al., 2005).

MBR technology has various advantages including a smaller footprint, better effluent water quality, disinfection, increased volumetric loading, and the generation of less sludge material compared to conventional treatment systems (Nagano et al., 1992; Knoblock et al., 1994; Adham et al., 2001). MBRs effectively overcome problems associated with poor settling of sludge in conventional activated sludge processes, and they permit the operation of bioreactors with considerably higher mixed liquor solids concentration, enabling the maintenance of extended retention times. Membrane processes commonly used belong to four main categories: microfiltration (MF), ultrafiltration (UF), nanofiltration (NF), and reverse osmosis (RO), and the main attributes of each are summarized in Table 5.2. NF is a more preferable economic option compared to RO because it requires lower operating pressures. UF mostly successfully eliminates suspended solids and rejects HMWC such as proteins and suspended solids. It allows most LMWCs including mono- and disaccharides, salts, amino acids, as well as organic and inorganic acids to pass through. Nanofiltration mostly rejects positively charged ions and ions with more than one negative charge (such as phosphates and sulfates), whereas reverse osmosis, in principle, rejects all dissolved and suspended materials, allowing only water to pass through. The typical pore sizes and operating pressures for the respective filtration processes are summarized in Table 5.2 with a note of caution that the sizes indicated

TABLE 5.2 Membrane Filtration Processes and Characteristics

Process	Microfiltration (MF)	Ultrafiltration (UF)	Nanofiltration (NF)	Reverse Osmosis (RO)
		Process Characteristics		
Membrane shape	Symmetrical and asymmetrical	Asymmetrical	Asymmetrical	Asymmetrical
Membrane material(s)	Ceramic, polypropylene (PP), polysulfone (PSO), polyvinylidenedifluoride (PVDF)	Thin film of ceramic, PSO, PVDF, or cellulose acetate (CA[a])	Thin film of CA	Thin film of CA
Membrane module	Tubular, hollow fiber	Tubular, hollow fiber, spiral wound, plate and frame	Tubular, spiral wound, plate and frame	Tubular, spiral wound, plate and frame
Pore size (μm)	0.02–4	0.02–0.2	0.02–0.002	<0.002
Rejection of	Bacteria, clay, and other optical microscope-visible particles	All those by MF, proteins, viruses, and polysaccharides	All those by MF and UF, mono-, di-, and oligosaccharides, polyvalent negatively charged ions, and HMWC[b]	All those by MF, UF, NF, and glucose, LMWC[c], amino acids
Operating pressure range (kPa)	0.3–200	0.3–500	500–1000	1000–5000
Primary applications	Removal of particles and turbidity	Removal of dissolved nonionic solutes	Removal of divalent ions (softening) and dissolved organic matter	Removal of monovalent ions (desalination)
		Chemical Resistance Based on Membrane Materials Specified Above[c]		
pH 3–8	Ceramic (+), PP (?), PVDF (+), PSO (+)	Ceramic (+), PSO (+), PVDF (+), CA (+)	CA (+)	CA (+)
pH < 3 or >8	Ceramic (+), PP (?), PSO (+), PVDF (+)	PSO (+), PVDF (+), CA (−)	CA (−)	CA (−)
Temp >35	Ceramic (+), PP (−), PSO (+), PVDF (+)	PSO (+), PVDF (+), CA (−)	CA (−)	CA (−)

(Continued)

223

TABLE 5.2 Continued

Process	Microfiltration (MF)	Ultrafiltration (UF)	Nanofiltration (NF)	Reverse Osmosis (RO)
Humic acid	Ceramic (−), PP (?), PSO (−), PVDF (−)	PSO (−), PVDF (−), CA (+)	CA (+)	CA (+)
Proteins	Ceramic (+), PP (?), PSO (+), PVDF (±)	PSO (+), PVDF (±), CA (±)	CA (±)	CA (±)
Polysaccharides	Ceramic (+), PP (?), PSO (+), PVDF (−)	PSO (+), PVDF (−), CA (−)	CA (−)	CA (−)
Aliphatic hydrocarbon	Ceramic (+), PP (?), PSO (−), PVDF (±)	PSO (−), PVDF (±), CA (−)	CA (−)	CA (−)
Aromatic hydrocarbon	Ceramic (+), PP (?), PSO (−), PVDF (+)	PSO (−), PVDF (+), CA (−)	CA (−)	CA (−)
Oxidizers	Ceramic (+), PP (?), PSO (+), PVDF (+)	PSO (+), PVDF (+), CA (±)	CA (±)	CA (±)
Ketones, Esters	Ceramic (+), PP (?), PSO (−), PVDF (+)	PSO (−), PVDF (+), CA (−)	CA (−)	CA (−)
Alcohol	Ceramic (+), PP (?), PSO (+), PVDF (+)	PSO (+), PVDF (+), CA (−)	CA (−)	CA (−)

Source: compiled based on Wagner (2001) and Membrane filtration Technology for Sustainable Water Management (American Water).

[a]Cellulose acetate is often a di-acetate or tri-acetate.

[b]HMWC refers to high molecular weight compounds such as proteins whereas LMWC[a] refers to low molecular weight compounds such as sodium chloride.

[c]The bracketed symbol (+) means high resistance, (±) dubious, (−) poor resistance or (?) unknown to that trait.

for NF and RO sizes are theoretical as they fall in the molecular and ionic range, respectively.

Membrane Type and Pore Size   Many materials are currently used by the membrane manufacturing industry and sold under an array of trade names. However, the bulk of the functional ingredients in the membranes are a few materials that are known to work effectively. Common key ingredients in membranes include polyamide or cellulose acetate (CA). The former are negatively charged and are believed to operate by repelling the negatively charged compounds. Neutral compounds can also be repelled depending on their physical and chemical properties (van der Bruggen et al., 1999). Membrane configuration and pore size varies depending on the manufacturer of the unit. Mostly used in UF, NF as well as RO filtration processes are membranes that are primarily made of CA because this material is relatively cheap and is hydrophilic, a characteristic that makes it less prone to fouling. However, cellulose acetate has a number of limitations with respect to temperature and pH. It can also be degraded by microorganisms. By contrast, the cellulose membranes (as opposed to cellulose acetate membranes) have good resistance irrespective of the prevailing pH, can withstand temperatures higher than 35°C, humic acids, proteins, aliphatic hydrocarbons, and alcohols. As a matter of fact, cellulose rather than CA-based membranes show more promise to the rejection of these compounds of concern but have traditionally not been used as widely in the filtration industry, commanding a meager 3–5% of the worldwide membrane market share possibly because cellulose membranes have questionable resistance to aromatic hydrocarbons, oxidizers, ketones, and esters (Wagner, 2001). As a basis for comparison to cellulose and cellulose acetate membranes, polysulfones (PSO), which are either polyethersulfone or polyarylethersulfone, typically used in some MF and UF processes, can withstand high temperatures and pH extremes (Table 5.2). It is noted that most PSOs do not tolerate polar solvents, fat, grease, and oil. Some MF processes use polypropylene (PP) or polyvinylidenedifluoride (PVDF). PVDF is difficult to manufacture but has some good qualities of filtration consistence and is highly resistant to hydrocarbons and oxidizing environments. Other membrane materials, which are not used as widely as PP, PSO, PVDE, and CA, include polyacrylonitrile (PAN), ceramic (SiO_2), and cellulose (a form of hydrolyzed cellulose acetate).

Besides pore size restrictions, membrane molecular size, as well as type, resistance to temperature and pH are possibly two of the other most important characteristics that have to be considered in predicting membrane efficacy in removing PPCPs and other organic pollutants. As a matter of fact, the effectiveness of a bioreactor can be influenced by temperature, type of compounds, contact time, and treatment protocol. Implicit in this observation is the likelihood that bioreactor membrane performance will slightly vary across treatment plants and practices. Thus, it is apparent from the summary in Table 5.2 that the high efficacy of cellulose-acetate-based membranes commonly used in NF and RO can be negatively impacted at pH < 3 or >8 and temperatures >35°C. Their rejection of aliphatic hydrocarbons, aromatic hydrocarbons, oxidizers, and functional groups such as ketones, esters, and alcohols to which a majority of PPCPs belong can therefore be dubious at best or even poor in some instances. The temperature capability of the membrane is dictated by the limitation of the membrane configuration. Thus, the higher the temperature, the more the need to balance the physical stress on the membrane itself and its elements. All membranes are sensitive to pressure. However, there are no firm rules that specify maximum allowable pressure for different membranes. To maximize efficacy, pressure, and temperature have been characterized as Wagner units, with each Wagner unit being the product of pressure (in bars) and temperature (in degrees Celsius). Wagner units of <1200 are believed to be ideal for operating a membrane.

Most of the MBR membranes have effective pore size of 0.01–0.4 μm, and the filtration process of MBR would physically remove larger microorganisms such as bacteria (2–3 μm) and protozoan parasites (4–15 μm) for which membrane filtration technology was originally envisioned. A comparison of different MBRs from six different manufactures, the differences resulting from different technologies including membrane configuration and pore size is shown in Table 5.3. Despite inherent differences in permeation cycle times, nitrification/denitrification capabilities, required amount of operator attention, required membrane cleaning frequency, power requirements, and robustness of the systems, all five MBR technologies produced excellent quality effluents in terms of pathogens, BOD, and TSS removal. A similar comparison with regard to the removal of PPCPs is not yet available, but this summary is intended to display the range of configurations in which membranes are set up.

TABLE 5.3   Variation in Membrane Configuration and Pore Size Based on Five Different Manufacturers

Membrane Type	Configuration and Pore Size
Enviroquip	Flat-panel membranes with 0.4-μm pore size arranged vertically in aeration tanks. The membranes had the ability to relax.
Ionics	Microfiber membranes with 0.4-μm pore size. They were arranged in a horizontal position in an aeration tank. The membranes had the ability to relax.
Zenon	Microfiber membranes with 0.04-μm pore size. They were arranged in a vertical position in an aeration tank. The membranes were relaxing and back-pulsing.
U.S. Filter	Microfiber membranes with 0.4-μm pore size. The membranes were arranged in a vertical position in an off-line tank. The membranes relaxed and back-pulsing.
Huber	Flat-panel membranes with 0.025-mm pore size. They are arranged in a vertical position on a rotating shaft in an aeration tank. They were subjected to spray washing.
Koch	Hollow-fiber membrane of 0.1-μm pore size. They were arranged in a vertical position with intermittent air scouring and back-flushing.

Source: Based on information downloaded from http://www.wrrc.hawaii.edu/MBR.htm as posted by R. Babcock (University of Hawaii; accessed on 11-09-2007).

In terms of membrane architecture, the pore size, pore geometry, as well as the membrane charge are pertinent to the efficacy of membranes. For negatively charged compounds, the degree of electrostatic repulsion and rejection depends on the amount of charge in the surface of the membrane. Membrane surface charge is determined by measuring the zeta potential (mV^+). Xu et al. (2005) showed that membranes with a highly negative charge (e.g., NF-200 and NF-90 with zeta potential of -15.3 and $-21.6\,mV^+$, respectively) were quite effective in rejecting all negatively charged compounds even when the compound had MWCO values that were lower than the pore size of the membrane. The contact angle of the membrane is also important and has been used to determine whether the membrane is hydrophilic (contact angle $<50°$) or hydrophobic ($>50°$) (Xu et al., 2005). Hydrophobic membranes are particularly effective in rejecting compounds that have hydrophobic moieties, the rejection being a result of hydrophobic–hydrophobic interactions. The same analogy of rejection of hydrophilic compounds by membranes with hydrophilic surfaces also applies. Furthermore, the presence of natural organic matter can clog membrane pores, changing the membrane's surface charge and resulting in improved rejection by enhancing steric and electrostatic exclusion (Thanuttamavong et al., 2002).

The surface of the membrane is also critical in determining membrane efficacy, with more hydrophilic surfaces enabling water to permeate easily. As the surface becomes more and more hydrophobic, it also becomes less permeable to water molecules. With pathogens, studies have shown that a consideration of the volumes of water to be treated (e.g., liters per hour), concentration ratio, or volume of the permeate, which is typically 75–95% of the feed volume and the expected permeate flux (e.g., liters per hour), a reliable prediction about the efficacy of membranes can be made. Similar parameterization needs to be done to determine what works best when it comes to rejecting PPCPs since parameters that are effective for rejecting pathogens may not necessarily apply wholesale to PPCPs.

Molecular Size of the Compounds The size and molecular length of the PPCP compound can also influence the efficacy of the membrane. Molecular length is, in this instance, defined as the maximum length of the molecule, whereas molecular width refers to its cross-sectional diameter. Kimura et al. (2003) tested 11 neutral pharmaceutically active compounds that greatly vary in molecular weight, octanol–water partition coefficients (K_{ow}), and dipole moments for regeneration by two distinctly different RO membranes. All of the assays were conducted at 20°C (pH 7 ± 0.1) and filtration was done over a 24-h period. A key assumption from that experiment was that by the end of that duration, the membrane was saturated. The percent regeneration (R) of the compound by the membrane was calculated from

$$R = \left(1 - \frac{C_p}{C_f}\right) \times 100$$

where C_p is the concentration of the compound in permeate after 24 h and C_f is the concentration of the compound in the feedwater after 24 h. Both molecular length and width are suspected to have some effect on the rejection of compounds by membranes (Xu et al., 2005). Although this possibility has not been widely investigated with regard to its influence on the removal of PPCPs by membranes, there is a tendency of the higher molecular weight compounds to be rejected to a greater extent compared to those with lower molecular weight in polyamide membranes (Table 5.4) depicting a sieving effect. Rejection was compared with the rejection of sodium chloride. Salt rejection is widely used by the membrane manufacturing industry to describe membrane rejection properties of their product. In general, rejection was higher with the polyamide membrane compared to the cellulose acetate membrane, the only exception being with sulfamethaxazole (Table 5.4). As a matter of fact, cellulose acetate had no rejection for 2-naphthol and the NAC standard. Overall, the rejection percentages for primidone, isophylantipyrine, carbamazepine, and sulfamethaxazole were comparable irrespective of the membrane used. However, this trend did not hold true with the cellulose acetate membrane, suggesting that both membranes act based on different mechanisms. In the same regard, molecular weight cutoff (MWCO) was more in agreement with rejection trends for the compounds presented in Table 5.4. However, an MWCO greater than 200 is predicted to have more than 90% of the compound rejected by the polyamide membrane. This prediction was only true for carbamazepine, indicating a discrepancy with MWCO predictions as well. This discrepancy may be attributable to the fact that polyethylene glycol (PEG) used as the standard in determining MWCO may not be entirely representative of removing pharmaceuticals by membranes. Further

TABLE 5.4 Rejection of Several Neutral Pharmaceuticals and Surrogates by the Polyamide (XLE) and the Cellulose Acetate (SC-3100) Membrane[a]

| | | | | | % Rejection with | |
| | | | | | XLE | SC-3100 |
Compound	MW[b]	Solubility (mg L)$^{-1}$	log K_{ow}	Dipole Moment (Debye)	(MWCO <200)	(MWCO 200–300)
Salt(NaCl)[b]	58.4	>100^c	−0.46^c	?	90	94
2-Naphthol	144	0.0255	2.70	0.873	57	0
4-Phenylphenol	170	56.2	3.20	1.124	61	11
Phenacetine	179	0.0766	1.58	1.675	74	10
Caffeine	194	0.0002	−0.07	3.862	70	44
NAC Standard	201	0.01	2.36	1.989	79	0
Primidone	218	0.05	0.91	2.696	87	85
Bisphenol A	228	0.012	3.32	0.709	83	18
Isopropylantipyrine	231	3×10^{-6}	1.94	3.851	78	69
Carbamazepine	236	17.7	2.45	3.286	91	85
Sulfamethaxazole	253	0.061	0.89	6.318	70	82
17β-estradiol	272	3.6	4.01	0.798	83	29

[a]Unless specified otherwise, all data were from Kimura et al. (2003).
[b]The salt used was 1000 mg NaCl L^{-1}.
[c]Values obtained from Chemfinder.cambridgesoft.com.

evidence with pesticides also shows that their removal with RO and NF strongly correlated with log K_{ow} (Kiso et al., 2001), but this parameter does not appear to correlate with the percent amount of PPCPs rejected (Table 5.4), suggesting some unique features for PPCPs compared to pesticides. A closer examination of the percent rejection with dipole moments also shows that the higher dipole moment compounds are rejected better than those with a lower dipole moment. In general, polar compounds and charged compounds can interact with membrane surfaces and attain better removal compared to neutral or less polar compounds.

In follow-up work, Kimura et al. (2005) classified the removal of various pharmaceuticals, by a 0.4-µm PVDF membrane versus conventional activated sludge into three distinct categories (Table 5.5). A few other physical characteristics of those compounds are also included in Table 5.5 to give some additional information that may influence membrane performance. Note that all of the compounds in category 2 in Table 5.5 have Cl⁻ ion in their structure. The chloride ion is fairly common in PPCPs (Jjemba, 2006), and chloride-based compounds are usually quite recalcitrant (Eker and Kargi, 2006).

TABLE 5.5 Removal of Various Compounds from Wastewater by Conventional Activated Sludge and Membrane Bioreactor (mg L⁻¹)

Category	Compound	MW	Mol. Width (Å)[a]	Mol. Length (Å)[a]	Solubility (mg L)⁻¹	log K_{ow}	pK_a
1. Easily removed by both conventional activated sludge and membrane bioreactor technology	Ibuprofen	206.3	5.04	10.56	21	3.97	4.91
2. Not efficiently removed by conventional activated sludge and membrane bioreactor technology	Clofibric acid	214.6	?	?	50 (in ethanol)	2.84	?
	Dichlorprop	235.1	?	?	350	3.03	3.00
	Diclofenac	296.2	7.2	9.37	2.37	4.51	4.15
3. Not effectively removed by conventional activated sludge but removed by membrane bioreactor technology	Ketoprofen	254.3	6.9	9.28	51	3.12	4.45
	Mefenamic acid	241.3	?	?	20	5.28	4.2
	Naproxen	230.3	5.32	11.23	159	3.18	4.15

[a]Molecular width and length values were compiled from Xu et al. (2005).

Other Important Considerations with Membranes   Oxidizers such as sodium hypochlorite (i.e., ClO_2), bromine, iodine, and ozone, which are typically used in the disinfection of wastewater, are not well tolerated by thin-film membranes. Such disinfectants can thus influence the efficacy of membranes in removing contaminants such as PPCPs. Furthermore, membranes can become fouled by microorganisms that can metabolize the membrane material. Thus, microbial counts of ≥ 100 cells/mL can be problematic. Likewise, dead-cell debris can also cause fouling. Membranes can also be fouled by heavy metals such as chromium. Thus, if heavy metals are deemed a problem, they should be precipitated from the wastewater prior to the filtration with membranes.

As indicated earlier, wastewater may also contain grease, oils, and solvents. If not emulsified, these will affect the efficacy of the membranes in the removal of PPCPs. In general, membranes tend to be more sensitive to cationic compounds such as laundry softeners and a variety of flocculating agents, whereas neutral (i.e., nonionic) compounds such as detergents are effectively rejected by most membranes (Table 5.1). However, as is typical with almost everything with membranes, there are some exceptions to these generalizations.

Membrane bioreactors can maintain extremely long retention times coupled with a diverse microbial community, all of which facilitate the degradation of compounds in the system. From the discussion above, it is clear that the outcome with membrane filtration in the removal of PPCPs is quite promising, but further research and validation, particularly in environmental samples that contain a range of contaminants, is still needed. Overall, removal largely depends on the compound (and its properties), and the type of material from which the membrane is made. Snyder et al. (2003) contend that the membrane may be effective in removing a compound of interest, but over time starts releasing plasticizers that may have endocrine disrupting properties. Furthermore, both ultrafiltration and reverse osmosis are still expensive and only used by select water treatment plants. However, membrane filtration is becoming increasingly popular as an advanced wastewater treatment technology.

5.2.2   Oxidation Processes

Oxidation processes have traditionally been used to disinfect water so as to rid it of pathogens, the efficiency of disinfection being influenced by the concentration of the disinfectant, contact time, temperature, and pH. Two major oxidizing agents are commonly used in the water treatment industry, that is, chlorination and ozonation. From a practical perspective, the oxidizing agent is mixed into the main wastewater stream followed by residence in a contacting tank or channel, allowing time for the oxidation process. Some oxidizing disinfectants have also been found to oxidize various organic contaminants, including PPCPs. In general, oxidation and subsequent reaction of PPCPs increases with pH as dissociated acidic compounds are more reactive than their protonated counterpart. Similarly, nondissociated basic compounds are quite susceptible to oxidation, with aromatic compounds showing more reactivity compared to aliphatic ones. The chemistry involved in these oxidation processes following chlorination and ozonation is discussed below.

5.2.2.1 Chlorination Chlorine (i.e., HOCl and OCl—) and chlorine-based disinfectants such as chlorine dioxide (ClO$_2$) have some oxidizing potential (Hoigné and Bader, 1994), and chlorination is the most widely used form of disinfection. The oxidation say of phenolic PPCPs (and other organic compounds) by chlorine occurs by the deprotonation of the phenolate anion followed by a sequential addition of chlorine to the aromatic ring, ultimately cleaving the ring. However, in the presence of ammonia, free chlorine will generate chloramines [Eqs. (5.4) and (5.5)], which are less reactive than free chlorine. That possibility of forming chloramines has a lot of implications in some wastewaters where the concentrations of ammonia can be substantial. Substantial presence of ammonium occurs if nitrification is limited or completely inhibited, a situation that can be facilitated by the presence of antibiotics (e.g., through excretion) in the wastewater:

$$HOCl + NH_3 \;\rightarrow\; NH_2Cl + H_2O \tag{5.4}$$
$$\text{Monochloramine}$$

$$NH_2Cl + HOCl \;\rightarrow\; NHCl_2 + H_2O \tag{5.5}$$
$$\text{Dichloramine}$$

Oxidation of PPCPs by chlorination can also be greatly hampered by the presence of natural organic matter (NOM), such inhibition resulting from the formation of various complexes (Adams et al., 2002).

Boyd et al. (2003) reported good removal of naproxen in natural surface water by chlorination and ozonation. Follow-up work by that group in bench-scale studies using ultrapure water treated by filtering through activated carbon, followed by a mixed-bed deionization tank, ultrafiltration membrane system, and exposure to UV prior to chlorination, showed that naproxen is transformed by chlorination into several intermediates within 6 min and continues over time but is not completely degraded (Fig. 5.4). However, it is not clear to what extent such transformation occurs under nonexperimental conditions considering the likely interference from natural organic matter in the water as water is not ultrapure under natural conditions. The transformation of naproxen in those studies also depended on the dosage of chlorine used, with the 10 mg chlorine L^{-1} leading to faster naproxen transformation compared to 1 and 5 mg chlorine L^{-1}. By comparison, chlorination, which is routinely conducted purely to control pathogens, is often done for only a few minutes, after which the chlorine is quenched. If these intermediates are also formed under normal chlorination practices, their effects on the ecosystem and human health are still unknown.

5.2.2.2 Ozonation Ozone (O$_3$) is used in water treatment as a disinfectant. Besides disinfection, ozonation can remove organic micropollutants including colored and odorous substances. It is a strong oxidant that reacts with water, forming free hydroxyl radicals ($^\bullet$OH) that can attack organic molecules either directly or indirectly. It can also directly oxidize organic constituents more readily with organic compounds compared to chlorine and chlorine dioxide. As a matter of fact, it has high reaction rates with compounds that have double bonds, activated aromatic structures, and heteroatoms such as nitrogen or sulfur. Molecular O$_3$ is a

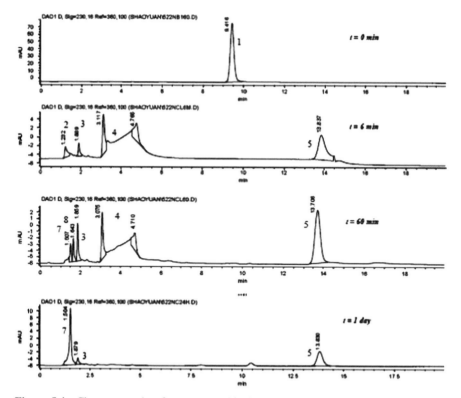

Figure 5.4 Chromatographs of naproxen and its intermediates in drinking water after chlorination at a molar ratio of naproxen:chlorine of 0.03 : 1 (i.e., 10 mg naproxen L^{-1} : 100 mg $Cl_2 L^{-1}$) at pH 6.23 (23°C). Naproxen (peak 1) is transformed in intermediates 2, 3, and 4 within 6 min. By one day, naproxen is gone and completely transformed into intermediates 7 (mainly) and 3. (From Boyd et al., 2005.)

selective electrophile that reacts readily and quickly with amines, phenols, and aliphatic compounds with unsaturated bonds, forming aldehydes, ketones, or carbonyl compounds (Langlais et al., 1991; Dantas et al., 2007). Additionally, ozone can participate in electrophilic reactions, particularly with aromatic compounds, or in nucleophilic reactions with many of the components of natural organic matter generating biodegrable dissolved organic carbon. Generally, ozonation increases the number of functional groups and the potency of the molecule, reactions that can distort the mode of action of pharmaceutical compounds (Ternes et al., 2003). When formed, the •OH radicals react less selectively compared to molecular O_3 but at even faster rates. This possibility has led to the combination of ozone with hydrogen peroxide to achieve even greater removal of organic contaminants (see Section 5.2.5).

With all of that oxidative potential, ozonation is capable of removing several types of PPCPs. For example, diclofenac was effectively removed by ozonation (Ternes et al., 2002) compared to the lower removal levels (i.e., 69%) obtained through conventional treatment (Ternes, 1998). The enhanced removal by ozone was possibly attributable to the presence of secondary amines in this compound. Carbamazepine, which has double bonds, and sulfamethoxazole, which has amines, are also effectively removed by ozone (Andreozzi et al., 2002). However, ozone reacts more slowly with clofibric acid, ibuprofen, ionpromide, and benzafibrate, despite their seemingly simple chemical structures. For example, in a study by Ternes et al. (2002), a low concentration of ozone ($0.5\,\mathrm{mg\,L^{-1}}$) removed 97% carbamazepine and diclofenac compared to a meager 15% of clofibric acid and primidone and 10% bezafibrate. To that effect, the removal by ozonation appears to be very product specific, but other environmental and physicochemical factors may also play an important role. For example, Dantas et al. (2007) recently reported effective removal of bezafibrate, with the initial concentrations of 0.5 mmol bezafibrate $\mathrm{L^{-1}}$ being diminished in 10 min following ozonation (Fig. 5.5). During that same period, approximately 73% of the initial chlorine content in the molecule was released into solution as chloride ions, resulting in a 30% increase in total organic carbon. Fibrates are some of the most used pharmaceuticals worldwide and have been found in sewage treatment plant effluents (Ternes, 1998). Dantas et al. (2007) also investigated the toxicity of the intermediates generated by this ozonation process using Microtox with *Vibrio fischeri* and found that the acute toxicity of the intermediates generated in the early stages of ozonating bezafibrate was higher than that of the parent compound.

Various antibiotics are also effectively removed from sewage effluents and river water by ozonation (Adams et al., 2003; Ternes et al., 2003). Other drugs such as

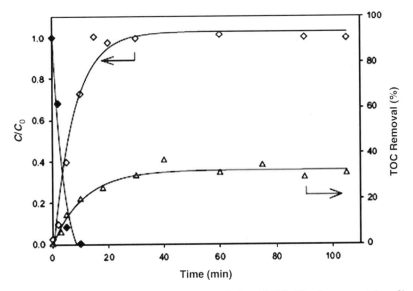

Figure 5.5 Ozonation of bezafibrate in water at pH 6 and 25°C. The ◆ represents bezafibrate concentration, ◇ represents the concentration of chlorine released from that degradation, whereas Δ represents the concentration of total organic carbon (TOC) removed during the ozonation process. (Adapted from Dantas et al., 2007.)

paracetamol (Andreozzi et al., 2003) and some estrogen steroids (Snyder et al., 2003) are also removed by ozonation. Huber et al. (2005) tested the removal of 11 pharmaceutical products from municipal wastewater effluents and showed that all of them, except the X-ray contrast medium iopromide were effectively oxidized with >2 mg ozone L^{-1}.

It should be noted that wastewater typically has a high total organic carbon (TOC) content under which the rate of ozone decay is increased. Like chlorination, ozonation also generates oxidation products (due to the deprotonation and double-bond substitution) whose nature and toxicity have rarely been determined. The disappearance of the primary pharmaceutical of interest during ozonation should not be regarded as a win-win situation all the time until the nature of products are determined to be harmless. A combination of both ozonation and GAC offer even better removal of PPCPs from water.

5.2.3 Ultraviolet Treatment

Several PPCPs have chromophores that absorb UV wavelengths, leading to some transformation. Thus, photodegradation is another process by which PPCPs can be removed from water. Buser et al. (1998b) and Tixier et al. (2003) documented photodegradation of diclofenac in lake water. Similarly, Latch et al. (2003) reported the photochemical conversion of triclosan to 2,8-dichlorodibenzo-p-dioxin in synthetic and natural waters. However, large doses of UV are usually required to cause any substantial transformation. Typical UV dosages in the water treatment industry are on the order of 30 mJ cm^{-2} (Adams et al., 2003). Adams et al. (2003) used UV dosages that were 100-to 300-fold higher than these typical doses and obtained removal rates that were only about 85% for several antibiotics except for carbadox and sulfathiazole where complete removal was registered with a dosage of about 7000 mJ cm^{-2}.

Photodegradation largely depends on the turbidity of the water, water depth, as well as the level of eutrophication of the water system under consideration. This observation has ecological implications especially for processes that depend on natural UV for treating their effluents of organic contaminants. These implications are even more significant in temperate regions where certain pharmaceutical compounds, such as those for respiratory ailments, are used more prominently during the winter when photodegradation is minimal due to much lower lux during that time of the year.

5.2.4 Electrolysis

Electrolysis as a means of removing contaminants from water works on the premise that the electrolyzed products generate hypochlorite (Morita et al., 2000; Nakajima et al., 2004; Kiura et al., 2002) in the presence of a low concentration of sodium chloride. Hirose et al. (2005) electrolyzed several antineoplastics, eliminating their biological activity and toxicity. The actual electrolysis was done with a cocktail of antineoplastic compounds in a 0.07 M NaCl solution using two platinum-irridium electrodes of 35 mm by 7 mm that were set at 5 mm apart and then introducing a 100-mA current (current density of $4\,A\,m^{-2}$) for a specific duration. Biological activity was determined by assaying the toxicity of the electrolyzed compounds (100 μL) against *Staphylococcus aureus* FDA209P incubated at 35°C, monitoring the turbidity of the bacterial cells over time. Cytotoxicity of the electrolyzed versus nonelectrolyzed water [expressed on a 50% cytotoxicity concentration (CC_{50}) basis] was determined using human lymphoblastoid (Molt-4) cells exposed to different concentrations of the test compound in the presence of a tetrazolium chromogenic marker enzyme that can form varying optical density at 450 nm, whose intensity depends on cell viability. Subsequent disappearance of the parent compound was monitored using HPLC. With electrolysis, the products were found to be noncytotoxic (Fig. 5.6).

More recently, Sirés et al. (2006) explored the electrochemical degradation of clofibric acid in water at various pH ranges (i.e., pH 2–12) using anodic oxidation with a cell containing a platinum or boron-doped diamond (BDD) anode and a stainless steel cathode. The platinum anode poorly decontaminated the solution of clofibric acid, whereas the BDD anode achieved the complete mineralization of the acid metabolite but at very low degradation rates and current efficiency. The high oxidizing power of the BDD anode compared to the platinum anode was attributed to its higher O_2 overpotential, which generates the strong oxidant hydroxyl radical [BDD(•OH)] adsorbed on its surface from water oxidation, that is,

$$BDD(H_2O) \rightarrow BDD(\text{•}OH) + H^+ + e^-$$

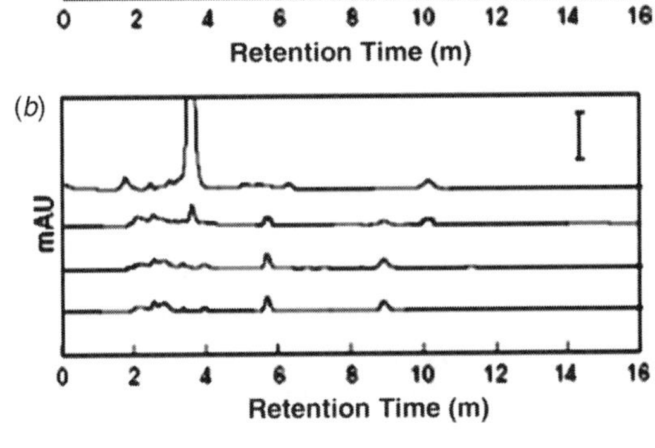

Figure 5.6 Chromatographs of electrolyzed epirubicin hydrochloride at (*a*) 480 nm and (*b*) 254 nm. The chromatographs were taken for samples after 0, 2, 4, and 6 h of electrolysis. (From Hirose et al., 2005.)

Follow-up work by Sirés et al. (2007) proposed combining BDD with Fenton processes as well as UV catalysts to favor the photodecomposition of complexes of Fe^{3+} with the carboxylic acids that are generated. The Fe^{3+} is regenerated, producing more •OH from the photoreduction of $Fe(OH)^{2+}$:

$$Fe(OH)^{2+} + hv \rightarrow Fe^{2+} + \bullet OH \tag{5.6}$$

5.2.5 Advanced Oxidation Processes

Section 5.2.2 focused on using oxidizing agents to remove PPCPs. Subsequent research by various groups has shown that combining these oxidizing agents (dubbed advanced oxidation processes; AOP) can provide even better removal rates in most instances (Zwiener and Frimmel, 2000; Huber et al., 2003; Snyder et al., 2006; Zweiner, 2007). Thus, combinations of several highly oxidizing agents, such as H_2O_2/ozone or UV/ozone, provide for nonselective hydroxyl radicals that react with many organic and inorganic constituents in fast reaction kinetics. A typical reaction vessel for conducting such advanced oxidation process is shown in Figure 5.7. For best results in bench-scale studies, one has to optimize the stirring speed, gas flow, and the concentration of ozone. Zweiner and Frimmel (2004) showed that increasing concentrations of O_3 and H_2O_2 increased the removal rates of clofribirc acid, ibuprofen, and diclofenac to 97.9, 99.4, and 99.9%, respectively, with O_3/H_2O_2 concentrations of 5 and 1.8 mg L^{-1} within 10 min of reaction time (Table 5.6). Similar enhancement of organic pollutant removal by ozone–hydrogen peroxide combinations have been reported by Kosaka et al. (2000) and more recently by Snyder et al. (2006). Combining UV with hydrogen peroxide also substantially removed nitro musks (musk ketone) from water compared to removals that have been reported with UV alone (Neamtu et al., 2000). However, the efficacy of such combined oxidants can be product specific under actual field conditions. For example, the removal of ibuprofen, dilantin, DEET, iopromide, and meprobamate

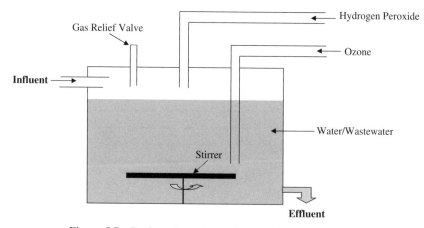

Figure 5.7 Design of an advanced oxidation process reactor.

TABLE 5.6 Enhanced Removal of Three Commonly Encountered Pharmaceuticals by Advanced Oxidation Process

	% of Compound Eliminated with		
	1 mg $O_3\,L^{-1}$ and 0.4 mg $H_2O_2\,L^{-1}$	3.7 mg $O_3\,L^{-1}$ and 1.4 mg $H_2O_2\,L^{-1}$	5 mg $O_3\,L^{-1}$ and 1.8 mg $H_2O_2\,L^{-1}$
Clofibric acid	21.8	92.7	97.9
Ibuprofen	29.2	94	99.4
Diclofenac	99.4	95.5	99.9

Source: Based on data from Zwiener and Frimmel (2000).

was enhanced by approximately 10% in Colorado River water, whereas that of androstenedione, pentoxifycline, testosterone, and progesterone in that same system was reduced by approximately 15% using O_3/H_2O_2 compared to O_3 alone.

The conventional techniques currently used in treating wastewater and biosolids by themselves are not very effective in significantly removing PPCPs. In most of those instances, the retention times may be shorter than the degradation half-lives of many PPCPs in wastewater, leading to discharge of undegraded PPCPs in the effluent (Xia et al., 2005). In general, the biodegradation of PPCPs increases with increasing hydraulic retention time (Kreuzinger et al., 2004; Khanal et al., 2006). For example, the removal of E1 increased from 64 to 94%, whereas that of E2 increased from 75 to 96% in MSW plants when the retention time was increased from 6 to 11 days (Ternes et al., 1999a). It is apparent that improving the removal of PPCPs from water and wastewater demands reliance on more than one treatment process at any particular treatment plant. The operation and efficacy of some of these methods, particularly those that are used for tertiary treatment, are seemingly promising in removing PPCPs, although the efficacy of some of them has been mostly based on laboratory or bench-scale studies. Testing for the presence of PPCPs in wastewater effluents not mandated by any regulatory agency at all and is still rare to routinely test for the presence of PPCPs in drinking water pre- and post-treatment at full-scale utilities because of the complexity of the methods that are used (see Chapter 2). However, where it occurs, its presence still poses a challenge and some of the methods discussed in this chapter can alleviate its presence.

6

FUTURE NEEDS

The volume of PPCPs that is used is expected to continue increasing worldwide as population density increases, per capita incomes rise, and new disease target groups as well as more potent compounds are identified. Appreciation of the relief of pain and suffering that pharmaceuticals provide, coupled with the sense of cleanliness and comfort that personal care products present to humanity, cannot be overemphasized. The drug development process has traditionally focused on designing, prescribing, regulating, and manufacturing PPCPs at acceptable high quality, with the ultimate goal of meeting the approval process and subsequent usage that totally excludes the compound getting into the environment. Similarly, health care providers have also been narrowly trained to focus on the disease without the recognition that most of what is prescribed can, through various pathways, be introduced into the environment. The unintended consequences of PPCPs in the environment cannot continue to be overlooked, and minimizing their impact in the environment is an enormous task that is not going to be completed overnight. It requires us to consider the entire life-cycle of PPCPs right from concept development through the designing and testing phases prior to approval. This holistic approach has been dubbed the cradle-to-cradle approach (Daughton, 2003a, 2003b) and offers an opportunity to exercise a good level of stewardship as opposed to the conventional cradle-to-grave approach that blindly assumes that the life span of the compounds end after they have been used for their intended purpose. One key component of this approach is making health care deliverers and patients aware of the medical and environmental consequences of our medication practices, including overprescribing. It also aims at

Pharma-Ecology: The Occurrence and Fate of Pharmaceuticals and Personal Care Products in the Environment. By P.K. Jjemba
Copyright © 2008 John Wiley & Sons, Inc.

minimizing pharmaceutical use by creating an awareness about the linkages between human health and ecological health. Some elements of this approach have been adapted to some extent with regard to the overuse of antibiotics because of the less subtle effects of antibiotic overuse, but has to be expanded to a larger audience and certainly to include other PPCPs. This approach should also be extended to veterinarians and farmers with regard to livestock management practices.

Drug development is a long, expensive, and complex process that currently involves collaboration between academics, clinicians, regulators, as well as industry (Fig. 6.1). Noticeably absent from this development process is a consideration that some of the compounds can get into the environment, ultimately impacting the ecosystem. Even where recognition of the potential problems of the presence of PPCPs in the environment has been made, addressing the issue has conventionally been looked at from separate lenses. It is apparent that a better understanding of this problem can be achieved using an integrated approach, and recognizing the fact that human and ecological health are intimately intertwined and deserve to be treated as one and the same (Daughton, 2004; Jjemba and Robertson, 2005). In this regard, ecosystem health is emerging as a discipline that is aimed at integrating our knowledge to assess and monitor ecosystems and related health problems in a more holistic fashion, linking public health and well-being with ecology and environmental degradation. It is important to note from the onset that high-quality health care and environmental

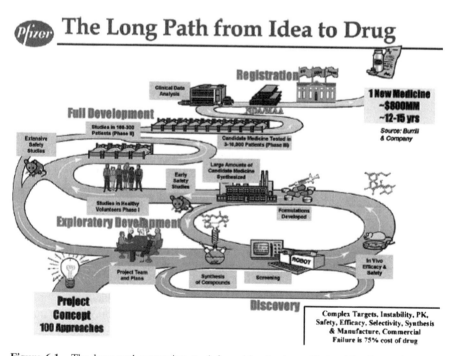

Figure 6.1 The long and expensive road from idea to drug. Noticeably absent from this lengthy path is the consideration that some of the compound will end up in the environment where its fate and impact are unknown. (From Kennedy and Bormann, 2006.)

protection should not have competing goals but rather be intimately linked to complement each other in a sustainable manner. The issue of PPCPs in the environment is more complex than it superficially seems, comprising many broad dimensions that are interrelated, deserving a multidisciplinary approach. That approach should ideally harmonize drug design, limit prescriptions only to cases that absolutely need them, ensure minimal PPCP "loss" that occurs through excretion, maximizing PPCP removal (from sewage/manure), and choose compounds that are more degradable in the environment over those that may have similar efficacy but with signs of higher persistence and/or potency in the environment. Other measures taken should include encouraging prudent use as well as proper disposal of unused PPCPs as to minimize exposure to the ecosystem. These ideas are expanded on below, without any intention to prioritize them, as future research and legislative needs.

6.1 IMPROVING ASSESSMENT OF THE RISKS FROM PPCPs IN THE ENVIRONMENT

There is still no agreeable procedure for assessing the risk of pharmaceuticals used in normal therapy on the environment. Predicated environmental concentration (PEC) values that trigger concern of pharmaceutical compounds in the environment have been established (EMEA, 2001; U.S. FDA, 1998) and were discussed in Chapter 4. The recently revised U.S. FDA guidelines require environmental risk assessment of new pharmaceutical compounds if the predicted dose is to exceed $1 \mu g L^{-1}$ (U.S. FDA, 1998). Despite its intentions, this is a value that was arbitrarily set, most possibly without any input from the U.S. EPA, the custodial agency for issues pertaining to environmental protection. The historical lack of interaction between these two important agencies even in the areas of drug effect monitoring in therapeutic settings has been well documented (Daughton, 2004). The U.S. National Research Council has called for considering nontraditional assessment and toxic end-point methods in determining the risk from these compounds (NRC, 1994). Therefore, future strategies to address this subject should not adapt a reductionistic approach but rather a comprehensive one that does not separate the individual components from one another. In the absence of a clear consensus about assessing the risks from PPCPs in the environment, some jurisdictions have adapted the precautionary principle to address concerns about these anthropogenic micropollutants getting into drinking water and other food products. Implicit in adapting this principle is the fact that some hazards from these compounds have been identified by various groups, with the scientific evaluations yet unable to determine the degree of harm with sufficient certainty.

6.2 EFFECT OF MIXTURES

When assessing the risk associated with mixtures, oftentimes the data for mixtures are not available. Thus, some scholars have assessed the risk from the individual components and used those to estimate risk using the dose-additivity model (for systemic toxicants) or a response-additivity model for carcinogens. Both models assume no interaction among chemicals and always leave us wondering as to what could be the interactive effects between different compounds in the environment. Such effects could be additive, synergistic, or antagonistic, increasing or decreasing the effects of the individual pharmaceutical compounds in the environment. It is also highly likely that the ecotoxicity of PPCPs is quite different from their clinically derived toxicity to humans and other organisms. Standardized toxicity tests may not adequately predict the functional and/or structural responses across various species based on data from only one species.

6.3 EFFECTS OF CHRONIC EXPOSURE TO LOW PPCP DOSES

The seemingly low PPCP concentrations typically encountered in the environments such as receiving waters are constant and could, therefore, have a cumulative effect. Such continuous exposure has implications on the long-term uptake by dwellers in those environments and defies studies that are done based on classical toxicology that are usually based on dose–response (acute exposure). What could be the toxicological effects of such low but sustained doses is still a poorly investigated toxicological question. The implications of this question can be illustrated by results from a preliminary experiment whereby increasing concentrations of three antimicrobial agents, that is, chloroquine, quinacrine, and metronidazole, were tested on soybean plants. While some extremely high doses of the compound were lethal to soybean plants, lower doses led to the development of stunted plants (Jjemba, 2002b), displaying an element of subtle effects. In that particular study, metronidazole also reduced the number of protozoa in the soil by one log unit. In another study, the exposure to concentrations of ethylestradiol as low as $2 \, \text{ng} \, \text{L}^{-1}$ inhibited the growth of testicles in fish, $10 \, \text{ng} \, \text{L}^{-1}$ caused mouthpart deformities in *Chiromonas riparius*, whereas $100 \, \text{ng} \, \text{L}^{-1}$ caused changes in sex ratios from $1:1$ females to males to $2:1$ in *Gammarus pulex* (Pascoe et al., 2003). In the same study, the regeneration of the digestive region in the cnidarian *Hydra vulgaris* to fully functional polyps was inhibited by a sustained exposure for 17 days to low doses $(10 \, \mu\text{g} \, \text{L}^{-1})$ of amlodipine, diazepam, and digoxin (Pascoe et al., 2003). Some of these effects at such low concentrations are transgenerational, an aspect that conventional toxicology does not easily quantify. However, it is worth noting that not all subtle effects are adverse or negative. Thus, the presence of low levels of PPCPs can cause some enhanced growth.

An advisory committee to study the biological effects of low-level exposures has been developed (see http:www.belleonline.com), but its focus is still almost entirely on clinical rather than ecological settings. Ecological risk assessment is less understood compared to human risk assessment, and better methods of quantifying and predicting the biological effects (ecotoxicity), particularly at low doses, are needed. The question of low but sustained doses also has relevance under soil and sediment environments, although even less work has been conducted to study the effects of PPCP contamination in those environments.

6.4 FORMULATION AND REGIMEN DESIGN AFFECTING PPCP BIOAVAILABILITY, EXCRETION, AND SORPTION

Chemists have been at the forefront of what is referred to as Green Chemistry with the aim of minimizing the use of ecologically hazardous materials and coming up with alternative ways of synthesizing ecofriendly compounds. However, that zeal has not yet been fully embraced in pharmaceutical chemistry. It is very desirable to investigate and develop drug formulations and regimens that have a high bioavailabity coupled with reduced proportions of the active ingredient being excreted. In animal systems, bioavailability has been linked to physicochemical characteristics such as the connectivity of the molecule, solubility, the shape and size of the molecule, as well as the electronic nature of the molecule (i.e., dielectric and conformational energy, number of aromatic rings, and electron affinity) (Turner et al., 2004). In the body, the bioavailability of a drug is defined as the rate at which the drug becomes available and the extent to which the administered dose is ultimately absorbed (Turner et al., 2004). It depends on the extent to which the drug dissolves, adsorbs, and/or gets metabolized at first pass in the body, among a variety of other factors. There is evidence that dosing regimen for various drugs need to be revisited as current dosages are often higher than what is medically necessary. Also related to this are the differences in drug absorption at different times of the day (i.e., circadian rhymes) and hunger status (i.e., empty stomach versus well fed individuals).

Under clinical settings, bioavailability changes in a cubic fashion with molecular reflectivity, which is based on molecular mass and density. Comparable or similar relationships need to be investigated in ecological settings. Bioavailability in the body also depends on the extent to which the drug dissolves and/or gets metabolized at first pass. Thus, some efforts have been devoted to controlled release of pharmaceuticals in the body as to increase the half-life by altering the molecular structure or using more stable carriers (Clarke et al., 2003; Kirsten et al., 1998). It is likely that such an approach reduces the percentage of the parent compound that is excreted, and by inference, also minimizes the amounts that are introduced in the environment. However, assessment to verify this contention and its likely implications on the occurrence of different PPCPs in the environment has not yet been conducted. Some of what might make such studies even more complicated is the fact that absorption, distribution, metabolism, and excretion pathways and profiles can greatly differ between organisms and even between compounds that belong to the same class (Daughton, 2004). Another ecofriendly approach could be investing in designing PPCPs that specifically have enhanced biodegradation and photolysis in the environment while maintaining their pharmaceutical activity.

6.5 USE OF QUANTITATIVE STRUCTURAL ACTIVITY RELATIONSHIPS IN ECOTOXICOLOGY

More often than not, the medicinal chemist/drug discovery worker has incomplete information about the structure of the target as the available structure may be from a different species. This is especially true of antimicrobial agents where one is unlikely to have structural information on all possible target species. The choice of multiple physicochemical properties, compared to only one or two properties to characterize a set of molecules provides a better description of the molecule. This argument is at the core of quantitative structural activity relationships (QASR) that have been used somewhat extensively to estimate drug activity and toxicity of chemicals (Auer et al., 1994; Lipnic, 1995). QSAR heavily relies on having a wide range of molecular descriptors available to provide predictive models. Such descriptors need to be properly identified and defined if QSAR is to be used to predict ecotoxicity of PPCPs in the environment as well. Attempts to use QSAR on a range of chemicals under ecological conditions has been published by the U.S. EPA in a computer program called ECOSAR (see http://www.epa.gov/opptintr/newchems/tools/21ecosar.htm). It has been used to estimate the toxicity of chemicals used in industry and discharged into water. It predicts the toxicity of these chemicals to aquatic organisms such as invertebrates, fish, and algae under chronic or acute conditions. Input parameters include the octanol–water partition coefficient (K_{ow}), molecular weight, charge density, and chemical structure (using the simplified molecular input line entry system, i.e., SMILES, notation). While this program shows some promise, it still needs more extensive testing with regard to PPCPs in the environment. From a clinical perspective, So and Karplus (1999) compared seven methods to predict binding energy. Five of the methods were QSAR based and the other two were structure based. The structure-based methods correlated quite poorly with the observed biological activity, while the QSAR-based methods correlated moderately but not strongly. These findings suggest that the methods used to determine/predict binding energy were not completely adequate and/or that binding energy alone may not entirely account for biological activity. By comparison, Aqvist et al. (1994) developed a thermodynamic cycle-based approach called linear interaction energy approach to predict binding energies. That method was not included in the comparisons by So and Karplus (1999), but it fit experimental results quite well. Considering that modification as a way of determining charge density inputs into the ECOSAR may improve on its predictability potential in future PPCP analyses.

6.6 USE OF GENOMICS IN BIOASSAY STUDIES

In the past, the potency of pharmaceuticals was determined from whole animals or whole tissue. Increasingly, the potency is being determined using data from isolated targets in vitro. While this approach is more efficient and economical in screening large quantities of candidate drugs, it has to be relied on with a bit of caution as a single bioassay cannot provide the full picture of the effects of that same compound in an organism in toxicological studies. Rather, several bioassays would be required to get a more comprehensive picture. Besides the more conventional bioassays such as cell-line culture and the use of indicator organisms such as *Vibrio* sp. in the MicroTox assay, gene expression bioassays promise a high sensitivity in determining the presence and impact of PPCPs in the environment. The exposure of organisms to pollutants can affect the expression of certain genes, the differences in expression patterns (versus those organisms that are not exposed) indicating how the pollutant/compound affects the organism. Genomics involves analyzing the changes in the expression of genes and can be applied after exposure to various PPCPs. If applied to toxicity, this aspect of genomics is referred to as toxico-genomics. Gene array technology enables us to study the differential expression of thousands of genes simultaneously. Typically, meaningful studies are conducted by comparing exposed versus nonexposed treatments, enabling the detection of patterns of the expressed genes with changes in mRNA levels (i.e., a snapshot of the genome) or matching the treatments with key genes or clusters of genes with expression patterns (i.e., gene profiling).

Gene arrays can allow us to screen a broad range of compounds, which helps in focusing more in-depth studies on the compounds that prove to be of further interest. In that sense then, contrary to its name, toxicogenomics is actually not analyzing the genome but rather the abundance of the mRNA in the biological sample. Messenger-RNA-based studies can enable us to determine whether exposure (and therefore response to a particular compound) is occurring even if the compound is at low concentrations that may not be detectable by chemical analysis. This approach can also be used to study the protein profiles, enabling long-term monitoring of the effects of exposure. Use of gene expression profiling to determine the effects of PPCPs on different nontarget organisms holds great potential in the ecotoxicity of these compounds. However, that technology is still relatively new, and it is increasingly becoming accepted that a several-fold change in gene expression levels by itself may not always be a result of statistically significant response to a particular stressor (Lucchini et al., 2001).

Microarrays offer a great opportunity in ecotoxicology as a bioassay tool. Sirisattha et al. (2004) used microarray analysis to study the toxicity of anionic detergents, for example, SDS and LAS on *Saccharomyces cerevisiae*. Both types of surfactants are well-characterized chemicals that accumulate in the environment due to their wide usage. *Saccharomyces cerevisiae* cells were grown at different concentrations of LAS and SDS, and the growth monitored using absorbance at 650 nm. Both LAS and SDS primarily exert toxicity by solubilizing proteins from membranes in living cells, although some oxidative stress in yeast cells by both detergents was reported (Sirisattha et al., 2004). Oxidative stress was expressed by indication of genes related to thioredoxin, glutathione, and glutaredoxin. Oxidative stress is known to damage DNA. Growth of *S. cerevisiae* was increasingly diminished with increasing concentrations of each compound, and, on extracting the RNA, its cDNA (complementary DNA) was subsequently labeled with Cy5. RNA from the control treatment, that is, without any LAS or SDS was labeled with Cy3. The two labeled cDNA pools were mixed and hybridized with a yeast DNA chip for 24–36 h at 65°C on an arrayer. The resulting images were quantified and the fluorescent intensity of each spot subtracted from the surrounding background to compute a Cy5/Cy3 intensity ratio. Intensity ratios >2.0 were considered up-regulated open reading frames (ORFs). Using this approach 104 ORFs (1.8%) were up-regulated with SDS treatment. Similarly, 1.6 and 2% ORFs were underregulated with LAS

TABLE 6.1   Summary of Total Number of Induced Genes in *Sacchromyces cerevisiae* with Two Detergents Categorized by Function

Function	With LAS		With SDS	
	Overregulated	Underregulated	Overregulated	Underregulated
Unclassified proteins	50	34	122	49
Cellular organization	25	33	93	39
Metabolism	26	12	76	14
Protein destination	9	3	31	2
Energy	3	2	29	2
Cell rescue, cell death, and aging	8	7	27	6
Cell growth, cell division, and DNA synthesis	7	9	25	11
Cellular transport and general transport mechanisms	5	4	18	5
Transport facilitation	6	4	16	7
Transcription	5	18	15	30
Still questionable classification	1	4	8	4
Signal transduction and communication	2	4	7	0
Ionic homeostasis	3	3	6	2
Protein synthesis	2	13	5	5
Cellular biogenesis	2	4	4	4
Percentage of total	1.8%	1.6%	5%	2%

and SDS, respectively (Table 6.1). It is important to clarify that merely finding an increase or decrease in gene expression should not necessarily be a basis for alarm but rather a cause for developing hypotheses, justifying further investigation.

6.7   SOCIAL RESPONSIBILITY, LEGISLATION, AND POLICY

While legislation can alleviate some of the problems associated with polluting the environment with PPCPs, it can only achieve limited success by itself. More inroads at reducing unnecessary usage and promoting efficient usage as well as developing better disposal programs can be attained through initiatives to educate the various professions involved in purveying PPCPs, as well as the general public (i.e., consumers). PPCPs need to be treated in a manner that is comparable to pesticides and biocides as they are pharmacologically active. Creating awareness about the dangers associated with herbicides and pesticides has had substantial success using programs that enlighten the general public about the dangers associated with those compounds. As indicated when introducing this chapter, health cannot be isolated from all the other aspects of sustainable living that environmentally conscious individuals are quite familiar with and to which they are already responsive. The general public is quite receptive to the broad theme of sustainability, especially if it aims at meeting society's needs in ways that do not adversely reduce the well-being of present and future generations.

In a recent survey, Seehusen and Edwards (2006) found that only 20% of patients at a pharmacy in the Pacific Northwest have ever been given advice by a medical provider on how to safely dispose of medication. The relatively few respondents who found it unacceptable to rinse medications down the sink also associated this unacceptable behavior with previous advice on the disposal of medication. Most of these disposal strategies can be achieved through voluntary programs once the awareness is created, with the notion of developing a more sustainable culture of using PPCPs. Educating patients to create the awareness that it is perfectly acceptable to return unused medications to the pharmacy or health care provider was identified by Seehusen and Edwards (2006) as a first step in getting patients to actually bring back the unwanted medications for proper disposal. To help in this cause, there is a need to include disposal guidelines in medical packages just like is currently the case with pesticides and other industrial chemicals (Okeke, 2002). Labeling proposed under the EU guidelines (EMEA, 2003) runs as follows: "*Medicines that are no longer required should not be disposed of in the wastewater or municipal drainage system but rather returned to a pharmacy or ask a pharmacist how to dispose of them in accordance with existing regulations.*" However, in reality most pharmacists are still uninformed about ecotoxicity issues and are not likely to have ready answers about ecologically acceptable ways of disposing PPCPs let alone how problematic these compounds can be once they get into the environment. Thus, issues pertaining to the occurrence of PPCPs in the environment have to be introduced into the curriculum of medical professionals, including nurses, public health workers, physicians, and pharmacists. An organization called H2E was initiated with the core mission of educating, motivating, and engaging health care professionals to adapt environmental practices that best increase operational efficiency in an environmentally sustainable system while improving the health of patients and the community (H2E, 2007). The mission of this and other similar organizations should be expanded to, besides emphasizing the reduction of waste in their practice, also specifically focus on practices that minimize the discharge of PPCPs into the environment.

The strategy of disposing pharmaceutical compounds through pharmacies that was introduced above should be actively supported by putting in place a viable recycling or disposal program at various pharmacies. Its success may involve changing some of the laws in some jurisdictions. For example, most states within the United States actually have laws in place that prohibit pharmacies from accepting returned medication from the public. Furthermore, federal laws prohibit the transfer of controlled substances under which prescription medications are included. Viability of this disposal service should ensure that there is no chance of other patients being reissued these otherwise expired or tainted drugs at the front end of the pharmacy. It should also ensure that the probability of selling the returned drugs through some other unscrupulous means is diminished. However, it is important to note that the disposal of pharmaceuticals in household waste and down the drain possibly contributes comparatively small amounts compared to those introduced through excretion. Thus, even more effort should be devoted to that mode of entry of these compounds into the environment.

6.8   NEED TO REVISIT THE DRUG APPROVAL AND ADVERTISING PROCESS

In 1962, the U.S. Senate passed a bill that mandated the pharmaceutical companies to ensure that the drugs they make have to be proven effective and that they do actually do work prior to approval by the U.S. FDA. During subsequent decades, the pharmaceutical industry made a very strategic decision to build strong relations with patient groups early in the process, prior to them applying for approval of candidate medications. That practice has proved quite helpful and politically influential in navigating the process of gaining approval of their products by the FDA. Increasing relationships with their clientele, which in essence includes the general public, has also included practices that directly market pharmaceutical products to the general public. That practice has, in part led to a steady increase in the sale of pharmaceutical products in general (Fig. 6.2). Coupled with this, the industry has also played its cards correctly and in some instances promoted some pharmaceutical products even prior to their approval, through preplacement branding, using those patient advocate groups. Preplacement branding has also been perpetuated by the industry "seeking" thought leaders who freely express their "opinion" in talks and peer review journals about the coming drug. As it happens, a number of these thought leaders are consultants that actively recruit/send patients into trials to these same companies (Critser, 2005). More recently, the Prescription Drug User Fee Act (PDUFA) of 1992 compelled the FDA to review new drug applications (NDA) over a short period of

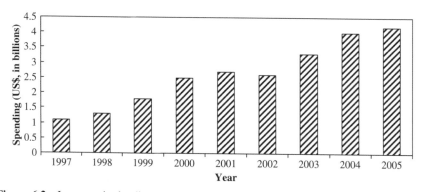

Figure 6.2   Increases in the direct-to-consumer spending on drugs within the United States. (Data from Government Accountability Office Report No. GAO-07-54 of 14th December, 2006.)

time. All of this mix of events has increasingly influenced the approval process, trumping the science behind the products in some instances. For example, some of the thought leaders are not entirely free of any conflict of interest from the pipeline drug about which they are expressing an opinion. Some of these and other calculated practices have led to the affirmation of pipeline drugs that may not always be a large improvement over the existing competitor (Avorn, 2004; Morgan and MacGibbon, 2007), leading to a cascade of compounds that may be even more persistent in the environment. Under the current approval requirements, the drug only needs to perform better than a placebo rather than show superiority to existing treatment alternatives. There is a need to revisit the approval process to, at the bear minimum, the extent that if a particular PPCP or class thereof exhibits potential risks to the environment, ways to address appropriate precautions and safety measures in terms of use and disposal should be proposed.

The message from health care professionals greatly influences health care outcomes. However, it is striking that the U.S. FDA does not preapprove advertisements that are presented by the drug companies. Rather, it is mandated to monitor their content when they are already in print or airing. This arrangement in some instances allows for misinformation and/or the transmission of partially true statements to the general public through sound bites and abbreviated messages. People feel more vulnerable when they are sick, and, when they get hold of such messages, they are likely to go along with their content to the extent that by the time the incorrect or questionable advertisement content is pulled, the misinformation would have been cultivated. There is no question that some subjects' opinions are shaped by what they have read or heard in these advertisements. Moyniham et al. (2002) refer to this practice as "disease mongering." This practice has surprisingly attracted little critical scrutiny, although its impact on medical practice, human health public consciousness, and national budgets has been substantial. Thus, cases of self-described ailments have steadily increased since the direct-to-consumer marketing concept in drug delivery was adapted. After watching the barrage of commercials about previously unknown and unrecognized conditions such as irritable bowel syndrome (IBS), GERD, restless leg syndrome, social phobia (shyness), and the like, individuals become convinced that they are sick and, as often suggested at the end of the commercial, are more than compelled to "ask their doctors about that pill." It is apparent that for some of these conditions, medication is not always the most logical answer. Related to this issue are the prevalent overmedication and polypharmacy practices, particularly in developed countries. The former refers to the unnecessary use of medication, whereas the latter is the simultaneous use of several medications to treat one or more conditions (Critser, 2005).

In the same regard, PPCP introductions through manure and biosolids that are applied to arable land is enhanced by the typically high application rates of these forms of crop fertilizers; rates that are usually higher than the crop needs (Kumar et al., 2005). It is recommended that manure and biosolids application rates be based on the nutrient requirements of the intended crop so as to minimize the unintended introduction of these micropollutants into the food chain at levels that can be avoided.

6.9   USE OF PRESCRIPTION RECORDS FOR MAPPING PPCPs

Prescription records are quite reliable and should, in an effort to link the environmental and medical community, be used to build a database for developing pharmaceutical use and consumption maps on a regular basis. As we saw in Chapter 2, the range of PPCPs is quite enormous and they cannot be identified by a single method. The availability of such information to water and wastewater utilities should enable them to focus on more product-specific target compounds in monitoring programs. Such an exercise should aim at maintaining the confidentiality of the individual consumers but provide enough information (e.g., Zip code and type of prescription as well as its quantity) to the utilities. Such a program could be modeled along databases such as The Prescriber Profiler (http://www.imshealth.com/ims/portal/front/indexC/0,2773,6599_1825_0,00) where one feeds in the physician's name and several parameters such as drug name, Zip code, sex (male/female), number of refills, disease category, and the like. Some maps have been developed to determine the increasing expense of medication in various localities, with emphasis on monetary aspects (Morgan et al., 2005). Using the same pharmacy records and survey approach, actual volumes of different pharmaceutical compounds consumed in specific municipalities, for example, can be determined and the sewage treatment process modified periodically as such information dictates.

In conclusion, while this is not an exhaustive list, it can serve as a stepping-stone for better multidisciplinary collaboration to combat the problem of PPCPs in the environment without compromising public health. The fact that PPCPs have been used for this long should not be a basis for complacency about their potential impact on the ecosystem and the well-being of organisms in the environment.

REFERENCES

Abraham M.H. and J. Lee (1999). The correlation and prediction of the solubility of compounds in water using an amemded salvation energy relationship. *Journal of Pharmaceutical Science* 88:868–880.

Acar J., M. Casewell, J. Freeman, H. Goossens, and C. Friis (2000). Avoparcin and virginia-mycin as animal growth promoters: A plea for science in decision making. *Clinical Microbiology and Infections* 6:477–482.

Adachi A., M. Kamide, R. Kawafume, N. Miki, and T. Kobuyashi (1990). Removal-efficiency of anionic and nonionic surfactants from chemical wastewater by a treatment plant using activated carbon adsorption and coagulation precipitation processes. *Environmental Technology* 11:133–141.

Adams C., Y. Wang, K. Loftin, and M. Meyer (2002). Removal of antibiotics from surface and distilled water in conventional water treatment processes. *Journal of Environmental Engineering* 128:253–260.

Adham, S., P. Gagliardo, L. Boulos, J. Oppenheimer, and R. Trussell (2001). Feasibility of the Membrane Bioreactor Process for Water Reclamation. *Water Science and Technology* 43:203–209.

Aerestrup F.M., Y. Agerso, P. Gerner-Smidt, M. Madsen, and L.B. Jensen (2000). Comparison of antimicrobial resistance phenotypes and resistance genes in *Enterococcus faecalis* and *Enterococci faecium* from humans in the community, broilers, and pigs in Denmark. *Diagnostic Microbiology and Infectious Disease* 37:127–137.

Aga D.S., R. Goldfish, and P. Kulshrestha (2003). Application of ELISA in determining the fate of tetracyclines in land-applied livestock wastes. *Analyst* 128:658–662.

Aga D.S. and E.M. Thurman (1997). Environmental immunoassays: Alternative techniques for soil and water analysis. *American Chemical Society Symposium Series* 657:1–20.

Agins A.P., L.R. Willis, and K. Gutierrez (2003). Drugs to Treat Asthma and Other Pulmonary Diseases. In: K. Gutierrez and S.F. Queener (Eds.). *Pharmacology for Nursing Practice*. Mosby, St. Louis, pp. 843–866.

Aherne G. and R. Briggs (1989). The relevance of the presence of certain synthetic steroids in the aquatic environment. *Journal of Pharmacy and Pharmacology* 41:735–736.

AHI (2002). Animal Health Institute. Available at: www.ahi.org.

AHR (2005). 2005 United States Animal Health Report. Available at: http://www.aphis. usda.gov/publications/animal_health/content/printable_version/AHR_2005B_508.pdf; accessed 2-14-2007.

Ahlborg U.G., L. Lipworth, L. Titus-Ernustoff, C-C. Hsieh, A. Hanberg, J. Baron, D. Trichopouls, and H.O. Adanie (1995). Organochlorine compounds in relation to breast cancer, endometrial cancer, and endometriosis: An assessment of the biological and epidemiological evidence. *Critical Reviews in Toxicology* 25:463–531.

Ahn K-H., K-G. Song, E. Cho, J. Cho, H. Yun, S. Lee, and J. Kim (2003). Enhanced biological phosphporus and nitrogen removal using a sequencing anoxic/anaerobic membrane bio-reactor (SAM) process. *Desalination* 157:345–352.

Al-Ahmad A., M. Wiedmann-Al-Ahmad, G. Schön, F.D. Daschner, and K. Kümmerer (2000). The role of Acinetobacter for biodegradability of quartenary ammonium compounds. *Bulletin of Environmental Contamination and Toxicology* 58:704–711.

Alder P., T. Steger-Hartmann, and W. Kalbfus (2001). Disinfection of natural and synthetic estrogen estrogenic steroid hormones in water samples from Southern and Middle Germany. *Acta Hydrochimica et Hydrobiologica* 29:227–241.

Alexander F.E., S.L. Patheira, A. Biondi, S. Brandalise, M.E. Chan, et al. (2001). Transplacental chemical exposure and risk of infant leukemia with MLL gene fusion. *Cancer Research* 61:2542–2546.

Alexander M. (1994). *Biodegradation and Bioremediation*. Academic, New York.

Alexander M. (2000). Aging, bioavailability, and overestimation of risk from environmental pollutants. *Environmental Science and Technology* 34:4259–4265.

Alimelkkila T., J. Kanto, and E. Iislo (1993). Pharmacokinetics and related pharmacodynamics of anticholinergic drugs. *Acta Anaethesiologica Scandinavica* 37:633–642.

Allred B. and G.O. Brown (1996). Boundary condition and soil attribute impacts on anionic surfactant mobility in unsaturated soil. *Groundwater* 34:964–972.

Altamura A.C., A.R. Moro, and M. Percudani (1994). Clinical pharmacokinetics of fluoxentine. *Clinical Pharmacokinetics* 26:201–214.

Altenburger R., T. Backhaus, W. Boedeker, M. Faust, M. Scholze, and L. H. Grimme (2000). Predictability of the toxicity of multiple chemical mixtures to *Vibrio fischeri*: Mixtures composed of similarly acting chemicals. *Environmental Toxicology and Chemistry* 19:2341–2347.

Ambrósio A.F., P. Soares-da-Silva, C.M. Carvalho, and A.P. Carvalho (2002). Mechanisms of action of carbamazepine and its derivatives, oxcarbamazepine, BIA2-093, and BIA2-024. *Neurochemical Research* 27:121–130.

Ames P.L. (1966). DDT residues in the eggs of the osprey in the northern United States and their relation to nesting success. *Journal of Applied Ecology* (Suppl.) 3:87–97.

Andersen H., H. Siegrist, B. Halling-Søresen, and T.A. Ternes (2003). Fate of estrogens in a municipal sewage treatment plant. *Environmental Science and Technology* 37:4021–4026.

Andersson M.I. and A.P. MacGowan (2003). Development of the quinolones. *Journal of Antimicrobial Chemotherapy* 51(Suppl. S1):1–11.

Andreozzi R., R. Marotta, G. Pinto, and A. Pollio (2002). Carbamazepine in water: Persistence, in the environment, on ozonation treatment and preliminary assessment on algal toxicity. *Water Research* 36:2869–2877.

Andreozzi R., M. Raffaele, and P. Nicklas (2003). Pharmaceuticals in STP effluents and their solar photodegradation in aquatic environment. *Chemosphere* 50:1319–1330.

Ankley G.T., K.M. Jensen, M.D. Kahl, J.J. Korte, and E.A. Makynen (2001). Description and evaluation of a short-term reproduction test with Fathead minnow (*Pinnephales promelas*). *Environmental Toxicology and Chemistry* 20:1276–1290.

Anonymous (2006). Causes of Type-1 diabetes. Available at: http://www.news-medical.net/?id=15965; accessed 10/01/2007.

Applebaum P.C. and P.A. Hunter (2000). The flouroquinolone antibacterial: Past, present and future perspectives. *International Journal of Antimicrobial Agents* 16:5–15.

Aqvist J., C. Medina, and J-E. Samuelsson (1994). A new method for binding affinity in computer-aided drug design. *Protein Engineering* 7:385–391.

Asbolt N.J., G.S. Grohmann, and C.S.W. Kueh (1993). Significance of polluted receiving waters in Sydney, Australia. *Water Science and Technology* 27:449–452.

Atkinson S., M.J. Atkinson, and A.M. Tarrant (2003). Estrogens from sewage in coastal marine environments. *Environmental Health Perspectives* 111:531–535.

Auer C.M., M. Zeeman, J.V. Nabholz, and R.G. Clements (1994). SAR. The US regulatory perspective. *SAR and QSAR in Environmental Research* 2:21–38.

Avorn J. (2004). *Powerful Medicines: The Benefits, Risks, and Costs of Prescription Drugs*. Knopf, New York.

AWWA (1995). *Water Sources: Principles and Practices of Water Supply Operations*. 2nd ed. American Water Works Association, Denver.

Azmitia E.C. (1999). Serotonin neurons, neuroplasticity and homeostasis of neural tissue. *Neuropsychopharmacology* 21:33S–45S.

Bakal R.S. and M.K. Stoskopf (2001). In vitro studies of the fate of sulfadimethoxine and ormetoprim in the aquatic environment. *Aquaculture* 195:95–102.

Baldi I., L. Filleul, B. Mohammed-Brahim, C. Fadrigoule, J-F. Dartigues, S. Schwall, J-P. Drevet, R. Salamon, and P. Brochard (2001). Neuropsychologic effects of longterm exposure to pesticides: Results from the French phytoner study. *Environmental Health Perspectives* 109:839–844.

Balk F. and R.A. Ford (1999). Environmental risk assessment for the polycyclic musks AHTN and HHCB in the EU I. Fate and exposure assessment. *Toxicology Letters* 111:57–79.

Ball P. (2000). Quinolone-induced QT interval prolongation: A not-so unexpected class effect. *Journal of Antimicrobial Chemotherapy* 45:557–559.

Barlaz M.A. (1996). The Microbiology of Municipal Solid Waste Landfills. In: A.C. Palminsano and M.A. Barlaz (Eds.). *Solid Waste Microbiology*. CRC Press, Boca Raton, FL, p. 31–70.

Barros M.P., M. Granbom, P. Colepicolo, and M. Pedersén (2003). Temporal mismatch between induction of superoxide dismutase and ascorbate peroxidase correlates with high

H_2O_2 concentration in seawater from clofibrate-treated red algae *Kappaphycus alvarezii*. *Archives of Biochemistry and Biophysics* 420:161–168.

Bedworth W.W. and D.L. Sedlak (1999). Sources and environmental fate of strongly complexed nickel in estruarine waters: The role of ethylenediaminetetraacetate. *Environmental Science and Technology* 33:926–931.

Bennie D.T. (1999). Review of the environmental occurrence of alkylphenols and alkylphenol ethoxylates. *Water Quality Research Journal of Canada* 34:79–122.

Berset J.D., T. Kupper, R. Etter, and J. Tarradellas (2004). Considerations about the enantioselective transformation of polycyclic musks in wastewater and sewage sludge and analysis of their fate in a sequencing batch reactor plant. *Chemosphere* 57:987–996.

Bertschy G., C. Bryois, G. Bondolfi, A. Velardi, P. Budry, D. Dascal, C. Martinet, D. Baettig, and P. Baumann (1997). The association of carbamazepine-mianserin? In opiate withdrawal: A double blind pilot study versus clonidine. *Pharmacological Research* 35:451–456.

Bewick M.W. (1978). Effect of tylosin and tylosin-fermentation waste on microbial activity of soil. *Soil Biology and Biochemistry* 10:403–407.

Bewick M.W. (1979). The adsorption and release of tylosin by clays and soils. *Plant and* 51:363–370.

Bindschedler M., P. Degen, G. Flesch, M. de Gasparo, and G. Preiswerk (1997). Pharmacokinetics and pharmacodynamic interaction of single oral doses of valsartan and furosemide. *European Journal of Clinical Pharmacology* 52:371–378.

Bingley P.J. and E.A.M. Gale (1989). Rising incidence of IDDM in Europe. *Diabetes Care* 12:289–295.

Birnbaum L.S. and S.E. Fenton (2003). Cancer and developmental exposure to endocrine disruptors. *Environmental Health Perspectives* 111:389–394.

Blacklock C.J., J.R. Lawrence, D. Wiles, E.A. Malcolm, I.H. Gibson, C.J. Kelly, and J.R. Paterson (2001). Salicylic acid in the serum of subjects not taking aspirin. Comparison of salicylic acid concentrations in the serum of vegetarians, non-vegetarians, and patients taking low dose aspirin. *Journal of Clinical Pathology* 54:553–555.

Blau K. and J.M. Halket (1993). *Handbook of Derivatives for Chromatography.* Wiley, New York.

Bodzek, M., Z. Debkowska, E. Lobos, and K. Konieczny (1996). Biomembrane wastewater treatment by activated sludge method. *Desalination* 107:83–95.

Bögi C., G. Levy, I. Lutz, and W. Kloas (2002). Functional genomics and sexual differentiation in amphibians. *Comparative Biochemistry and Physics B* 133:559–570.

Bond R.G. and F.A. Digiano (2004). Evaluating GAC performance using the ICR database. *Journal of the American Water Works Association* 96(6):96–104.

Borgen K., G.S. Simonsen, A. Sundsfjord, Y. Wasteson, Ø. Olsvik, and H. Kruse (2000). Continuing high prevalence of VanA-type vancomycin-resistant enterococci on Norwegian poultry farms three years after avoparcin was banned. *Journal of Applied Microbiology* 89:478–485.

Bound J.P. and N. Voulvoulis (2005). Household disposal of pharmaceuticals as a pathway for aquatic contamination in the United Kingdom. *Environmental Health Perspectives* 113:1705–1711.

Boxall A.B.A., P. Blackwell, R. Cavallo, P. Kay, and J. Tolls (2005). The Sorption and Transport of a Sulphonamide Antibiotic in Soil Systems. In: D.R. Dietrich, S.F. Webb, and T. Petry (Eds.). *Hot Spot Pollutants: Pharmaceuticals in the Environment.* Elsevier, New York, pp. 37–49.

Boxall A.B.A., P. Johnson, E.J. Smith, C.J. Sinclair, E. Stutt, and L.S. Levy (2006). Uptake of veterinary medicines from soils into plants. *Journal of Agricultural and Food Chemistry* 54:2288–2297.

Boxall A.B.A., D.W. Kolpin, B. Halling-Sørensen, and J. Tolls (2003). Are veterinary medicines causing environmental risks? *Environmental Science Technology* 37:286A–294A.

Boyd G.R. and D.A. Grimm (2001). Occurrence of pharmaceutical contaminants and screening of treatrment alternatives for southeastern Louisiana. *Annals of the New York Academy of Sciences* 948:80–89.

Boyd G.R., H. Reemtsma, D.A. Grime, and S. Mitra (2003). Pharmaceuticals and personal care products (PPCPs) in surface and treated waters of Louisina, USA and Ontario, Canada. *Science of the Total Environment* 311:135–149.

Boyd G.R., S.Y. Zhang, and D.A. Grimm (2005). Naproxen removal from water by chlorination and biofilm processes. *Water Research* 39:668–676.

Boyle M. and E.A. Paul (1989). Carbon and nitrogen mineralization kinetics in soil previously amemded with sewage sludge. *Soil Science Society of America Journal* 53:99–103.

Bozdogan B., R. Leclerq, A. Lozniewski, and W. Weber (1999). Plasmid-mediated coresistance to streptogramins and vancomycin in *Enterococcus faecium* HM1032. *Antimicrobial Agents and Chemotherapy* 43:2097–2098.

Brady J.F., J. Turner, and D.H. Skinner (1995). Application of a triasulfuron enzyme immunoassay to the analysis of incurred residues in soil and water samples. *Journal of Agricultural and Food Chemistry* 43:2542–2547.

Breda M., M.S. Benedetti, M. Bani, C. Pellizzoni, I. Poggesi, G. Brianceschi, M. Rocchetti, D. Sassela, and R. Rimoldi (1999). Effect of rifabutin on ethambutol pharmacokineticsa in healthy volunteers. *Pharmacology Research* 40:351–356.

Brighty K.E. and T.D. Gootz (1997). The chemistry and biological profile of trovafloxacin. *Journal of Antimicrobial Chemotherapy* 39(Suppl. B):1–14.

Brooks B.W., S.M. Richards, J.J. Weston, P.K. Turner, J.K. Stanley, T.W. La Point, R. Brain, E.A. Glidewell, A.R.D. Massengale, W. Smith, C.L. Blank, K.R. Solomon, M. Slattery, and C.M. Foran (2005). Aquatic Ecotoxicology of Fluoxentine: A Review of Recent Research. In: D.R. Dietrich, S.F. Webb, and T. Petry (Eds.). *Hot Spot Pollutants: Pharmaceuticals in the Environment.* Elsevier, New York, pp. 165–187.

Brooks B.W., P.K. Turner, J.K. Stanley, J.J. Weston, E.A. Glidewell, C.M. Foran, M. Slattery, T.W. La Point, and D.B. Huggett (2003). Waterborne and sediment toxicity of fluoxetine to select organisms. *Chemosphere* 52:135–142.

Brown K.D. (2004). Pharmaceutically active compounds in residential and hospital effluent, municipal wastewater, and the Rio Grande in Albuquerque, New Mexico. Water Resource Program, University of New Mexico, Albuquerque, NM. Publication No. WRP-9.

Brunström B., J. Axelsson, and K. Halldin (2003). Effects of endocrine modulators on sex differentiation in birds. *Ecotoxicology* 12:287–295.

Bull D.L., G.W. Ivie, J.G. MacConnell, V.F. Gruber, C.C. Ku, B.H. Arison, J.M. Stevenson, and W.J.A. VandenHeuvel (1984). Fate of avermectin B1a in soil and plants. *Journal of Agricultural and Food Chemistry* 32:94–102.

Bumpus J.A., M. Tien, D. Wright, and S.D. Aust (1985). Oxidation of persistent environmental pollutants by a white rot fungus. *Science* 228:1434–1436.

Burhenne J., M. Ludwig, and Spiteller M. (1997). Photolytic degradation of fluoroquinolone carboxylic acids in aqueous solution. Part II: Isolation and structural elucidation of polar photometabolites. *Environmental Science and Pollution Research* 4:61–67.

Burhenne J., M. Ludwig, and M. Spiteller (1999). Polar photodegradation products of quinolones determined by HPLC/MS/MS. *Chemosphere* 38:1279–1286.

Burnette T.C. and P. Demiranda (1994). Metabolic disposition of the acyclovir prodrug valaciclovir in the rat. *Drug Metabolism and Disposition* 22:60–64.

Buser H-R., M.D. Müller, and N. Theobald (1998a). Occurrence of the pharmaceutical drug clofibric acid and the herbicide mecoprop in various Swiss lakes and in the North Sea. *Environmental Science and Technology* 32:188–192.

Buser H.-R., T. Poiger, and M.D. Müller (1998b). Occurrence and fate of the pharmaceutical drug diclofenac in surface waters: Rapid photodegradation in a lake. *Environmental Science and Technology* 32:3449–3456.

Buser H.-R., T. Poiger, and M.D. Müller (1999). Occurrence and environmental behavior of the chiral pharmaceutical drug ibuprofen in surface waters and in wastewater. *Environmental Science and Technology* 33:2529–2535.

Butcher S.S., R.J. Charlson, G.H. Orians, and G.V. Wolfe (Eds.) (1992). *Introduction. Global Biogeochemical Cycles.* Academic, New York, pp. 1–7.

Cadby P.H., W.R. Troy, and M.G.H. Vey (2002). Consumer exposure to fragrance ingredients: Providing estimates for safety evaluation. *Regulatory Toxicology and Pharmacology* 36:246–252.

Calamari D., E. Zuccato, S. Castiglioni, R. Bagnati, and R. Fanelli (2003). Strategic survey of therapeutic drugs in the Rivers Po and Lambro in Northern Italy. *Environmental Science and Technology* 37:1241–1248.

Campagnolo E.R., K.R. Johnson, A. Karpati, C.S. Rubin, D.W. Kolpin, M.T. Meyer, J.E. Esteban, R.W. Currier, K. Smith, K.M. Thug, and M. McGeehin (2002). Antimicrobial residues in animal waste and water resources proximal to large-scale swine and poultry feeding operations. *Science of the Total Environment* 299:89–95.

Campbell J. (2007). Drugs on tap: Pharmaceuticals in our drinking water. *Pharmacy Practice.* Available at: http://www.cielap.org/pdf/news.DrugsOnTap.pdf; accessed 5/8/2007.

Campoli-Richards D.M., J.P. Monk, and A. Price (1988). Ciprofloxacin. A review of its antibacterial activity, pharmacokinetic properties and therapeutic use. *Drugs* 35:373–447.

Caraffini, Assalve, Stingeni, and Lisi (1994). Tylosin, an airborne contact allergen in veterinarians. *Contact Dermatitis* 5:327–328.

Carlsen E., A. Ginerchman, N. Keidling, and N.E. Skakkebaek (1995). Declining sperm quality and increasing incidence of testicular cancers: Is there a common cause? *Environmental Health Perspectives* 130:137–139.

Carson R. (1962). *Silent Spring.* Houghton Mufflin, Boston.

Casanova L.M., C.P. Gerba, and M. Karpiscak (2001). Chemical and microbial characteristics of household greywater. *Journal of Environmental Science and Health Part A* 36:395–401.

Casey F.X.M., G.L. Larsen, H. Hakk, and J. Simunek (2003). Fate and transport of 17β-estradiol in soil-water system. *Environmental Science and Technology* 37:2400–2409.

Chambers H.P. (1997). Methicillin resistance in staphylococci: Molecular and biochemical basis and clinical implications. *Clinical Microbiology Reviews* 10:781–791.

Chander K. and P.C. Brookes (1993). Residual effects of zinc, copper and nickel in sewage sludge on microbial biomass in a sandy loam. *Soil Biology and Biochemistry* 25:1231–1239.

Chander Y., K. Kumar, S.M. Goyal, and S. Gupta (2005). Antibacterial activity of soil-bound antibiotics. *Journal of Environmental Quality* 34:1952–1957.

Chee-Sanford J.C., R.I. Aminov, I.J. Krapac, N. Garrigues-Jeanjean, and R.I. Mackie (2001). Occurrence and diversity of tetracycline resistance genes in lagoons and groundwater underlying two swine production facilities. *Applied and Environmental Microbiology* 67:1494–1502.

Chen H.J., K.J. Bloch, and J.A. Maclean (2000). Acute eosinophilic hepatitis from trovafloxacin. *New England Journal of Medicine* 342:359–360.

Chen Y., J.P.N. Rosazza, C.P. Reese, H-Y. Chang, M.A. Nowakowski, and J.P. Kiplinger (1997). Microbial models of soil metabolism: Biotransformations of danofloxacin. *Journal of Industrial Microbiology and Biotechnology* 19:378–384.

Chiemchaisri, C., K. Yamamoto, and S. Vigneswaran (1993). Household membrane bioreactor in domestic wastewater treatment. *Water Science and Technology* 27:171–178.

Choquet-Kastylevsky G., T. Vial, and J. Descotes (2002). Allergic adverse reactions to sulfonamides. *Current Allergy and Asthma Report* 2:16–25.

Christensen F.M. (1998). Pharmaceuticals in the environment—a human risk? *Regulatory Toxicology and Pharmacology* 28:212–221.

Christian T., R.J. Schneider, H.A. Färber, D. Skutlarek, M. Meyer, and H.E. Goldbach (2003). Determination of antibiotic residues in manure, soil, and surface waters. *Acta Hydrochimica et Hydrobiologica* 31:36–44.

Christofi N., E. Aspichueta, D. Dalzell, A. De la Sota, J. Etxebarria, T. Fernandes, M. Gutierrez, J. Morton, U. Obst, and P. Schmellenkamp (2003). Congruence in the performance of model nitrifying activated sludge plants located in Germany, Scotland and Spain. *Water Research* 37:177–187.

Chu D.T.W. (1999). Recent developments in macrolides and ketolides. *Current Opinion in Microbiology* 2:467–474.

Ciani A. (2003). Photolysis processes in soils and other porous media. Ph.D. dissertation, Swiss Federal Institute of Technology, Zürich, Switzerland. Available at: http://e-collection.ethbib.ethz.ch/ecol-pool/diss/fulltext/eth15306.pdf.

Cicek, N., H. Winnen, M.T. Suidan, B.E. Wrenn, V. Urbain, and J. Manem (1998). Effectiveness of the membrane bioreactor in the degradation of high molecular weight compounds. *Water Research* 32:1553–1563.

Clark D.E. (1999a). Rapid calculation of polar molecular surface area and its application to the prediction of transport phenomena. 1. Prediction of intestinal absorption. *Journal of Pharmaceutical Sciences* 88:807–814.

Clark D.E. (1999b). Rapid calculation of polar molecular surface area and its application to the prediction of transport phenomena. 2. Prediction of blood-brain barrier penetration. *Journal of Pharmaceutical Sciences* 88:815–821.

Claycamp H.G. and B.H. Hooberman (2004). Risk Assessment of Streptogramin Resistance in *Enterococcus faecium* Attributable to the Use of Streptogramins in Animals. Virginiamycin Risk Assessement. Draft Report. Available at: http://www.fda.gov/cvm/Documents/SREF_RA_FinalDraft.pdf; accessed 2-15-2007.

Cleuvers M. (2003). Aquatic ecotoxicity of pharmaceuticals including the assessment of combination effects. *Toxicology Letters* 142:185–194.

Cleuvers M. (2004). Mixture toxicity of the anti-inflammatory drugs diclofenac, ibuprofen, naproxen, and acetylsalicylic acid. *Ecotoxicology and Environmental Safety* 59:309–315.

Cleuvers M. (2005). Aquatic Ecotoxicity of Pharmaceuticals Including the Assessment of Combination Effects. In: Dietrich D.R., S.F. Webb, and T. Petry (Eds.). *Hot Spot Pollutants: Pharmaceuticals in the Environment*. Elsevier, New York, pp. 189–202.

Colborn T., F.S. Vom Saal, and A.M. Soto (1993). Developmental effects of endocrine disrupting chemicals in wildlife and humans. *Environmental Health Perspectives* 101:378–384.

Collier C.A. (2007). Pharmaceutical contaminants in potable water: Potential concern for pregnant women and children. *EcoHealth* 4:164–171.

Colucci M.S., H. Bort, and E. Topp (2001). Persistence of estrogenic hormones in agricultural soils: I. 17β-estradiol and estrone. *Journal of Environmental Quality* 30:2070–2076.

Colucci M.S. and E. Topp (2001). Persistence of estrogenic hormones in agricultural soils: II. 17β-estradiol. *Journal of Environmental Quality* 30:2077–2080.

Comerton A.M., R.C. Andrews, and D.M. Bagley (2005). Evaluation of an MBR-RO system to produce high quality reuse water: Microbial control, DBP formation and nitrate. *Water Research* 39:3982–3990.

COMPAS Inc. (2002). F&DA Environmental Assessment Regulations Project Benchmark Survey. A Report to Health Canada (POR-02–13) Office of Regulatory and International Affairs. Available at http://www.hc-sc.gc.ca/ewh-semt/alt_formats/hpfb-dgpsa/pdf/contaminants/rep-rap_por-02-03_e.pdf; accessed 4/25/2007.

Court M.H., S.X. Duan, L.M. Hesse, K. Venkatakishnan, and D.J. Greenblett (2001). Cytochrome P-450 is responsible for interindividual variability of propofol hydroxylation by human liver microsomes. *Anesthesiology* 94:110–119.

Critchley I.A., R.S. Blosser-Middleton, M.E. Jones, J.A. Karlowsky, E.A. Karginova, C. Thornsberry, and D.F. Sahm (2002). Phenotypic and genotypic analysis of levofloxacin-resistant clinical isolates of *Streptococcus pneumoniae* collected from 13 countries during 1999–2000. *International Journal of Antimicrobial Agents* 20:100–107.

Critser G. (2005). *Generation Rx: How Prescription Drugs are Altering American Lives, Minds and Bodies*. Houghton Mifflin, New York.

Crittenden J.C., P.S. Reddy, H. Arora, J. Trynoski, D.W. Hand, D.L. Perram, and R.S. Summers (1991). Predicting GAC performance with rapid small-scale column tests. *Journal of the American Water Works Association* 83(1):77–87.

Crittenden J.C., R.R. Trussell, D.W. Hand, K.J. Howe, and G. Tchobanoglous (2005). *Water Treatment: Principles and Design*. Wiley, Hoboken, NJ.

Crittenden J.C., K. Vaitheeswaran, D.W. Hand, E.W. Howe, E.M. Aieta, C.H. Tate, M.J. McGuire, and M.K. Davis (1993). Dissolved organic carbon using granular activated carbon. *Water Research* 27:715–721.

Cserháti T., E. Forgács, and G. Oros (2002). Biological activity and environmental impact of anionic surfactants. *Environment International* 28:337–348.

Cuthbert R., R.E. Green, S. Ranade, S. Saravanan, D.J. Pain, V. Prakash, and A.A. Cunningham (2006). Rapid population declines of Egyptian vultures (*Neophron percnopterus*) and red-headed culture (*Sarcogyps calvus*) in India. *Animal Conservation* 9:349–354.

Czaker R. (2006). Serotonin immunoreactivity in a highly enigmatic metazoan phylum, the prenervous Dicymida. *Cell and Tissue Research* 326:843–850.

Czech P., K. Weber, and D.R. Dietrich (2001). Effects of endocrine modulating substances on reproduction in the hermaphroditic snail *Lymnaea stagnalis* L. *Aquatic Toxicology* 53:103–114.

Danese P., A. Zanca, and M.A. Bertazzoni (1994). Occupational contact dermatitis from tylosin. *Contact Dermatitis* 30:122–123.

Dantas R.F., M. Cantarino, R. Marotta, C. Sans, S. Esplugas, and R. Andreozzi (2007). Bezafibrate removal by means of ozoanation: Primary intermediates, kinetics, and toxicity assessment. *Water Research* 41:2525–2532.

Das B.S., L.S. Lee, P.S.C. Rao, and R.P. Hultgren (2004). Sorption and degradation of steroid hormones in soils during transport: Column studies and model evaluation. *Environmental Science and Technology* 38:1460–1470.

Daughton C.G. (2003a). Craddle-to-craddle stewardship of drugs for minimizing their environmental disposition while promoting human health. I. Rationale for and avenues towards a green pharmacy. *Environmental Health Perspectives* 111:757–774.

Daughton C.G. (2003b). Craddle-to-craddle stewardship of drugs for minimizing their environmental disposition while promoting human health. II. Drug disposal, waste reduction, and future directions. *Environmental Health Perspectives* 111:775–785.

Daughton C.G. (2004). PPCPs in the Environment: Future Research—Beginning with the End in Mind. In: Kümmerer K. (Ed.). *Pharmaceuticals in the Environment*, 2nd ed. Springer, New York, pp. 463–495.

Daughton C.G. and T.A. Ternes (1999). Pharmaceutical and personal care products in the environment: Agents of subtle changes? *Environmental Health Perspectives* 107:907–938.

Davies J. and C.F. Amábile-Cuevas (2003). The Rise of Antibiotic Resistance. In: C.F. Amábile-Cuevas (Ed.). *Multiple Drug Resistant Bacteria*. Horizon Scientific Press, Wymondham, UK, pp. 1–7.

Dazzo F., P. Smith, and D. Hubbel (1973). The influence of manure slurry irrigation on the survival of fecal organisms in Scranton fine sand. *Journal of Environmental Quality* 2:470–473.

de Groot A.N., P.W. van Dongen, and T.B. Vree (1998). Ergot alkaloids: Current status and review of clinical pharmacology and therapeutic use compared with other oxytocics in obstetrics and gynaecology. *Drugs* 56:523–535.

de la Cruz F. and J. Davies (2000). Horizontal gene transfer and the origin of species: Lessons from bacteria. *Trends in Microbiology* 8:128–133.

Delanghe B., F. Nakamura, H. Myoga, Y. Magara, and E. Guibal (1994). Drinking water denitrification in a membrane bioreactor. *Water Science and Technology* 30:157–160.

Del Estal J.L., A.I. Alvarez, C. Villaverde, and J.G. Prieto (1993). Comparative effects of anionic, natural bile acid surfactants and mixed micelles on the intestinal absorption of the anthelmintic albendazole. *International Journal of Pharmaceutics* 91:105–109.

De Liguoro M., V. Cibin, F. Capolongo, B. Halling-Sørensen, and C. Montesissa (2003). Use of oxytetracycline and tylosin in intensive calf farming: Evaluation of transfer to manure and soil. *Chemosphere* 52:203–212.

Denslow N.D. et al. (2001). Induction of gene expression in sheephead minnows (*Cyprinodon variegatus*) treated with 17β-estradiol, diethylstilbestrol, or ethinylestradiol: The use of mRNA fingerprints as an indicator of gene regulation. *General and Comparative Endocrinology* 121:250–260.

Desbrow C., E.J. Routledge, G.C. Brighty, J.P. Sumpter, and M. Waldock (1998). Identification of estrogenic chemicals in STW effluents. I. Chemical fraction and in vitro biological screening. *Environmental Science and Technology* 32:1549–1558.

Devic E., L. Paquereau, R. Steinberg, D. Caput, and Y. Audigier (1997). Early expression of a beta 1-adrenergic receptor and catecholamines in *Xenopus* oocytes and embryos. *FEBS Letter* 417:184–190.

Díaz-Cruz M.S., M.J.L. de Alda, and D. Barceló (2003). Environmental behavior and analysis of veterinary and human drugs in soils, sediments and sludge. *Trends in Analytical Chemistry* 22:340–351.

DiMicco J.A. and K. Gutierrez (2003). Drugs to Treat Diabetes Mellitus. In: K. Gutierrez and S.F. Queener (Eds.). Pharmacology for Nursing Practice. Mosby, St. Louis, pp. 955–976.

Dimitrov S.D., N.C. Dimitrova, J.D. Walker, G.D. Veith, and O.G. Mekenyan (2003). Bioconcentration potential predictions based on molecular attributes—an early warning approach for chemicals found in humans, birds, fish and wildlife. *QSAR and Combinatorial Science* 22:58–68.

Doll T.E. and F.H. Frimmel (2003). Fate of pharmaceuticals—photodegradation by simulated solar UV-light. *Chemosphere* 52:1757–1769.

Dollery C.T. (Ed.) (1991). Therapeutic Drugs, Vol. 1 and 2. Churchill Livingstone, Edinburg.

Donaldson E.M. and G.A. Hunter (1982). Sex control in fish with particular reference to salmonids. *Canadian Journal of Fisheries and Aquatic Sciences* 39:99–110.

Donoghue J. and T.R. Hylan (2001). Antidepressant use in clinical practice: Efficacy v effectiveness. *British Journal of Psychiatry* 179(Suppl. 42):s9–s17.

Donoho A.L. (1984). Biochemical studies on the fate of monensin in animals and in the environment. *Journal of Animal Science* 58:1528–1539.

Drewes J.E., T. Heberer, Rauchand, and K. Reddersen (2003). Fate of pharmaceuticals during groundwater recharge. *Ground Water Monitoring and Remediation* 23:64–72.

Drewes J.E., T. Heberer, and K. Reddersen (2002). Fate of pharmaceuticals during indirect potable reuse. *Water Science and Technology* 46:73–80.

Drillia P., K. Stamatelatou, and G. Lyberatos (2005). Fate and mobility of pharmaceuticals in solid matrices. *Chemosphere* 60:1034–1044.

Duft M., U. Schulte-Oehlmann, M. Tillman, B. Markert, and J. Oehlmann (2003). Toxicity of triphenyltin and tributyltin to the freshwater mudsnail *Potamopyrgus antipoderum* in a new sediment biotest. *Environmental Toxicology and Chemistry* 22:145–152.

Eckles W.P., B. Ross, and R.K. Isensee (1993). Pentobarbital found in groundwater. *Ground Water* 31:801–804.

Edison R.J. and M. Muenke (2004). Central nervous system and limb anomalies in case reports of first trimester statin exposure. *New England Journal of Medicine* 350:1579–1582.

Eker S. and F. Kargi (2006). Impacts of COD and DCP loading rates on biological treatment of 2,4-dichlorophenol (DCP) containing wastewater in a perforated tubes biofilm reactor. *Chemosphere* 64:1609–1617.

Emblidge J.P. and M.E. DeLorenzo (2006). Preliminary risk assessment of the lipid-regulating pharmaceutical clofibric acid, for three estuarine species. *Environmental Research* 100:216–226.

EMEA (2001). Draft CPMP Discussion Paper on Environmental Risk Assessment of Non-Genetically Modified Organisms (Non-GMO) Containing Medicinal Products for Human Use. CPMP/SWP/4447/00. Final. European Agency for the Evaluation of Medicinal Products, London.

Emmerson A.M. and A.M. Jones (2003). The quinolones: Decades of development and use. *Journal of Antimicrobial Chemotherapy* 51(Suppl S1):13–20.

Esiobu N., L. Armenta, and J. Ike (2002). Antibiotic resistance in soil and water environments. *International Journal of Environmental Health Research* 12:133–144.

EU (1994). Assessment of Potential Risks for the Environment Posed by Medical Products for Human Use, Excluding Products Containing Live Genetically Modified Organisms. EU Ad Hoc Working Party. III/5504/94. Draft 4. European Agency for the Evaluation of Medicinal Products, London.

Falkiner F.R. (1998). The consequences of antibiotic use in horticulture. *Journal of Antimicrobial Chemotherapy* 41:429–431.

FEDESA (1999). Antibiotics use in farm animals does not threaten human health. Press Release by the European Federation of Animal Health. Brussels, July 13 (website details).

FEDESA (2003). Annual veterinary antibiotic use in the EU in 1999. Available at: http://www.fedesa.be; accessed on 10/05/2003.

Fent K., A.A. Weston, and D. Caminada (2006). Ecotoxicology of human pharmaceuticals. *Aquatic Toxicology* 76:122–159.

Ferber D. (2003). Antibiotic resistance. WHO advises kicking the livestock antibiotic habit. *Science* 301:1027.

Ferguson P.L., C.R. Iden, and B.J. Brownawell (2000). Analysis of alkylophenol ethoxylate metabolites in the aquatic environment using liquid chromatography-electrospray mass spectrometry. *Analytical Chemistry* 72:4322–4330.

Ferrari B., N. Paxéus, R. Lo Giudice, A. Pollio, and J. Garric (2003). Ecotoxicological impact of pharmaceuticals found in treated wastewaters: Study of carbamazepine, clofibric acid, and diclofenac. *Ecotoxicology and Environmental Safety* 55:359–370.

Fey P.D., T.J. Safranek, M.E. Rupp, E.F. Dunne, E. Ribot, P.C. Iwen, P.A. Bradford, F.J. Angulo, and S.H. Hinrichs (2000). Ceftriaxone-resistant Salmonella infection acquired by a child from cattle. *New England Journal of Medicine* 342:1242–1249.

Field J.A. (2003). Biodegradation of Chlorinated Compounds by White Rot Fungi. In M.M. Häggblom and I.D. Bossert (Eds.). *Dehalogenation: Microbial Processes and Environmental Applications.* Kluwer Academic, pp. 159–204.

Finlay-Moore O., P.G. Hartel, and M.L. Cabrera (2000). 17β-Estradiol and testosterone in soil and runoff from grasslands amended with broiler litter. *Journal of Environmental Quality* 29:1604–1611.

Folmer L.C., N.D. Denslow, V. Rao, M. Chow, D.A. Crain, J. Enblom, J. Marcino, and L.J. Guillette Jr. (1996). Vitellogenin induction and reduced serum testosterone concentrations in feral male carp (*Cyprinus carpio*) captured near a major metropolitan sewage treatment plant. *Environmental Health Perspectives* 104:1096–1101.

Folmer L.C., M. Hemmer, R. Hemmer, C. Bowman, K. Kroll, and N.D. Denslow (2000). Comparative estrogenicity of estradiol, ethynylestradiol and diethylstilbestrol in an in vivo, male sheephead minnow (*Cyprinodon variegates*), vitellogenin bioassay. *Aquatic Toxicology* 49:77–88.

Fong P.P. (1998). Zebra mussels spawning is induced in low concentrations of putative serotonin re-uptake inhibitors. *Biology Bulletin* 194:143–149.

Fong P.P. (2001). Antidepressants in Aquatic Organisms: A Wide Range of Effects. In: C.G. Daughton and T.L Jones-Lepp (Eds.). *Pharmaceuticals and Personal Care Products in the Environment—Scientific and Regulatory Issues.* ACS Symposium Series 791. American Chemical Society, pp. 264–281.

Fong P., P. Huminski, and L. D'Urso (1998). Induction and potentiation of parturition? In Fingernail clams by selective serotonin re-uptake inhibitors. *Journal of Experimental Zoology* 280:260–264.

Fridkin S.K., J. Hageman, L.K. McDougal, et al. (2003). Epidemiological and microbiological characterization of infections caused by *Staphylococcus aureus* with reduced susceptibility to vancomycin, United States, 1997–2001. *Clinical Infectious Diseases* 36:429–439.

Frimmel F.H. (1994). Photochemical aspects related to humic substances. *Environment International* 20:373–384.

Frimmel F.H. (1998). Impact of light on the properties of aquatic natural organic matter (NOM). *Environment International* 24:559–571.

Frimmel F.H. and D.P. Hessler (1994). Photochemical Degradation of Triazine and Anilide Pesticides in Natural Waters. In: G.R. Helz, R.G. Zepp, and D.G. Crosby (Eds.). *Aquatic and Surface Photochemistry.* pp. 137–147.

Fromme H., T. Otto, K. Pilz, and F. Neugebauer (1999). Levels of synthetic musks; Bromocyclene and PCBs in eel (*Anguilla anguilla*) and PCBs in sediment samples from some waters of Berlin/Germany. *Chemosphere* 39:1723–1735.

Fry D.M and C.K. Toone (1981). DDT-induced feminization of gull embryos. *Science* 213:922–924.

Fry D.M., C.K. Toone, S.M. Speich, and R.J. Peard (1987). Sex ratio skew and breeding patterns of gulls: Demographic and toxicological considerations. *Studies in Avian Biology* 10:26–43.

Fujioka Y. (2002). Effects of hormone treatments and temperature on sex-reversal of Nigorobuna *Carassius carassius grandoculis. Fish Science* 68:889–893.

Gagné F., C. Blaise, and C. André (2006). Occurrence of pharmaceutical products in a municipal effluent and toxicity to rainbow trout (Oncorhychus mykiss) hepatocytes. *Ecotoxicology and Environmental Safety* 64:329–336.

Gatermann R., S. Biselli, H. Huhnerfuss, G.G. Rimkus, M. Hecker, and L. Karbe (2002). Synthetic musks in the environment. Part 1: Species-dependent bioaccumulation of polycyclic and nitro musk fragrances in freshwater fish and mussels. *Archives of Environmental Contamination and Toxicology* 42:437–446.

Gavalchin J. and S.E. Katz (1994). The persistence of fecal-borne antibiotics in soil. *J. AOAC International* 77, 481–485.

Gesell M., E. Hammer, M. Specht, W. Francke, and F. Schauer (2001). Biotransformation of biphenyl by Paecilomyces lilacinus and characterization of ring cleavage products. *Applied and Environmental Microbiology* 67:1551–1557.

Ghosal D.N. and S.K. Mukherjee (1970). Studies of sorption and desorption of two basic antibiotics by and from clays. *Journal of the Indian Society of Soil Science* 18:243–247.

Gilbertson T.J., R.E. Hornish, P.S. Jaglan, K.T. Koshy, J.L. Nappier, G.L. Stahl, A.R. Cazers, J.M. Nappier, M.F. Kubicek, G.A. Hoffman, and P.J. Hamlow (1990). Environmental fate of ceftiofur sodium, a cephalosporin antibiotic. Role of animal excreta in its decomposition. *Journal of Agricutlural and Food Chemistry* 38:890–894.

Glass B.D., M.E. Brown, S. Daya, M.S. Worthington, P. Drummond, E. Antunes, M. Lebete, S. Anoopkumar-Dukie, and D. Maharaj (2001). Influence of cyclodextrins on the photostability of selected drug molecules in solution and the solid-state. *International Journal of Photoenergy* 3:205–211.

Glassmeyer S.T., E.T. Furlong, D.W. Kolpin, J.D. Cahill, S.D. Zaugg, S.L. Werner, M.T. Meyer, and D.D. Kryak (2005). Transport of chemical and microbial compounds from known wastewater discharges: Potential for use as indicators of human fecal contamination. *Environmental Science and Technology* 39:5157–5169.

Golet E.M., A.C. Alder, and W. Giger (2002a). Environmental exposure and risk assessment of flouroquinolone antibacterial agents in wastewater and river water of the Glatt Valley watershed, Switzerland. *Environmental Science and Technology* 37:3645–3651.

Golet E.M., A.C. Alder, A. Hartmann, T.A. Ternes, and W. Giger (2001). Trace determination of fluoroquinolone antibacterial agents in urban wastewater by solid-phase extraction and liquid chromatography with fluorescence detection. *Analytical Chemistry* 73:3632–3638.

Golet E.M., A. Strehler, A.C. Alder, and W. Giger (2002b). Determination of fluoroquinolone antibacterial agents in sewage sludge and sludge-treated soil using accelerated solvent extraction followed by solid-phase extraction. *Analytical Chemistry* 74:5455–5462.

Gomez J., R. Mendez, and J.M. Lemna (1996). The effect of antibiotics on nitrification processes. *Applied Biochemistry and Biotechnology* 57/58:869–876.

Gonsalves D. and D.P.H. Tucker (1977). Behaviour of oxytetracycline in Florida citrus soils. *Archives of Environmental Contamination and Toxicology* 6:515–523.

Goole J., F. Vanderbist, and K. Amighi (2006). Development and evaluation of new multiple-unit levodopa sustained-release floating dosage forms. *International Journal of Pharmaceutics* 334:35–41.

GSK (2006). Imitrex (sumatriptan) nasal spray. Prescribing information. GlaxoSmithKline, Research Triangle Park, NC. Available at: http://us.gsk.com/products/assets/us_imitrex_nasal_spray.pdf; accessed 1/29/2007.

Guillette, Jr., L.J. (2006). Endocrine disrupting contaminants—beyond the dogma. *Environmental Health Perspectives* 114(Suppl. 1):9–12.

Guitton J., T. Burronfosse, M. Sanchez, and M. Desage (1997). Quantification of propofol. *Analytical Letters* 30:1369–1378.

Gulland F.M.D. and A.J. Hall (2007). Is marine health deteriorating? Trends in the Global reporting of marine mammal disease. *EcoHealth* 4:135–150.

Gunn A. and J.W. Sadd (1994). The effect of ivermectin on the survival, behavior and cocoon production of the earthworm *Eisenia foetida*. *Pedobiologia* 38:327–333.

Gutierrez K. and S.F. Queener (2003). *Pharmacology for Nursing Practice*. Mosby St. Louis.

Hagenstein P.R., R.G. Flocchini, J.C. Bailar, et al. (2002) Air Emission from Animal Feeding Operations: Current Knowledge, Future Needs. National Research Council of the National Academies. Available at: http://www.epa.gov/ttn/chief/ap42/ch09/related/nrcanimalfeed_dec2002.pdf; accessed 9/27/2007.

Halden R.U. and D.H. Paull (2004). Co-occurrence of tricarban and triclosan in US water resources. *Environmental Science and Technology* 39:1420–1326.

Halley B.A., W.J.A. VandenHeuvel, and P.G. Wislocki (1993). Environmental effects of the usage of avermectins in livestock. *Veterinary Parasitology* 48:109–125.

Halling-Sørensen, B. (2001). Inhibition of aerobic growth and nitrification of bacteria in sewage sludge by antibacterial agents. *Archives of Environmental Contamination and Toxicology* 40:451–460.

Halling-Sørensen B., H.C. Holten Lützhøft, H.R. Andersen, and F. Ingerslev (2000). Environmental risk assessment of antibiotics: Comparison of mecillinam, trimethoprim and ciprofloxacin. *Journal of Antimicrobial Chemotherapy* 46(Suppl. S1):53–58.

Halling-Sørensen B., A.M. Jacobsen, J. Jensen, G. Sengelov, E. Vaclavik, and F. Ingerslev (2005). Dissipation and effects of chlortetracycline and tylosin in two agricultural soils: A field-scale study in Southern Denmark. *Environmental Toxicology and Chemistry* 24:802–810.

Halling-Sørensen B., S.N. Nielsen, P.F. Lanzky, F. Ingerster, H.C.H. Lützhøff, and S.E. Jørgensen (1998). Occurrence of pharmaceutical substances in the environment—a review. *Chemosphere* 36:357–393.

Halling-Sørensen, B., G. Sengeløv, F. Ingerslev, and L.B. Jensen (2003). Reduced antimicrobial potencies of oxytetracycline, tylosin, sulfadiazin, streptomycin, ciprofloxacin and olaquindox due to environmental processes. *Archives of Environmental Contamination and Toxicology* 44:7–16.

Halling-Sørensen B., G. Sengeløv, and J. Tjørnelund (2002). Toxicity of tetracyclines and tetracycline degradation products to environmentally relevant bacteria, including selected tetracycline-resistant bacteria. *Archives of Environmental Contamination and Toxicology* 42:263–271.

Hammerum A.H., L.B. Jensen, and F.M. Aarestrup (1998). Detection of the SatA gene and transferability of virginiamycin resistance in Enterococcus faecium from food animals. *FEMS Microbiology Letters* 168:145–151.

Hamscher G., H.T. Pawelzick, S. Sczesny, H. Nau, and J. Hartung (2003). Antibiotics in dust originating from a pig fattening farm: A new source of health hazard for farmers? *Environmental Health Perspectives* 111:1590–1594.

Hamscher G., S. Sczesny, H. Höper, and H. Nau (2002). Determination of persistent tetracycline residues in soil fertilized with liquid manure by high-performance liquid chromatography with electron spray ionization tandem mass spectrometry. *Analytical Chemistry* 74:1509–1518.

Hansch C. and A. Leo (1979). *Substituent Constants for Correlation Analysis in Chemistry and Biology*. Wiley, New York.

Hansch C., P.P. Maloney, T. Fujita, and R.M. Muir (1962). Correlation of biological activity of phenoxyacetic acids with Hammett substituent constants and partition coefficients. *Nature (London)* 194:178–180.

Hansen P.K., B.T. Lunestad, and O.B. Samuelsen (1992). Effects of oxytetracycline, oxalinc acid, and flumequine on bacteria in an artificial marine fish farm. *Canadian Journal of Microbiology* 39:1307–1312.

Harnagea-Theophilus E. and M.R. Miller (1999). Acetaminophen alters estrogenic responses in vitro: Stimulation of DNA synthesis in estrogen-responsive human breast cancer cells. *Toxicological Sciences* 46:38–44.

Hartmann A., A.C. Alder, T. Koller, and R.W. Widmer (1998a). Identification of fluoroquinolone antibiotics as the main source of umuc genotoxicity in native hospital wastewater. *Environmental Toxicology and Chemistry* 17:383–393.

Hatzinger P.B. and M. Alexander (1995). Effect of aging of chemicals in soil on their biodegradability and extractability. *Environmental Science and Technology* 29:537–545.

Hawkey P. (2003). Mechanisms of quinolone action and microbial response. *Journal of Antimicrobial Chemotherapy* 51(Suppl S1):29–35.

Hayes J.R., A.C. McIntosh, S. Qaiyumi, J.A. Johnson, L.L. English, L.E. Carr, D.W. Wagner, and S.W. Joseph (2001). High-frequency recovery of quinupristin-dalfopristin-resistant *Enterococcus faecium* isolates from poultry production environment. *Journal of Clinical Microbiology* 39:2298–2299.

H2E (2007). Hospitals for a Healthy Environment. Available at: http://cms.h2e-online.org/about/.

Heberer T. (2002a). Occurrence, fate, and removal of pharmaceutical residues in the aquatic environment: A review of recent research data. *Toxicology Letters* 131:5–17.

Heberer T. (2002b) Tracking persistent pharmaceutical residues from municipal sewage to drinking water. *Journal of Hydrology* 266:175–189.

Heberer T. and M. Adam (2005). Occurrence, Fate and Removal of Pharmaceutical Residues in the Aquatic Environment: An Extended Review of Recent Research Data. In: D.R. Dietrich, S.F. Webb, and T. Petry (Eds.). Hot Spot Pollutants: Pharmaceutical in the Environment. Elsevier, New York, pp. 11–36.

Heberer T. and D. Feldmann (2004). In: K. Kümmerer (Ed.). *Pharmaceuticals in the Environment: Sources, Fate, Effects and Risks.* Springer, New York, pp. 391–410.

Heberer T., K. Reddersen, and A. Mechlisnki (2002). From municipal sewage to drinking water: Fate and removal of pharmaceutical residues in the aquatic environment in urban areas. *Water Science and Technology* 46:81–88.

Hennion M.-C. (1998). Applications and validation of immunoassays for pesticides analysis. *Analusis Maganize* 26:M149–M155.

Henschel K.P., A. Wenzel, M. Diderick, and A. Fliedner (1997). Environmental hazard assessment of pharmaceuticals. *Regulatroy Toxicology and Pharmacology* 25:220–223.

Hensiek A.E. and M.R. Trimble (2002). Relevance of new psychotropic drugs for the neurologist. *Journal of Neurology, Neurosurgery and Psychiatry* 72:285–286.

HERA (2004). Human and environmental risk assessment on ingredients of household cleaning products. Polycyclic musks. Available at: http://www.heraproject.com/files/28-E-36551E10F8EF-E807-E4199B9BB0076A9F.pdf.

Hernandez-Rauda R., G. Rozas, P. Rey, J. Otero, and M. Aldegunde (1999). Changes in the pituitary metabolism of monoamines (dopamine, norepinephrine, and serotonin) in female and male rainbow trout (*Oncorhynchus mykiss*) during gonadal recrudescence. *Physiological and Biochemical Zoology* 72:353–359.

Herwig R.P. and J.P. Gray (1997). Microbial response to antibacterial treatment in marine microcosms. *Aquaculture* 152:139–154.

Hess R., W. Staubli, and W. Reiss (1965). Nature of hepatomegalic effect produced by ethylchlorophenoxyl-isobutyrate in rat. *Nature* 208(5013):856–858.

Heuther S.E. (2000). Mechanisms of Hormonal Regulation. In: S.E. Heuther and K.L. McCance (Eds.). *Understanding Pathophysiology*, 2nd ed. Mosby, St. Louis, pp. 445–504.

Hiemke C. and S. Härther (2000). Pharmacokinetics of selective serotonin reuptake inhibitors. *Pharmacology and Therapeutics* 85:11–28.

Hinz B. and K. Brune (2002). Cyclooxygeanse-2–10 years later. *Journal of Pharmacology and Experimental Therapeutics* 300:367–375.

Hiramatsu K. (1998). The emergence of *Staphylococcus aureus* with reduced susceptibility to vancomycin in Japan. *American Journal of Medicine* 104:7S–10S.

Hiramatsu K., L. Cui, M. Kuroda and T. Ito (2001). The emergence and evolution of methicillin-resistant *Staphylococcus aureus*. *Trends in Microbiology* 9:486–493.

Hiramatsu K., H. Hanaki, T. Ino, K. Yabuta, T. Oguri, and F.C. Tenover (1997). Methicillin-resistant *Staphylococcus aureus* clinical strains with reduced vancomycin susceptibility. *Journal of Antimicrobial Chemotherapy* 40:135–136.

Hirose J., F. Kondo, T. Nakano, T. Kobayashi, N. Hiro, Y. Ando, H. Takenaka, and K. Sano (2005). Inactivation of antineoplastics in clinical wastewater by electrophoresis. *Chemosphere* 60:1018–1024.

Hirsch R., T. Ternes, K. Haberer, and K.L. Kratz (1999). Occurrence of antibiotics in the aquatic environment. *Science of the Total Environment* 225:109–118.

Hoigné J. and H. Bader (1994). Characterization of water quality criteria for ozonation processes. 2. Lifetime of added ozone. *Ozone Science and Engineering* 16:121–134.

Holakoo L., G. Nakhla, E.K. Yanful and A.S. Bassi (2005). Simultaneous nitrogen and phosphorus removal in a continuously fed and aerated membrane bioreactor. *Journal of Environmental Engineering* 131:1469–1472.

Holland W., T. Morrison, Y. Chang, N. Wiernsperger, and B.J. Stith (2004). Metaformin (glucophage) inhibits tyrosine phosphatase activity to stimulate the insulin receptor tyrosine kinase. *Biochemical Pharmacology* 67:2081–2091.

Holm J.V., K. Rügge, P.L. Bjerg, and T.H. Christensen (1995). Occurrence and distribution of pharmaceutical organic compounds in the groundwater downgradient of a landfill (Grindsted, Denmark). *Environmental Science and Technology* 29:1415–1420.

Holt D., D. Harvey, and R. Hurley (1993). Chloramphenicol toxicity. *Adverse Drug Reactions and Toxicological Reviews* 12:83–95.

Holthaus et al. (2002) The potential for estradiol and ethinylestradiol to sorb to suspended and bed sediments in some English rivers. *Environmental Toxicology and Chemistry* 21:2526–2535.

Howden B.P., P.B. Ward, P.G. Charles, T.M. Korman, A. Fuller, P. du Cros, E.A. Grabsch, S.A. Roberts, J. Robson, K. Read, N. Bak, J. Hurley, P.D.R. Johnson, A.J. Morris, B.C. Mayall, and M.L. Grayson (2004). Treatment outcomes for serious infections caused by methicillin-resistant *Staphylococcus aureus* with reduced vancomycin susceptibility. *Clinical Infectious Diseases* 38:521–528.

Howe R.A., A. Monk, M. Wootton, T.R. Walsh, and M.C. Enright (2004). Vancomycin susceptibility within methicillin-resistant *Staphylococcus aureus* lineages. *Emerging Infectious Diseases* 10:855–857.

Hua W., E.R. Bennett, and R.J. Letcher (2006). Ozone treatment and the depletion of detectable pharmaceuticals and atrazine herbicide in drinking water sourced from the upper Detroit River, Ontario, Canada. *Water Research* 40:2259–2266.

Huang C-H. and D.L. Sedlak (2001). Analysis of estrogenic hormones in municipal wastewater effluent and surface water using enzyme-linked immunosorbent assay and gas chromatography/tandem mass spectrometry. *Environmental Toxicology and Chemistry* 20:133–139.

Huber M.M., A. Göbel, A. Joss, N. Hermann, D. Löffler, C.S. McArdell, A. Reid, H. Siegrist, T.A. Ternes, and U. von Gunten (2005). Oxidation of pharmaceuticals during ozonation of municipal wastewater effluents: A pilot study. *Environmental Science and Technology* 39:4290–4999.

Huchon G.K., G. Gialdroni-Grassi, P. Leophonte, F. Manresa, T. Schaberg, and M. Woodhead (1996). Initial antibiotic therapy for lower respiratory infection in the community: A European survey. *European Respiratory Journal* 9:1590–1595.

Hümpel M., U. Tauber, W. Kuhnz, M. Pfeffer, K. Brill, R. Heithecker et al. (1990). Comparison of serum ethinyl stradiol, sex-hormone-binding globulin, corticoid-binding globulin and cortisol levels in women using two low-dose combined contraceptives. *Hormone Research* 33:35–39.

Hunt E., N. Pearson, T.M. Willson, and A. Takle (2002). Intracellular Targets. In: F.D. King (Eds.). Medicinal Chemistry: Principles and Practice. Royal Society of Chemistry, Cambridge, UK, pp. 42–63.

Hurd, H.S., S. Doores, D. Hayes, A.G. Mathew, J. Maurer, P. Silley, R. Singer, and R.N. Jones (2004). Semi-quantitative risk assessment of the human health impact attributable to the use of macrolides in food animals. *Journal of Food Protection* 67:980–992.

Ibabe A., E. Bilbao, and M.P. Cajaraville (2005). Expression of peroxisome proliferator-activated receptors in zebrafish (*Danio rerio*) depending on gender and developmental stage. *Histochemistry and Cell Biology* 123:75–87.

Ibabe A., M. Grabenbauer, E. Baumgart, H.D. Fahimi, and M.P. Cajaraville (2002). Expression of peroxisome proliferator-activated receptors in zebrafish (*Danio rerio*). *Histochemistry and Cell Biology* 118:231–239.

IFRA (2006). International Fragrance Association. Available at: http://www.ifraorg.org/News.asp.

Ignatavicius D.D. and M.L. Workman (2002). *Medical Surgical Nursing: Critical Thinking for Collaborative Care*. W.B. Saunders, Philadelphia.

Ingerslev, F. and B. Halling-Sørensen (2000). Biodegradability properties of sulfonamides in activated sludge. *Environmental Toxicology and Chemistry* 19:2467–2473.

Ingerslev F. and B. Halling-Sørensen (2001). Biodegradability of metronidazole, olaquindox, and tylosin and formation of tylosin degradation products in aerobic soil-manure slurries. *Ecotoxicology and Environmental Safety* 48:311–320.

Ingerslev F. and B. Halling-Sørensen (2003). Evaluation of analytical chemical methods for detection of estrogens in the environment. Working Report No. 44. Danish Environmental Protection Agency, Danish Minsitry of the Environment. Available at: http://www2.mst.dk/udgiv/publications/2003/87-7972-968-1/pdf/87-7972-p970-3.pdf; accessed 4/29/2007.

Ingerslev F., L. Torang, M.L. Loke, B. Halling-Sørensen, and N. Nyholm (2001). Primary bio-degradation of veterinary antibiotics in aerobic and anaerobic surface water simulation systems. *Chemosphere* 44:865–872.

Iwai T., K. Tanonaka, S. Kasahara, R. Inoue, and S. Takeo (2002). Protective effect of propranolol on mitochondrial function in the ischaemic heart. *British Journal of Pharmacology* 136:472–480.

Jacobsen A.-M., A. Lorenzen, R. Chapman, and E. Topp (2005). Persistence of Testosterone and 17β-Estradiol in Soils Receiving Swine Manure or Municipal Biosolids. *Journal of Environmental Quality* 34:861–871.

Jafvert C.T. (1991). Sediment- and saturated-soil-associated reactions involving an anionic surfactant (dodecyl sulfate). 2. Partition of PAH compounds among phases. *Environmental Science and Technology* 25:1039–1045.

Jafvert C.T. and J.K. Heath (1991). Sediment- and saturated-soil-associated reactions involving an anionic surfactant (dodecylsulfate). 1. Precipitation and micelle formation. *Environmental Science and Technology* 25:1031–1039.

Jeffrey P. (2002). Pharmacokinetics. In: F.D. King (Eds.). *Medicinal Chemistry: Principles and Practice.* The Royal Society of Chemistry, Cambridge, UK, pp. 118–137.

Jensen J. (2001). Veterinary Medicines and Soil Quality: The Danish Situation as an Example. In: C.G. Daughton and T. Jones-Lepp (Eds.). *Scientific and Regulatory Issues, Symposium Series*, Vol. 791. American Chemical Society, Washington, D.C., pp. 282–302.

Jensen J., P.H. Krogh, and L.E. Sverdrup (2003). Effects of the antimicrobial agents tiamulin, olanquindox and metronidazole and the anthelmintic ivermectin on the soil invertebrate species *Folsomia fimetaria* (Collembola) and *Enchytraeus crypticus* (Enchytraeidae). *Chemosphere* 50:437–443.

Jeon H-K., Y. Chung, and J-C. Ryu (2006). Simulatenous determination of benzophenone type UV filters in water and soil by gas chromography-mass spectrometry. *Journal of Chromatography A* 1131:192–202.

Jin S. (1997). Regulation Realities and Recommendation on Antimicrobial Use in Food Animal Production in China. In: The Medical Impact of the Use of Antimicrobials in Food Animals. WHO, Geneva, Switzerland.

Jjemba P.K. (2002a). The potential impact of veterinary and human therapeutic agents in manure and biosolids on vegetation: A review. *Agriculture, Ecosystems and Environment* 93:267–278.

Jjemba P.K. (2002b). The effect of chloroquine, quinacrine, and metronidazole on both soybean (*Glycine max*) plants and soil microbiota. *Chemosphere* 46:1019–1025.

Jjemba P.K. (2004). *Environmental Microbiology: Principles and Applications.* Science, Enfield, NH.

Jjemba P.K. (2006). Excretion and ecotoxicity of pharmaceutical and personal care products in the environment. *Ecotoxicity and Environmental Safety* 63:113–130.

Jjemba P.K., B.K. Kinkle, and J.R. Shann (2006). *In situ* enumeration and probing of pyrene-degrading soil bacteria. *FEMS Microbiology Ecology* 55:287–298.

Jjemba P.K. and B.K Robertson (2003). The Fate and Potential Impact of Pharmaceutical Compounds to Non-Target Organisms in the Environment. *Proceedings of the Third International Conference on Pharmaceutical and Endocrine Disrupting Chemicals in Water.* Minneapolis MN (3/19/2003). National Groundwater Association. NGWA Press, pp. 184–194.

Jjemba P.K. and B.K. Robertson (2005). Antimicrobial agents with improved clinical efficacy versus their persistence in the environment: Synthetic 4-quinolone as an example. *EcoHealth* 2:171–182.

Jobling S., M. Nolan, C.R. Tyler, G. Brighty, and J.P. Sumpter (1998). Widespread sexual disruption in wild fish. *Environmental Science and Technology* 32:2498–2506.

Jones R. (2002). Contemporary pathogens of antibiotic resistance in humans. Available at: http://www.fda.gov/ohrms/dockets/ac/03/briefing/3919B2_03_Handout%20from%20Carnevale.pdf; accessed 12/05/2007.

Jones R.N. and D.J. Biedenbach (2003). Comparative activity of gerenoxacin (BMS 284756), a novel desfluoroquinolone tested against 8,331 isolates from community-acquired respiratory tract infections: North American results from the SENTRY Antimicrobial Surveillance Program (1999–2001). *Diagnostic Microbiology and Infectious Disease* 45:273–278.

Joquevial C., R. Martino, V. Gilard, M. Malat-Martino, P Canal, and U. Niemeyer (1998). Urinary excretion of cyclophosphamide in humans determined by phosphorus-31 nuclear magnetic resonance spectroscopy. *Drug Disposition and Metabolsim* 26:418–428.

Jørgensen, S.E. and B. Halling-Sørensen (2000). Drugs in the environment. *Chemosphere* 40:691–699.

Jürgens M.D., K.I.E. Holthaus, A.C. Johnson, J.J.L. Smith, M. Hetheridge, and R.J. William (2002). The potential for estradiol and ethinylestradiol degradation in English rivers. *Environmental Toxicology and Chemistry* 21:480–488.

Kannan K., J.L. Reiner, S.H. Yun, E.E. Perrotta, L. Tao, B. Johnson-Restrepo, and B.D. Rodan (2005). Polycyclic musk compounds in higher trophic level aquatic organism and humans in the United States. *Chemosphere* 61:693–700.

Kantin R., M.G.Z. Baumgarten, M. Cabeda, A.C. Beaumord, and T.L. De Almeida (1981). Concentration of anionic detergents in Rio Grande water (South Brazil). *Marine Pollution Bulletin* 12:50–53.

Kaplowitz N. (2003). Drug-Induced Liver Disorders: Introduction and Oveview. In: N Kaplowitz and L.D. Deleve (Eds.). *Drug-Induced Liver Disease*. Marcel Dekker, New York, pp. 1–9.

Kasim N.A., M. Whitehouse, C. Ramachandran, M. Bermejo, H. Lennernas, A.S. Hussain, H.E. Junginger, S.A. Stavchansky, K.K. Midha, V.P. Shah, G.L. Amidon, G.L. (2004). Molecular properties of WHO essential drugs and provisional biopharmaceutical classification. *Molecular Pharmaceutics* 1:85–96.

Kay P., P.A. Blackwell, and A.B.A. Boxall (2004). Fate of veterinary antibiotics in a macroporous tile drained clay soil. *Environmental Toxicology and Chemistry* 23:1136–1144.

Keller B.J., H. Yamanaka, and R.G. Thurman (1992). Inhibition of mitochondrial respiration and oxygen-dependent hepatoxicity by six structurally dissimilar peroxisomal proliferating agents. *Toxicology* 71:49–61.

Kendell R.J., T.A. Anderson, R.J. Baker, C.M. Bens, J.A. Carr, L.A. Chiodo, G.P. Cobb III, R.L. Dickerson, K.R. Dixon, L.T. Frame, M.J. Hooper, C.F. Martin, S. T .McMurry, R. Patino, E.E. Smith, and C.W. Theodorakis (2001). Ecotoxicology. In: C.D. Klaassen (Eds.). *Casarett and Doull's Toxicology: The Basic Science of Poisons*. McGraw-Hill, New York, pp. 1013–1045.

Kennedy S.P. and B.J. Bormann (2006). Effective partnering of academic and physician scientists with the pharmaceutical drug development industry. *Experimental Biology and Medicine* 231:1690–1694.

Khanal S.K., B. Xie, M.L. Thompson, S. Sung, S-K. Ong, and J. van Leeuwen (2006). Fate, transport and biodegradation of natural estrogens in the environment and engineered systems. *Environmental Science and Technology* 40:6537–6546.

Kieke A.L., M.A. Borchardt, B.A. Kieke, S.K. Spencer, M.F. Vandermause, K.E. Smith, S.L. Jawahir, and E.A. Belongia (2006). Use of streptogramin growth promoters in poultry and isolation of streptogramin-resistant *Enterococcus faecium* from humans. *Journal of Infectious Diseases* 194:1200–1208.

Kirk L.A., C.R. Tyler, C.M. Lye, and J.P. Sumpter (2002). Changes in estrogenic and androgenic activities at different stages of treatment in wastewater treatment works. *Environmental Toxicology and Chemistry* 21:972–979.

Kim K.S., B.S. Oh, J.W. Kang, D.M. Chung, W.H. Cho, and Y. Choi (2005). Effect of ozone and GAC process for the treatment of micropollutants and DBPs control in drinking water. Pilot scale evaluation. *Ozone Science and Engineering* 27:69–79.

Kimura K., G. Amy, J.E. Drewes, T. Heberer, T-U. Kim, and Y. Watanabe (2003). Rejection of organic micro-pollutants (disinfection by-products, endocrine disrupting compounds, and pharmaceutically active compounds) by NF/RO membranes. *Journal of Membrane Science* 227:113–121.

Kimura K., S. Toshima, G. Amy, and Y. Watanabe (2004). Rejection of neutral endocrine disrupting compounds (EDCs) and pharmaceutical active compounds (PhACs) by RO membranes. *Journal of Membrane Science* 245:71–78.

Kinney C.A., E.T. Furlong, S.D. Zaugg, M.R. Burkhardt, S.L. Werner, J.D. Cahill, and G.R. Jorgesen (2006). Survey of organic wastewater contaminants in biosolids destined for land application. *Environmental Science and Technology* 40:7207–7215.

Kiso Y., Y. Sigiura, T. Kitao, and K. Nishimura (2001). Effects of hydrophobicity and molecular size on rejection of aromatic pesticides with nanofiltration membranes. *Journal of Membrane Science* 192:1–10.

Kiura H., K. Sano, S. Morimatsu, T. Nakano, C. Morita, M. Yamaguchi, T. Maeda, and Y. Katsuoka (2002). Bactericidal activity of electrolyzed acid water from solution containing sodium chloride at low concentration, in comparison with that at high concentration. *Journal of Microbiological Methods* 49:285–293.

Klaver A.L. and R.A. Matthews (1994). Effects of oxytetracycline on nitrification in a model aquatic system. *Aquaculture* 123:237–242.

Knoblock, M.D., P.M. Sutton, P.N. Mishra, K. Gupta, and A. Jason (1994). Membrane biological reactor system for treatment of oily wastewaters. *Water and Environment Research* 66:133–139.

Kobayashi K., T. Ishizuka, N. Shimada, Y. Yoshimura, K. Kamijima, and K. Chiba (1999). Sertraline N-demethylation is catalyzed by multiple isoforms of human cytochrome P-450 in vitro. *Drug Metabolism and Disposition* 27:763–766.

Koester C.J., S.C. Simmonich, and B.K. Esser (2003). Environmental analysis. *Analytical Chemistry* 75:2813–2829.

Kolodzieg E.P., J.L. Gray, and D.L. Sedlak (2000). Quantification of steroid hormones with phenomenal properties in municipal wastewater effluent. *Environmental Toxicology and Chemistry* 22:2622–2629.

Kolpin D.W., E.T. Furlong, M.T. Meyer, E.M. Thurman, S.D. Zaugg, L.B. Barber, and H.T. Buxton (2002). Pharmaceuticals, hormones, and other organic wastewater contaminants in U.S. streams 1999–2000; a national reconnaissance. *Environmental Science and Technology* 36:1202–1211.

Kolpin D.W., M. Skopec, M.T. Meyer, E.T. Furlong, and S.D. Zaugg (2004). Urban contribution of pharmaceuticals and other organic wastewater contaminants to streams during differing flow conditions. *Environmental Science and Technology* 38:119–130.

Kolwzan B., T. Traczewska, and M. Pawlaczyk-Szipilowa (1991). Examination of resistance of bacteria isolated from drinking water to antibacterial agents. *Environmental Protection Engineering* 17:53–60.

Konstantinou I.K., A.K. Zarkadis, and T.A. Albanis (2001). Photodegradation of selected herbicides in various natural waters and soils under environmental conditions. *Journal of Environmental Quality* 30:121–130.

Korte J.J, M.D. Kahl, K.M. Jensen, M.S. Pasha, L.G. Parks, G.A. LeBlanc, and G.T. Ankley (2000). Fathead minnow vitellogenin: Complementary DNA sequence and messenger RNA and protein expression after 17β-estradiol treatment. *Environmental Toxicology and Chemistry* 19:972–981.

Kosaka K., H. Yamada, S. Matsui, and K. Shishida (2000). The effects of the co-existing compounds on the decomposition of micropollutants using the ozone/hydrogen peroxide process. *Water Science and Technology* 42:353–361.

Kreuzinger N., M. Clara, B. Strenn, and H. Kroiss (2004). Relevance of the sludge retention time (SRT) as design criteria for wastewater treatment plants for the removal of endocrine disruptors and pharmaceuticals from wastewater. *Water Science and Technology* 50:149–156.

Kuch H.M. and K. Ballschmitter (2001). Determination of endocrine-disrupting phenolic compounds and estrogens in surface and drinking water by HRGC-(NCI)-MS in picograms per liter range. *Environmental Science and Technology* 35:3201–3206.

Kücken D., H-H. Feucht, and P-M. Kaulfers (2000). Association of qacE and qacEΔ1 with multiple resistance to antibiotics and antiseptics in clinical isolates of gram-negative bacteria. *FEMS Microbiology Letters* 183:95–98.

Kudoh A., H. Isihara, and A. Matsuki (1998). Effects of carbamazepine on pain scores of unipolar depressed patients with chronic pain: A trial of off-on-off-on design. *Clinical Journal of Pain* 14:61–65.

Kühn and Müller (2000). Riverbank filtration—an overview. *Journal of the American Water Works Association* 92:60–69.

Kühne M., D. Ihnen, G. Moller, and O. Agthe (2000). Stability of tetracycline in water and liquid manure *Journal of Veterinary Medicine Series A—Physiology, Pathology, Clinical Medicine* 47:379–384.

Kumar K., S.C. Gupta, Y. Chander, and A.K. Singh (2005). Antibiotic use in agriculture and its impact on the terrestrial environment. *Advances in Agronomy* 87:1–54.

Kumar K., A. Thompson, A.K. Singh, Y. Chandler, and S.C. Gupta (2004). Enzyme-linked immunosorbent assay for ultratrace determination of antibiotics in aqueous samples. *Journal of Environmental Quality* 33:250–256.

Kümmerer K. (2004a). Resistance in the environment. *Journal of Antimicrobial Chemotherapy* 54:311–320.

Kümmerer K. (2004b). Emission from Medical Care Units. In: K. Kümmerer (Ed.). *Pharmaceuticals in the Environment: Sources, Fate, Effects and Risks*. Springer, New York, pp. 27–44.

Kümmerer K., A. Al-Ahmad, and V. Mersch-Sundermann (2000). Biodegradability of some antibiotics, elimination of the genotoxicity and affection of wastewater bacteria in a simple test. *Chemosphere* 40:701–710.

Kümmerer K., A. Eitel, U. Braun, P. Hubner, F. Daschner, G. Mascart, M. Milandri, F. Reinthaler, and J. Verhoef (1997a). Analysis of benzalkonium chloride in the effluent from European hospitals by solid-phase extraction and high-performance liquid chromatography with post-column ion-pairing and fluorescence detection. *Journal of Chromatography A* 774:281–286.

Kümmerer K. and E. Helmers (2000). Hospitals as a source of gadolinium in the aquatic environment. *Environmental Science and Technology* 34:573–577.

Kümmerer K. and A. Henninger (2004). Promoting resistance by the emission of antibiotics from households into effluent. *Clinical Microbiology and Infection* 9:1203–1214.

Kümmerer K., T. Steger-Hartmann, and M. Meyer (1997b). Biodegradability of the anti-tumor agent ifosfomide and its occurrence in hospital effluents and sewage. *Water Research* 31:2705–2710.

Kuperstein R. and Z. Sasson (2000). Effects of antihypertensive therapy on glucose and insulin metabolism and on left ventricular mass. *Circulation* 102:1802–1806.

Kuspis D.A. and E.P. Krenzelok (1996). What happens to expired medications? A survey of community medication disposal. *Veterinary and Human Toxicology* 38:48–49.

Kwok M., S.L. Chow, J. Holts, and M.S.S. Chow (2003). Panic disorders: A pharmacological armamentarium. *Formulary* 38:431–438.

Lacasse J.R. and J. Leo (2005). Serotonin and depression: A disconnect between the advertisements and the scientific literature. *PLoS Medicine* 2:1211–1216.

La Du B.N., H.G. Mandel, and E.L. Way (Eds.) (1979). *Fundamental of Drug Metabolism and Drug Disposition*. Robert E. Kreiger, Huntington New York.

Lai K.M., K.L. Johnson, M.D. Scrimshaw, and J.N. Lester (2000). Binding of waterborne steroid estrogens to solid phases in rivers and estuarine system. *Environmental Science and Technology* 34:3890–3894.

Lange I.G., A. Daxenberger, B. Schiffer, H. Witters, D. Ibarreta, and H.H.D. Meyer (2002). Review—Sex hormones originating from different livestock production systems: Fate and potential disrupting activity in the environment. *Analytica Chimica Acta* 473:27–37.

Lange R. and D. Dietrich (2002). Environmental risk assessment of pharmaceutical drug substances—conceptual considerations. *Toxicology Letters* 31:97–104.

Länge R., T.H. Hutchinson, C.P. Croudace, F. Siegmund, H. Schweinfurth, P. Hampe, G.H. Panter, and J.P. Sumpter (2001). Effects of the synthetic estrogen 17α-ethinylestradiol on the life-cycle of the fathead minnow (*Pimephales promelas*). *Environmental Toxicology and Chemistry* 20:1216–1227.

Langlais B., D. Recknow, and D.R. Brink (1991). *Ozone in Water Treatment, Applications and Engineering*. Lewis, Chelsea, MI.

Lanzky P.F. and B. Halling-Sørensen, (1997). The toxic effects of the antibiotic metronidazole on aquatic organisms. *Chemosphere* 35:2553–2561.

Latch D.E., J.L. Packer, W.A. Arnold, and K. McNeill (2003). Photochemical conversion of triclosan to 2,8-dichlorodibenzo-*p*-dioxin in aqueous solution. *Journal of Photochemistry and Photobiology A* 158:63–66.

Laukova A. (2000). Vancomycin-resistant enterococci isolates from the rumen content of deer. *Microbios* 97:95–101.

Laurie D., A.J. Manson, F. Rowell, and J. Seviour (1989). A rapid qualitiative ELISA test for the specific detection of morphine in serum or urine. *Clinica Chemica Acta* 183:183–196.

Lautier J., J.L. Chanal, and A. Delhon (1986). Comparative hydrolipidemic effects of clofibric acid and itanoxone on the metabolism and distribution of lipids in the crab *Pachygraspus marmoratus* (Decopoda, Brachyura). *Comparative Biochemistry and Physiology C* 85:269–274.

Laville N., S. Ait-Aissa, E. Gomez, C. Casellas, and J.M. Porcher (2004). Effects of human pharmaceuticals on cytotoxicity, EROD activity and ROS production in fish hepatocytes. *Toxicology* 196:42–55.

Layton A.C., B.W. Gregory, J.R. Seward, T.W. Schultz, and G.S. Sayler (2000). Mineralization of steroidal hormones by biosolids in wastewater treatment systems in Tennessee U.S.A. *Environmental Science and Technology* 34:3925–3931.

Leaver M.J., J. Wright, and S.G. George (1998). A peroxisomal proliferator-activated receptor gene from the marine flatfish, the plaice (*Pleuronectes platessa*). *Marine Environmental Research* 46:75–79.

Lee W.M. (2003). Drug-induced hepatoxicity. *New England Journal of Medicine* 349:474–485.

Lee H.B. and D. Liu (2002). Degradation of 17β-estradiol and its metabolites by sewage bacteria. *Water, Air and Soil Pollution* 134:351–366.

Lerch O. and P. Zinn (2003). Derivatization and gas chromatography-chemical ionization mass spectrometry of selected synthetic and natural endocrine disruptive chemicals. *Journal of Chromatography A* 991:77–97.

Lever R. (1996). Infection in atopic dermatitis. *Dermatology Therapy* 1:32–37.

Levine D.P. (2006). Vancomycin: A history. *Clinical Infectious Diseases* 42:S5–S12.

Levy S.B. (1992). *The Antibiotic Paradox—How Miracle Drugs are Destroying the Miracle*. Plenum, New York.

Lewis J.A., M.J. Clemens, and J.R. Tata (1976). Morphological and biochemical changes in the hepatic endoplasmic reticulum and golgi apparatus of male *Xenopus laevis* after induction of egg yolk protein synthesis by oestradiol-17β. *Molecular and Cellular Endocrinology* 4:311–329.

Li S., C.A. Wagner, J.A. Freisen, and D.W. Borst (2003). 3-Hydroxy-3-methylglutaryl-coenzyme A reductase in the lobster mandibular organ: Regulation by the eyestalk. *General and Comparative Endocrinology* 134:147–155.

Lin C., S. Gupta, D. Loebenberg, and M.N. Cayen (2000). Pharmacokinetics of an everninomycin (SCH 27899) in mice, rats, rabbits, and cynomolgus monkeys following intravenous adminstration. *Antimicrobial Agents and Chemotherapy* 44:916–919.

Lindsey M.E., M. Meyer, and E.M. Thurman (2001). Analysis of trace levels of sulfonamide and tetracycline antimicrobials in groundwater and surface water using solid-phase extraction and liquid chromatography/mass spectrometry. *Analytical Chemistry* 73:4640–4646.

Lipnic R.L. (1995). Structural Activity Relationships. In: G.M. Rand and S.R. Petrocelli (Eds.). *Fundamentals of Toxicology*. Taylor and Francis, Washington, D.C., pp. 609–655.

Litz N., H.W. Doering, M. Thiele, and H.-P. Blume (1987). The behavior of linear alkylbenzenesulfonate in different soils: A comparison between field and laboratory studies. *Ecotoxicology and Environmental Safety* 14:103–116.

Liu M. and D. Roy (1995). Surfactant-induced interactions and hydraulic conductivity changes in soil. *Waste Management* 15:463–470.

Lo M.W., J.B. McCrea, C.R. Shadle, M. Hesney, R. Chiou, D. Cylc, A.S. Yuan, and M.R. Goldberg (2000). Enalapril in RAPIDISC (wafer formulation): Pharmacokinetic evaluation of a novel, convenient formulation. *International Journal of Clinical Pharmacology and Therapeutics* 38:327–332.

Loftin K., C. Henry, C. Adams, and M. Mormile (2005). Inhibition of microbial metabolism in anaerobic lagoons by selected sulfonamides, tetracyclines, lincomycin, and tylosin tartrate. *Environmental Toxicology and Chemistry* 24:782–788.

Lorenzen A., R. Chapman, J.G. Hendel, and E. Topp (2005). Persistence and pathways of testosterone dissipation in agricultural soil. *Journal of Environmental Quality* 34:854–860.

Lorenzen A., J.G. Hendel, K.L. Conn, S. Bittman, A.B. Kwabiah, G. Lazarovitz, D. Massé, T.A. McAllister, and E. Topp (2004). Survey of hormone activities in municipal biosolids and animal manures. *Environmental Toxicology* 19:216–225.

Lovelock J.E. (1988). *The Ages of Gaia: A Biography of our Living Earth.* W.W. Norton, New York.

Lucchini S., A. Thompson, and J.C.D. Hinton (2001). Microarrays for microbiologists. *Microbiology* 147:1403–1414.

Lundholm C.E. (1997). DDE-induced eggshell thinning in birds: Effects of p,p-DDE on the calcium and prostaglandin metabolism of the eggshell gland. *Comparative Biochemistry and Physics C* 118:113–128.

Lunestad B.T. and J. Goksøyr (1990). Relaxation in the antibacterial effect of oxytetracycline in sea water by complex formation with magnesium and calcium. *Diseases of Aquatic Organisms* 9:67–72.

Lunn G. and L.C. Hellwig (1998). *Handbook of Derivatization Reactions for HPLC.* Wiley, New York.

Malekani K., J.A. Rice, and J.-S. Lin (1997). The effect of sequential removal of organic matter on the surface morphology of humin. *Soil Science* 162:333–342.

Malintan N.T. and M.A. Mohd (2006). Determination of sulfonamides in selected Malaysian swine wastewater by high-performance liquid chromatography. *Journal of Chromatography A* 1127:154–160.

Mannaerts G.P. et al., (1979). Mitochondrial and peroxisomal fatty acid oxidation in liver homogenates and isolated hepatocytes from control and clofibrate-treated rats. *Journal of Biological Chemistry* 254:4585–4595.

Mansell J., J. Drewes, and T. Rauch (2004). Removal mechanism of endocrine disrupting compounds (steroids) during soil aquifer treatment. *Water Science and Technology* 50:229–234.

Marchesi J.R., W.A. House, G.F. White, N.J. Russell, and I.S. Farr (1991). A comparative study of the adsorption of linear alkyl sulphates and alkylbenzene sulphonates on river sediments. *Colloids and Surfaces* 53:63–78.

Marchesi J.R., S.A. Owen, G.F. White, W.A. House, and N.J. Russell (1994). SDS-degrading bacteria attach to riverine sediment in response to the surfactant or its primary biodegradation product dodecan-1-ol. *Microbiology* 140:2999–3006.

Marcus A.J., M.J. Broekman, and D.J. Pinsky (2002). Cox inhibitors and thromboregulation. *New England Journal of Medicine* 347:1025–1026.

Marengo, J.R., R.A. Kok, K. O'Brien, R.R. Velagaleti, and J.M. Stamm (1997). Aerobic degradation of (^{14}C)-sarafloxacin hydrochloride in soil. *Environmental Toxicology and Chemistry*: 16:462–471.

Marieb E.N. (2001). *Human Anatomy and Physiology.* Benjamin Cummings, NewYork.

Martens R., H-G. Wetzstein, F. Zadrazil, M. Capelari, P. Hoffmann, and N. Schmeer (1996). Degradation of the fluoroquinolone enrofloxacin by wood-rotting fungi. *Applied and Environmental Microbiology* 62:4206–4209.

Martindale (1999). In: K. Parfitt (Ed.). *Martindale: The Complete Drug Reference*, 33rd ed. Pharmaceutical Press, London.

Matsuoka S., M. Kikuchi, S. Kimura, Y. Kurokawa, and S. Kawai (2005). Determination of estrogenic substances in the water of Muko river using in vitro assays, and the degradation of natural estrogens by aquatic bacteria. *Journal of Health Science* 51:178–184.

McAllister T.A., L.J. Yanke, G.D. Inglis, and M.E. Olson (2001). Is antibiotic use in diary cattle causing antibiotic resistance? *Advances in Diary Technology* 13:229–242.

McArdell C.S., E. Molnar, M.J. Suter, and W. Giger (2003). Occurrence and fate of macrolide antibiotics in wastewater treatment plants and in the Glatt Valley watershed, Switzerland. *Environmental Science & Technology* 37:5479–5486.

McAvoy D.C., B. Schatowitz, M. Jacob, A. Hauck, and W.S. Eckoff (2002). Measurement of triclosan in wastewater treatment systems. *Environmental Toxicology and Chemistry* 21:1323–1329.

McCann S.J., O. White, and B. Keevil (2002). Assay of teicoplanin in serum: Comparison of high-performance liquid chromatography and fluorescence polarization immunoassay. *Journal of Antimicrobial Chemotherapy* 50:107–110.

McGuire J.M., W.S. Boniece, C.E. Higgins, M.M. Hoehn, W.W. Stark, J. Westhead, and R.N. Wolfe (1961). Tylosin, a new antibiotic. I. Microbiological studies. *Antibiotics and Chemotherapy* 11:320–327.

McKay L.D. (1998). Hydrogeology and Groundwater Contamination. In: S.D. Pillai (Ed.). *Microbial Pathogens within Aquifers*. Springer, New York, pp. 1–23.

McKeller Q.A. (1997). Ecotoxicology and residence of antihelmitic compounds. *Veterinary Parasitology* 77:413–435.

McMurray L.M., M. Oethinger, and S.B. Levy (1998). Overexpression of *marA, soxS*, or *acrAB* produces resistance to triclosan in laboratory and clinical strains of *Escherichia coli*. *FEMS Microbiology Letters* 166:305–309.

Medinsky M.A. and J.L. Valentine (2001). Toxicokinteics. In: C.D. Klaassen (Ed.). *Casarett and Doull's Toxicology: The Basic Science of Poisons*. McGraw-Hill, New York, pp. 225–237.

Meinert R., J. Schuz, P. Kaatsch, and J. Michaelis (2000). Leukemia and non-Hodgkin's lymphoma in childhood and exposure to pesticides: Results of a register-based case-control study in Germany. *American Journal of Epidemiology* 7:639–646.

Méndez E., J.V. Planas, J. Castillo, I. Navarro, and J. Gutiérrez (2001). Identification of a Type II insulin-like growth factor receptor in fish embryos. *Endocrinology* 142:1090–1097.

Merck (2005). Fosamax. Available at: http://www.fda.gov/cder/foi/label/2005/020560s038_021575s007lbl.pdf#search=%22fosamax%20merck%20whitehouse%20station%20oral%20solution%22; accessed 10/04/2006.

Metcalfe C., X-S. Miao, W. Hua, R. Letcher, and M. Servos (2004). Pharmaceuticals in the Canadian Environment. In: K. Kümmerer (Ed.). *Pharmaceuticals in the Environment: Sources, Fate, Effects and Risks*. Springer, New York, pp. 67–90.

Metcalfe C.D., X. Miao, B.G. Koenig, and I. Struger (2003). Distribution of acidic and neutral drugs in surface waters near sewage treatment plants in the lower Great Lakes, Canada. *Environmental Toxicology Chemistry* 22:2881–2889.

Meteyer C.U., B.A. Rideout, M. Gilbert, H.L. Shivaprasad, and J.L. Oaks (2005). Pathology and proposed pathophysiology of diclofenac poisoning in free-living and experimentally exposed oriental white-backed vultures (*Gyps bengalensis*). *Journal of Wildlife Diseases* 41:707–716.

Meyer F.P. (1996). Minocycline for acne. Food reduces minocycline's bioavailability. *British Journal of Medicine* 312:1101.

Miao X. and C.D. Metcalfe (2003). Determination of pharmaceuticals in aqueous samples using positive and negative voltage switching microbore liquid chromatography/electrospray ionization tandem mass spectrometry. *Journal of Mass Spectrometry* 38:27–34.

Miège C., M. Favier, C. Brosse, J-P. Canler, and M. Coquery (2007). Concentrations and fluxes of β blockers in effluents of wastewater treatment plants (WWTPs) for the Lyon area (France). Available at: http://www.cid.csic.es/emco/pdf%20oral/Miege.pdf; accessed 5/8/2007.

Migliore L., E. Alessi, L. Busani, and A. Caprioli (2002). Effects of the use of flumequine in aquaculture: Microbial resistance and sediment contamination. *Fresenius Environmental Bulletin* 11:1–5.

Migliore, L., G. Brambilla, P. Casoria, C. Civitareale, S. Cozzolino, and L. Gaudio (1996). Effect of sulphadimethoxine on barley (*Hordeum distichum* L., Poaceae, Liliopsida) in laboratory terrestrial models. *Agriculture, Ecosystem, and Environment* 60:121–128.

Migliorie L., G. Brambilla, S. Cozzolino, and L. Gaudio (1995). Effect on plants of sulfadimethoxine used in intensive farming (*Panicum miliaceum, Pisum sativum* and *Zea mays*). *Agriculture Ecosystems and Environment* 52:103–110.

Migliore, L., C. Civitareale, S. Cozzolino, P. Casoria, G. Brambilla, and L. Gaudio (1998). Laboratory models to evaluate phytotoxicity of sulphadimethoxine on terrestrial plants. *Chemosphere* 37:2957–2961.

Migliore, L., S. Cozzolino, and M. Fiori (2000). Phytotoxocity to and uptake of flumequine used in intensive aquaculture on the aquatic weed, *Lythrum salicaria* L. *Chemosphere* 40:741–750.

Migliore L., S.C. Sivitareale, G. Brambilla, and G. Dojmi Di Delupis (1997). Toxicity of several important agricultural antibiotics to Artemia. *Water Research* 31:1801–1806.

Millam J.R., C.B. Craig-Veit, A.E. Quaglino, A.L. Erichsen, T.R. Famula, and D.M. Fry (2001). Post-hatch oral estrogen exposure impairs adult reproductive performance of zebra finch in a sex-specific manner. *Hormones and Behavior* 40:542–549.

Miller D., V. Urdaneta, and A. Weltman (2002). Vancomycin-resistant *Staphylococcus aureus*—Pennsylvania, 2002. *Morbidity and Mortality Weekly Report* 51:509.

Miller M.R., E. Wentz, and S. Ong (1999). Acetaminophen alters estrogenic responses in vitro: Inhibition of estrogen-dependent vitellogenin production in trout liver cells. *Toxicological Sciences* 48:30–37.

Mitchell A.A. (2003). Systematic identification of drugs that cause birth defects—a new opportunity. *New England Journal of Medicine* 349:2556–2559.

Mitema E.S., G.M. Kikuvu, H.C. Wegener, and K. Stohr (2001). An assessment of antimicrobial consumption in food producing animals in Kenya. *Journal of Veterinary Pharmacology and Therapeutics* 24:385–390.

Moellering R.C. (2003). Linezolid: The first oxazolidinone antimicrobial. *Annals of Internal Medicine* 138:135–142.

Moellering R.C. (2006). Vancomycin: A 50-year reassessment. *Clinical Infectious Diseases* 42:S3–S4.

Mojaverian P., E. Radwanski, M.B. Affrime, M.N. Cayen, and C.C. Lin (1994). Pharmacokinetics of the triazole antifungal agent genaconazole in healthy men after oral and intravenous administration. *Antimicrobial Agents and Chemotherapy* 38:2758–2762.

Möller P., P. Dulski, M. Bau, A. Knappe, A. Pekdeger, and C. Sommer-von Jarmersted (2000). Anthropongeinc gadolinium as a conservative tracer in hydrology. *Journal of Geochemical Exploration* 69/70:409–414.

Mölstad S., C.S. Lundborg, A-K. Karlsson, and O. Cars (2002). Antibiotic prescription rates vary markedly between 13 European countries. *Scandinavian Journal of Infectious Diseases* 34:366–371.

Montgomery-Brown J., J.E. Drewes, P. Fox, and M. Reinhard (2003). Behavior of alkylphenol polyethoxylate metabolites during soil aquifer treatment. *Water Research* 37:3672–3681.

Montgomery-Brown J. and M. Reinhard (2003). Occurrence and behavior of alkylphenol polyethoxylates in the environment. *Environmental Engineering Science* 20:471–486.

Morgan S. and B. MaćGibbon (2007). Pharmaceutical use and outcomes: Always a need for a sober second look. *HealthCare Policy* 3:10–15.

Morgan S., M. McMahon, J. Lam, D. Mooney, and D. Raymond (2005). Canadian Rx Atlas. Center for Health Services and Policy Research, Vancover, BC, Canada. Available at: http://www.chspr.ubc.ca/files/publications/2005/chspr05-35R.pdf; accessed 11/15/2007.

Morita C., K. Sano, S. Morimatsu, H. Kiura, T. Goto, T. Kohno, W. Hong, H. Miyoshi, A. Iwasawa, Y. Nakamura, M. Tagawa, O. Yokosuka, H. Saisho, T. Maeda, and Y. Katsuoka (2000). Disinfection potential of electrolyzed solutions containing sodium chloride at low concentrations. *Journal of Virologial Methods* 85:163–174.

Moyniham R., I. Heath, and D. Henry (2002). Selling sickness: The pharmaceutical industry and disease mongering. *British Medical Journal* 324:886–890.

Muñoz-Bellido J.L., S. Muñoz-Criado, and J.A. Garcìa-Rodriguez (1996). In vitro activity of psychiatric drugs against *Corynebacterium urealyticum* (*Corynebacterium* group D2). *Journal of Antimicrobial Chemotherapy* 37:1005–1009.

Murdoch D. and D. McTavish (1992). Sertraline: A review of its pharmacodynamic properties and therapeutic potential in depression and obsessive compulsive disorder. *Drugs* 44:604–624.

Murray B.E. (1997). Vancomycin-resistant enterococci. *American Journal of Medicine* 101:284–293.

Murray M.D. and D.C. Brater (1993). Renal toxicity of the nonsteroidal anti-inflammatory drugs. *Annual Reviews of Pharmacology and Toxicology* 32:435–465.

Musavi S. and P. Kakker (2003). Effect of diazepam treatment and its withdrawal on pro/antioxidative in rat brain. *Molecular and Cellular Biochemistry* 245:51–56.

Nagano, A., E. Arikawa, and H. Kobayashi (1992). The treatment of liquor wastewater containing high-strength suspended solids by membrane bioreactor system. *Water Science and Technology* 26:887–895.

Nakada N., T. Tanishima, H. Shinohara, K. Kiri, and H. Takada (2006). Pharmaceutical chemicals and endocrine disrupters in municipal wastewater in Tokyo and their removal during activated sludge treatment. *Water Research* 40:3297–3303.

Nakajima N., T. Nakano, F. Harada, H. Taniguchi, I. Yokoyama, J. Hirose, E. Daikoku, and K. Sano (2004). Evaluation of disinfective potential of reactivated free chlorine in pooled tap water by electrolysis. *Journal of Microbiological Methods* 57:163–173.

Nakata H. (2005). Occurrence of synthetic musk fragrances in marine mammals and sharks from Japanese coastal waters. *Environmental Science and Technology* 39:3430–3434.

Nash J.P., D.E. Kime, L.T.M. Van der Ven, P.W. Wester, F. Brion, G. Maack, P. Stahlschmidt-Allner, and C.R. Tyler (2004). Long-term exposure to environmental concentrations of the pharmaceutical ethynylestradiol causes reproductive failure in fish. *Environmental Health Perspectives* 112:1725–1733.

Nation J.L. (2002). *Insect Physiology and Biochemistry.* CRC Press, Boca Raton, FL.

Neamtu M., I. Siminiceanu, and A. Kettrup (2000). Kinetics of nitromusk compounds degradation in water by ultraviolet radiation and hydrogen peroxide. *Chemosphere* 40:1407–1410.

Nentwig G., M. Oetken, and J. Oehlmann (2004). Effects of Phamaceuticals on Aquatic Invertebrates—The Example of Carbamazepine and Clofibric Acid. In: K. Kümmerer (Ed.). *Pharmaceuticals in the Environment: Sources, Fate, Effects and Risks.* Springer, New York, pp. 195–208.

Newcombe G., M. Drikas, S. Assemi, and R. Beckett (1997). Influence of characterized natural organic material on activated carbon adsorption: I. Characterization of concentrated reservoir water. *Water Research* 31:965–972.

Nice H.E., D. Morritt, M. Crane, and M. Thorndyke (2003). Long-term and transgenerational effects of nonylphenol exposure at a key stage in the development of *Crassostrea gigas*. Possible endocrine disruption? *Marine Ecology Progress Series* 256:293–300.

Nickerson J.G., S.G. Dugan, G. Drouin, and T.W. Moon (2001). A putative beta-adrenoceptor from the rainbow trout (*Oncorhynchus mykiss*). Molecular characterization and pharmacology. *European Journal of Biochemistry* 268:6465–6472.

Nishijima S., S. Namura, S. Kawai, H. Hosokawa, and Y. Asada (1995). *Staphylococcus aureus* on hand surface and nasal carriage in patients with atopic dermatitis. *Journal of the American Academy of Dermatology* 32:677–679.

Noble W.C. (1998). Skin bacteriology and the role of *Staphylococcus aureus* in infection. *British Journal of Dermatology* 139(Suppl. 53):9–12.

Noble W.C., Z. Virani, and R.G. Cree (1992). Co-transfer of vancomycin and other resistance genes from *Enterococcus faecalis* NCTC to *Staphylococcus aureus*. *FEMS Microbiology Letters* 72:195–198.

Novartis (2002). Miacalcin (Calcitonin-Salmon) T2002–84 89018101. Available at: www. http://www.pharma.us.novartis.com/product/pi/pdf/miacalcin_injection.pdf#search = %22 Miacalcin%20(Calcitonin-Salmon)%20T2002-84%2089018101%22; downloaded 10/04/2006.

Nowack B. (2002). Environmental chemistry of aminopolycarboxylate chelating agents. *Environmental Science and Technology* 36:4009–4016.

Nowara A., J. Burhenne, and M. Spiteller (1997). Binding of fluoroquinolone carboxylic acid derivatives to clay minerals. *Journal of Agriculture and Food Chemistry* 45:1459–1463.

NRC (1994). *Science and Judgement in Risk Assessment.* National Academy Press, Washington, D.C.

Nunes B., F. Carvalho, and L. Guilhermino (2006). Effects of widely used pharmaceuticals and a detergent on oxidative stress biomarkers of the crustacean *Artemia parthenogenatica*. *Chemosphere* 62:581–594.

Nygaard B.T., B.T. Lunestead, H. Hektoen, J.A. Berge, and V. Hormazabel (1992). Resistance to oxytetracyline, oxolinic acid and furazolidone in bacteria from marine sediments. *Aquaculture* 104:31–36.

Oaks J.L., M. Gilbert, M.Z. Virani, R.T. Watson, C.U. Meteyer, B.A. Rideout, H.L. Shivaprasad, S. Ahmed, M.J.I. Chaudhry, M. Arshad, S. Mahmood, A. Ali, and A.A. Khan (2004). Diclofenac residues as the cause of vulture population decline in Pakistan. *Nature* 427:630–633.

Ochman H., J.G. Lawrence, and E.A. Groisman (2000). Lateral gene transfer and the nature of bacterial innovation. *Nature* 405:299–304.

Odum E.P. (1983). *Basic Ecology*. Saunders College, Philadelphia, PA.

Ohlsen K., T. Ternes, G. Werner, U. Wallner, D. Löffler, W. Ziebuhr, W. Witte, and J. Hacker (2003). Impact of antibiotics on conjugational resistance gene transfer in *Staphylococcus aureus* in sewage. *Environmental Microbiology* 5:711–716.

Oka H., Y. Ito, and H. Matsumoto (2000). Chromatographic analysis of tetracycline antibiotics in foods. *Journal of Chromatography* A882:109–133.

Okeke C.C. (2002). Current USP requirements and proposals on packaging systems. *American Pharmaceutical Outsourcing* 3:34–38.

Onan L.J. and T.M. LaPara (2003). Tylosin-resistant bacteria cultivated from agricultural soil. *FEMS Microbiology Letters* 220:15–20.

Orvos D.R., D.J. Versteeg, J. Inauen, M. Capdevielle, A. Rothenstein, and V. Cunningham (2002). Aquatic toxicity of triclosan. *Environmental Toxicology and Chemistry* 21:1338–1349.

O'Sullivan A.J., M.C. Kennedy, J.H. Casey, R.O. Day, B. Corrigan, and A.D. Wodak (2000). Anabolic androgenic steroids: Medicinal assessment of present, past and potential users. *Medical Journal of Australia* 173:323–327.

Palm K., K. Luthman, A-L. Ungell, G. Strandlund, F. Beigi, P. Lundahl, and P. Artursson (1998). Evaluation of dynamic polar molecular surface area as predictor of drug absorption: Comparison with other computational and experimental predictors. *Journal of Medicinal Chemistry* 41:5382–5392.

Panter G.H., R.S. Thompson, N. Beresord, and J.P. Sumpter (1999). Transformation of a non-oestrogenic steroid metabolite to an oestrogenically active substance by minimal bacterial activity. *Chemosphere* 38:3579–3596.

Parker V.G., R.M. Mayo, B.N. Logan, B.J. Holder, and P.T. Smart (2002). Toxins and diabetes mellitus: An environmental connection? *Diabetes Spectrum* 15:109–112.

Parnetti L. (1995). Clinical pharmacokinetics of drugs for Alzheimers disease. *Clinical Pharmacokinetics* 29:110–129.

Pascoe D., W. Karntanut, and C.T. Müller (2003). Do pharmaceuticals affect freshwater invertebrates? A study with the cnidarian *Hydra vulgaris*. *Chemosphere* 51:521–528.

Paul C., S.M. Rhind, C.E. Kyle, H. Scott, C. McKinell, and R.M. Sharpe (2005). Cellular and hormonal disruption of fetal testis development in sheep reared on pasture treated with sewage sludge. *Environmental Health Perspectives* 113:1580–1587.

Peck A.M. and K.C. Hornbuckle (2004). Synthetic musk fragrances in Lake Michigan. *Environmental Science and Technology* 38:367–372.

Peitsaro N., O.V. Anichtchik, and P. Panula (2000). Identification of a histamine H3-like receptor in zebrafish (*Danio rerio*) brain. *Journal of Neurochemistry* 75:718–724.

Penrose L.S. and G.F. Smith (1966). *Down's Anomaly*. Little, Brown, Boston.

Perkins E. and D. Schlenk (1998). Immunochemical characterization on hepatic cytochrome P450 isozymes in the channel catfish: Assessment of sexual, development and treatment-related effects. *Comparative Biochemistry and Physiology* C 121:305–310.

Peschka M., J.P. Eubeler, and T.P. Knepper (2006). Occurrence and fate of barbiturates in the aquatic environment. *Environmental Science and Technology* 40:7200–7206.

Petersen A., J.S. Andersen, T. Kaewmak, T. Somsiri, and A. Dalsgaard (2002). Impact of integrated fish farming on antimicrobial resistance in a pond environment. *Applied and Environmental Microbiology* 68:6036–6042.

Peterson E.W., R.K. Davis, and H.A. Orndroff (2000). 17β-estradiol as an indicator of animal waste contamination in mantled karst aquifers. *Journal of Environmental Quality* 29:826–834.

Peterson S.M., G.E. Batley, and M.S. Scammell (1993). Tetracycline in antifouling paints. *Marine Pollution Bulletin* 26:96–100.

Pfluger P. and D.R. Dietrich (2001). Pharmaceuticals in the Environment—An Overview and Principle Considerations. In: K. Kümmerer (Ed.). *Pharmaceuticals in the Environment: Sources, Fate, Effects and Risks*. Springer, New York, pp. 11–17.

Piferrer F. (2001). Endocrine sex control strategies for the feminization of teleost fish. *Aquaculture* 197:229–281.

Pinck L.A., W.F. Holton, and F.E. Allison (1961a). Antibiotics in soil: 1. Physico-chemical studies of antibiotic-clay complexes. *Soil Science* 91:22–28.

Pinck L.A., D.A. Soulides, and F.E. Allison (1961b). Antibiotics in soils: 2. Extent and mechanism of release. *Soil Science* 91:94–99.

Planas J.V., E. Méndez, N. Baños, E. Capilla, I. Navarro, and J. Gutiérrez (2000). Insulin and IGF-I receptors in trout dipose tissue are physiologically regulated by circulating hormone levels. *Journal of Experimental Biology* 203:1153–1159.

Poiger T., H.-R. Buser, et al. (2004). Occurrence of UV filter compounds from sunscreens in surface waters: Regional mass balance in two Swiss Lakes. *Chemosphere* 55:951–963.

Porras A.G., S.D. Holland, and B.J. Gertz (1999). Pharmacokinetics of alendronate. *Clinical Pharmacokinetics* 36:315–328.

Prokvová M., J. Augustin, and A. Vrbanova (1997). Enrichment, isolation and characterization of dialkyl sulfosuccinate degrading bacteria *Comamonas terrigena* N3H and *Comamonas terrigena* N1C. *Folia Microbiologica* 42:635–639.

Prueksaritanont T., J.J. Zhao, B. Ma, B.A. Roadcap, C.Y. Tang, Y. Qui, L.D. Liu, J.H. Lin, P.G. Pearson, and T.A. Baillie (2002). Mechanistic studies on metabolic interactions between gemfibrozil and statins. *Journal of Pharmacology and Experimental Therapeutics* 301:1042–1051.

Purdom C.E., V.J. Bye, N.C. Eno, P.A. Hardiman, J.P. Sumpter, and C.R. Tyler (1994). Estrogenic effects of effluents from sewage treatment works. *Chemical Ecology* 8:275–285.

Pursell L., T. Dineen, J. Kerry, S. Vaughan, and P. Smith (1996). The biological significance of breakpoint concentrations of oxytetracycline in media for the examination of marine sediment microflora. *Aquaculture* 145:21–30.

Qu B., Q.-T. Li, K.P. Wong, T.M.C. Tan, and B. Halliwell (2001). Mechanisms of clofibrate hepatoxicity: Mitochondrial damage and oxidative stress in hepatocytes. *Free Radical Biology and Medicine* 31:659–669.

Quaglino A.E., C.B. Craig-Veit, M.R. Viant, A.L. Erichsen, D.M. Fry, and J.R. Millam (2002). Oral estrogen masculinizes female zebra finch song system. *Hormones and Behavior* 41:236–241.

Queener S.F. and K. Gutierrez (2003). Ophthalmic Drugs. In: K. Gutierrez and S.F. Queener (Eds.). *Pharmacology for Nursing Practice*. Mosby, St. Louis, MO, pp. 1075–1094.

Quiroga J.M., D. Sales, and A. Gómez-Parra (1989). Experimental evaluation of pollution potential of anionic surfactants in the marine environment. *Water Research* 23:801–808.

Raloff J. (1998). Drugged waters. *Science News* 153:187–189.

Raloff J. (2000). More waters test positive for drugs. *Science News* 157:212–??

Raloff J. (2002). Pharm pollution: Excreted antibiotics can poison plants. *Science News* 161:406–408.

Ranganathan R., E.R. Sawin, C. Trent, and H.R. Horvitz (2001). Mutation in the *Caenahabditis elegans* serotonin reuptake transporter MOD-5 reveal serotonin-dependent and independent activities of fluoxatine. *Journal of Neuroscience* 21:5871–5884.

Ranny R.E. (1977). Comparative metabolism of ethinylestradiol in man. *Journal of Toxicology and Environmental Health* 3:139–166.

Rapport D.J., J.M. Howard, R. Lannigan, C.M. Anjema, and W. McCauley (2001). Strange bedfellows: Ecosystem health in the medical curriculum. *Ecosystem Health* 7:155–162.

Renner D. (2002). Do cattle growth hormones pose an environmental risk? *Environmental Science and Technology* 36:194A–197A.

Renshaw P.F, A.R. Guimaraes, and M. Fava (1992). Accumulation of fluoxentine and norflouxentine in human brain during therapeutic administration. *American Journal of Psychiatry* 149:1592–1594.

Retsema J. and W. Fu (2001). Macrolides: Structures and microbial targets. *International Journal of Antimicrobial Agents* 18:S3–S10.

Reynolds J.E.F. (Ed.) (1996). *Martindale—The Extra Pharmacopoeia*. Pharmaceutical Press, London.

Rhodes G., G. Huys, J. Swings, P. Mcgann, M. Hiney, P. Smith, and R.W. Pickup (2000). Distribution of oxytetracycline resistance plasmids between aeromonads in hospital and aquaculture environments: Implication of Tn*1721* in dissemination of the tetracycline resistance determinant Tet A. *Applied and Environmental Microbiology* 66:3883–3890.

Richards S.M., C.J. Wilson, D.J. Johnson, D.M. Castle, M. Lam, S.A. Mabury, P.K. Sibley, and K.R. Solomon. (2004). Effects of pharmaceutical mixtures in aquatic microcosms. *Environmental Toxicology and Chemistry* 23:1035–1042.

Richardson M.L. and J.M. Bowron (1985). The fate of pharmaceutical chemicals in the aquatic environment. *Journal Pharmacy and Pharmacology* 37:1–12.

Ricking M., J. Schwarzbauer, J. Hellon, A.S. Svenson, and V. Zitko (2003). Polycyclic aromatic musk compounds in sewage treatment plant effluents of Canada and Sweden— first results. *Marine Pollution Bulletin* 46:410–417.

Rimkus G.G. and M. Wolf (1996). Polycyclic musk fragrances in human adipose tissue and human milk. *Chemosphere* 33:2033–2043.

Ring B.J., J.A. Eckstein, J.S. Gillespie, S.N. Binkley, M. VandenBranden, and S.A. Wrighton (2001). Identification of the human cytochrome P450 responsible for in vitro formation of r- and s-norfloxentine. *Journal of Pharmacology and Experimental Therapeutics* 297:1044–1050.

Ripa S., L. Ferrante, and M. Prenna (1993). Pharmacokineitcs of fluconazole in normal volunteers. *Chemotherapy* 39:6–12.

Roberts M.C., J. Sutcliffe, P. Courvalin, L.B. Jesen, J. Rood, and H. Seppala (1999). Nomenclature for macrolide and macrolide-lincosamide-streptogramin B resistance determinants. *Antimicrobial Agents and Chemotherapy* 43:2823–2830.

Robertson J.R., L.K. Fox, D.D. Hancock, J.M. Gay, and T.E. Besser (1994). Ecology of *Staphylococcus aureus* isolated from various sites on diary farms. *Journal of Diary Science* 77:3354–3364.

Rodriguez I., J. B. Quintana, J. Carpinteiro, A. M. Carro, R. A. Lorenzo, and R. Cela (2003). Determination of acidic drugs in sewage water by gas chromatography–mass spectrometry as tert.-butyldimethylsilyl derivatives. *Journal of Chromatography A* 985:265–274.

Rooklidge S.J., E.R. Burns, and J.P. Bolte (2005). Modeling antimicrobial contaminant removal in slow sand filtration. *Water Research* 39:331–339.

Routaledge E.J., D. Sheahan, C. Desbrow, G.C. Brighty, M. Waldock, and J.P. Sumpter (1998). Identification of estrogenic chemicals in STW effluents. 2. In vivo responses in trout and roach. *Environmental Science and Technology* 32:1559–1565.

Rowland M. and T.N. Tozer (1995). *Clinical Pharmacokinetics: Concepts and Applications*, 3rd ed. Williams and Wilkins, Baltimore, MD.

Ruoff K.L., D.R. Kuritzkes, J.S. Wolfson, and M.J. Ferraro (1988). Vancomycin-resistant Gram-positive bacteria isolated from human sources. *Journal of Clinical Microbiology* 26:2064–2068.

Russell A.D. (2000). Do biocides select for antibiotic resistance? *Journal of Pharmacy and Pharmacology* 52:227–233.

Ruyter B., O. Andersen, A. Dehli, A.-K. Ostlund Farrants, T. Gjoen, and M.S. Thomassen (1997). Peroxisome proliferator activated receptors in Atlantic salmon (*Salmo salar*): Effects on PPAR transcription and acyl-CoA oxidase activity in hepatocytes by peroxisome proliferators and fatty acids. *BBA-Lipids and Lipid Metabolism* 1348:331–338.

Salvatore M.J. and K.E. Katz (1993). Solubility of antibiotics used in animal feeds in selected solvents. *Journal of AOAC International* 76:952–956.

Samuelson O.B. (1994). Degradation of oxytetracylcine in seawater at two different temperatures and light intensities, and the persistence of oxytetracycline in the sediment from a fish farm. *Aquaculture* 83:7–16.

Samuelson O.B., B.T. Lunestad, A. Ervik, and S. Fjelde (1994). Stability of antibacterial agents in an artificial marine aquaculture sediments studied under laboratory conditions. *Aquaculture* 126:283–290.

Samuelsen O.B., V. Torsvik, and A. Ervik (1992). Long-range changes in oxytetracycline concentration and bacterial resistance toward oxytetracycline in a fish farm sediment after medication. *Science of the Total Environment* 114:25–36.

Saperstein G.L., S. Hinckley, and J.E. Post (1988). Taking the team approach to solving staphylococcal mastitis infection. *Veterinary Medicine* 83:940–947.

Sarmah A.K., M.T. Meyer, and A.B.A. Boxall (2006). A global perspective on the use, sales, exposure pathways, occurrence, fate and effects of veterinary antibiotics (VAs) in the environment. *Chemosphere* 65:725–759.

Sassman S.A. and L.S. Lee (2005). Sorption of three tetracyclines by several soils: Assessing the role of pH and cation exchange. *Environmental Science and Technology* 39:7452–7459.

Schlumpf M., B. Cotton, M. Conscience, V. Haller, B. Steinmann, and W. Lichtensteiger (2002). *In vitro* and *in vivo* estrogenicity of UV screens. *Environmental Health Perspectives* 109:239–244.

Schlüsener M.P., K. Bester, and M. Spiteller (2003a). Determination of antibiotics from soil by pressurized liquid extraction and liquid chromatography tandem mass spectrometry. *Journal of Chromatography A* 1003:21–28.

Schlüsener M.P., K. Bester, and M. Spiteller (2003b). Determination of antibiotics such as macrolides, ionophores and tiamulin in liquid manure by HPLC-MS/MS. *Analytical and Bioanalytical Chemistry* 375:942–947.

Schmid T., J. Gonzales-Valero, H. Rufli, and D.R. Dietrich (2005). Determination of Vitellogen Kinetics in Male Fathead Minnows (*Pimephales promelas*). In: D.R. Dietrich, S.F. Webb, and T. Petry (Eds.). *Hot Spot Pollutants: Pharmaceuticals in the Environment*. Elsevier, New York, pp. 115–129.

Schneider, Kuch, Braun, and J.W. Met (2004). Pharmaceuticals in landfill leachates and receiving WWTP influents. SETAC. Available at: http://www.iswa.unistuttgart.de/ch/poster/setac04_pharmaceuticals_in_landfill_leachates.pdf; accessed 4/23/2007.

Schowanek D. and S. Webb (2005). Exposure Simulation for Pharmaceuticals in European Surface Waters with GREAT-ER. In: D.R. Dietrich, S.F. Webb, and T. Petry (Eds.). *Hot Spot Pollutants: Pharmaceuticals in the Environment*. Elsevier, New York, pp. 63–78.

Schrag S.J., V. Perrot, and B.R. Levin (1997). Adaptation to fitness costs of antibiotic resistance in *Escherichia coli*. *Proceedings of the Royal Society of London Series B—Biological Sciences* 264:1287–1291.

Schreurs R.H.M.M., J. Legler, E. Artola-Garicano, T.L. Sinnige, P.H. Lanser, W. Seinen, and B. van der Burg (2004). In vitro and in vivo antiestrogenic effects of polycyclic musks in Zebrafish. *Environmental Science and Technology* 38:997–1002.

Schulte-Oehlmann U., M. Oetken, J. Bachmann, and J. Oehlmann (2004). Effects of Ethinylestradiol and Methyltestosterone in Prosobranch Snails. In: K. Kümmerer (Ed.). *Pharmaceuticals in the Environment: Sources, Fate, Effects and Risks*. Springer, New York, pp. 233–247.

Schumann W. (1991). Pharmacokinetics of oral-turinabol in man. *Pharmazie* 46:650–654.

Schwaiger J., O. H. Spieser, C. Bauer, H. Ferling, U. Mallow, W. Kalbfus, and R. D. Negele (2000). Chronic toxicity of nonylphenol and ethinyl-estradiol: Haematological and histopathological effects in juvenile common carp (*Cyprirus carpio*). *Aquatic Toxicology* 51:69–78.

Schwartz T., T. Kohnen, B. Jansen, and U. Obst (2003). Detection of antibiotic-resistant bacteria and their resistance genes in wastewater, surface water and drinking water biofilms. *FEMS Microbiology Ecology* 43:325–335.

Sedlak D.L., C.-H. Huang, and K. Pinkston (2004). Strategies for Selecting Pharmaceuticals to Assess Attenuation During Indirect Potable Water Reuse. In: K. Kümmerer (Ed.). *Pharmaceuticals in the Environment: Sources, Fate, Effects and Risks*. Springer, New York, pp. 107–120.

Sedlak D.L. and K.E. Pinkston (2001). Factors affecting the concentrations of pharmaceuticals released to the aquatic environment. *Water Research Update* 120:56–64.

Seehusen D.A. and J. Edwards (2006). Patient practices and beliefs concerning disposal of medications. *Journal of American Board of Family Medicine* 19:542–547.

Selvaratnam S. and J.D. Kunberger (2004). Increased frequency of drug-resistant bacteria and fecal coliforms in an Indiana Creek adjacent to farmland amended with treated sludge. *Canadian Journal of Microbiology* 50:653–656.

Sengeløv G., Y. Agersø, B. Halling-Sørensen, S.B. Baloda, J.S. Andersen, and L.B. Jensen (2003). Bacterial antibiotic resistance levels in Danish farmland as a result of treatment with pig manure slurry. *Environment International* 28:587–595.

Setchell K.D.R., S.J. Gosselin, M.B. Welsh, J.O. Johnston, W.F. Balistreri, L.W. Kramer, B.L. Dresser, and M.J. Tarr (1987). Dietary estrogens—a probable cause of infertility and liver disease in captive cheetahs. *Gastroenterology* 93:255–233.

Seveno N.A., D. Kallifidas, K. Smalla, J.D. van Elsas, J-M. Collard, A.D. Karagouni, and E.M.H. Wellington (2002). Occurrence and reservoirs of antibiotic resistance genes in the environment. *Reviews in Medical Microbiology* 13:15–27.

Sharpe R.M. and N.E. Shakkebaek (2003). Male reproductive disorders and the role of endocrine disruption: Advances in understanding and identification of areas for future research. *Pure and Applied Chemistry* 75:2023–2038.

Shore L.S., Y. Kapulnik, B. Ben-Dov, Y. Fridman, S. Wininger, and M. Shemesh (1992). Effects of estrone on 17β-estradiol on vegetative growth of *Medicago sativa*. *Physiologia Plantarium* 84:217–222.

SICOR (2005). Adrucil®. (Fluorouracil injection, USP) Package insert Y36-102-103. Available at: newsicor.com/products/pi/adrucil_6-2005_y36-102-103.pdf; accessed 5/20/2007.

Sievert D.M., M.L. Boulton, G. Stolzman, et al. (2002). *Staphylococcus aureus* resistant to vancomycin—United States, 2002. *Morbidity and Mortality Weekly Report* 51:565–567.

Simonich S.L., T.W. Federle, W.S. Eckoff, A. Rottiers, S. Webb, D. Sabaliunas, and W. de Wolf (2002). Removal of fragrance materials during U.S. and European wastewater treatment. *Environmental Science and Technology* 36:2839–2847.

Simonsen G.S., H. Haaheim, K.H. Dahl, H. Kruse, A. Lovseth, O. Olsvik, and A. Sundsfjord (1998). Transmisssion of vanA-type vancomycin-resistant enterococci and vanA resistance elements between chicken and humans at avoparcin-exposed farms. *Microbial Drug Resistance—Mechanisms Epidemiology and Disease* 4:313–318.

Sirés I., P.L. Cabot, F. Centellas, J.A. Garrido, R.M. Rodríguez, C. Arias, and E. Brillas (2006). Electrochemical degradation of clofibric acid in water by anodic oxidation: Comparative study with platinum and boron-doped diamond electrodes. *Electrochimica Acta* 52:75–85.

Sirisattha S., Y. Momose, E. Kitagawa, and H. Iwahashi (2004). Toxicity of anionic detergents determined by *Saccharomyces cerevisiae* microarray analysis. *Water Research* 38:61–70.

Sischo W.M., L.E. Hieder, G.Y. Miller, and D.A. Moore (1993). Prevalence of contagious pathogens of borne mastitis and use of mastitis control practices. *Journal of the American Veterinary Medical Association* 202:595–500.

Sithole B.B. and R.D. Guy (1987a). Models for tetracylcine in aquatic environments. 1. Interaction with bentonite clay systems. *Water Air and Soil Pollution* 32:303–314.

Sithole B.B. and R.D. Guy (1987b). Models for tetracycline in aquatic environments. 2. Interaction with humic substances. *Water Air and Soil Pollution* 32:315–321.

Skov T. et al (1990). Risks for physicians handling antineoplastic drugs. *Lancet* 336:1446–?

Smith K.E., J.M. Besser, C.W. Hedberg, F.E. Leano, J.B. Bender, J.H. Wicklund, B.P. Johnson, K.A. Moore, M.T. Osterholm, et al. (1999a). Quinolone-resitant *Campylobacter jejuni* infections in Minnesota, 1992–1998. *New England Journal of Medicine* 340:1525–1532.

Smith S.R. (1996). *Agricultural Recycling of Sewage Sludge and the Environment*. CAB International, Wallingford, Oxon, UK.

Smith T.L., M.L. Pearson, K.R. Wilcox, C. Cruz, M.V. Lancaster, B. Robinson-Dunn, F.C. Tenover, M.J. Zervos, J.D. Dand, E. White, W.R. Jarvis, et al. (1999b). Emergence of vancomycin reistiance in *Staphylococcus aureus*. *New England Journal of Medicine* 340(7):493–501.

Smith-Kielland A., B. Skuterud, K.M. Olsen, and J. Morland (2001). Urinary excretion of diazepam metabolites in healthy volunteers and drug users. *Scandinavian Journal of Clinical and Laboratory Investigations* 61:237–246.

Snyder S.A., P. Westerhoff, Y. Youn, and D.L. Sedlak (2003). Pharmaceuticals, personal care products and endocrine disruptors in water: Implications for water treatment. *Environmental Engineering Science* 20:449–469.

So S-S. and M. Karplus (1999). A comparative study of ligand-receptor complex binding affinity prediction methods based on glycogen phosphorylase inhibitors. *Journal of Computer Aided Molecular Design* 13:243–258.

Sorensen T.L., M. Blom, D.L. Monnet, N. Frimodt-Moller, R.L. Poulsen, and F. Espersen (2001). Transient intestinal carriage after ingestion of antibiotic-resistant *Enterococcus faecium* from chicken and pork. *New England Journal of Medicine* 345:1161–1166.

Spaepen K.R.I., L.J.J. Van Leemput, P.G. Wislocki, and C. Verschueren (1997). A uniform procedure to estimate the predicted environmental concentration of the residues of veterinary medicines in soil. *Environmental Toxicology and Chemistry* 16:1977–1982.

Spika J.S., S.H. Waterman, G.W. Hoo, M.E. St. Louis, R.E. Pacer, S.M. James, M.L. Bissett, L.W. Mayer, J.Y. Chiu, B. Hall, et al. (1987). Chloramphenicol-resistant *Salmonella newport* traced through hamburger to diary farms. A major persisting source of human salmonellosis in California. *New England Journal of Medicine* 316:565–570.

Staels B., J. Dallongeville, J. Auwerx, K. Schoonjans, E. Leitersdorf, and J.-C. Fruchart (1998). Mechanism of action of fibrates on lipid and lipoprotein metabolism. *Circulation* 98:2088–2093.

Stanislavska J. (1979). Communities of organisms during treatment of sewage containing antibiotics. *Polish Archives of Hydrobiologia* 26:221–229.

Stass H. and D. Kubitza (1999). Pharmacokinetics and elimination of moxifloxacin after oral and intravenous administration in man. *Journal of Antimicrobial Chemotherapy* 43(Suppl. B):83–90.

Steger-Hartmann T., K. Kümmerer, and A. Hartmannl (1997). Biological degradation of cyclophosphamide and its occurrence in sewage water. *Ecotoxicology and Environmental Safety* 36:174–197.

Steger-Hartmann T., K. Kümmerer, and J. Schecker (1996). Trace analysis of the antineoplastics ifosfamide and cyclophosphamide in sewage water by two-step solid-phase extraction and gas chromatography-mass spectrometry. *Journal of Chromatography A* 726:179–184.

Steinbrook R. (2002). Testing medications in children. *New England Journal of Medicine* 347:1462–1469.

Stene L.C., D. Hongve, P. Magnus, K.S. Rnningen, and G. Joner (2002). Acidic drinking water and risks of childhood-onset type 1 diabetes. *Diabetes Care* 25:1534–1538.

Sternebring B., A. Linden, K. Anderson, and A. Melander (1992). Carbamazepine kinetics and adverse effects during and after ethanol exposure in alcoholics and healthy volunteers. *European Journal of Clinical Pharmacology* 43:393–397.

Stewart D.J., D. Grewaal, R.M. Green, S. Verma, J.A. Maroun, D. Redmond, L. Robillard, and S. Gupta (1991). Bioavailability and pharmacology of oral idarubicin. *Cancer Chemotherapy and Pharmacology* 27:308–314.

Straub J.O. (2005). Concentrations of the UV Filter Ethylhexyl Methyoxycinnamate in the Aquatic Compartment: A Comparison of the Modeled Concentrations for Swiss Surface Waters with Empirical Monitoring Data. In: D.R. Dietrich, S.F. Webb, and T. Petry (Eds.) *Hot Spot Pollutants: Pharmaceutical in the Environment.* Elsevier, New York, pp. 51–62.

Strong L. (1993). Overview: The impact of avermectins on pastureland ecology. *Veterinary Parasitology* 48:3–17.

Strong L. and T.A. Brown (1987). Avermectins in insect control and biology: A review. *Bulletin of Entomological Research* 77:357–389.

Stuer-Lauridsen F., M. Birkved, L.P. Hansen, H-C.H. Lützhøft, and B. Halling-Sørensen (2000). Environmental risk assessment of human pharmaceuticals in Denmark after normal therapeutic use. *Chemosphere* 40:783–793.

Stumm-Zollinger E. and G.M. Fair (1965). Biodegradation of steroid hormones. *Journal of the Water Pollution Control Federation* 37:1506–1510.

Stumpf M., T.A. Ternes, K. Haberer, and W. Baumann (1998). Isolation of ibuprofen-metabolites and their importance as pollutants of the aquatic environment. *Von Wasser* 91:291–303.

Swan G., V. Naidoo, R. Cuthbert, R.E. Green, D.J. Pain, D. Swarup, V. Prakash, M. Taggart, L. Bekker, D. Das, J. Diekmann, M. Diekmann, E. Killian, A. Meharg, R.C. Patra, M. Saini, and K. Wolter (2006). Removing the threat of diclofenac to critically endangered Asian vultures. *PLOS Biology* 4:395–402.

Swartz R.D., B. Starmann, A.M. Hovarth, S.C. Olson, and E.L. Posvar (1990). Pharmacokinetics of quinaprilat during continuous ambulatory peritoneal-dialysis. *Journal of Clinical Pharmacology* 30:1136–1141.

Tacket C.O., L.B. Dominguez, H.J. Fisher, and M.L. Cohen (1985). An outbreak of multiple-drug-resistant *Salmonella enteritis* from raw milk. *Journal of the American Medical Association* 253:2058–2060.

Takai H., S. Pedersen, J.O. Johnson, J.H.M. Metz, P.W.G. Groot Koerkamp, G.H. Uenk et al. (1998). Concentrations and emissions of airborne dust in livestock buildings in northern Europe. *Journal of Agricultural Engineering Research* 70:59–77.

Tang J., H-H. Liste, and M. Alexander (2002). Chemical assays of availability to earthworms of polycyclic aromatic hydrocarbons in soil. *Chemosphere* 48:35–42.

Tenover F.C., L.M. Weigel, P.C. Appelbaum, L.K. McDougal, J. Chaitram, S. McAllister, N. Clark, G. Killgore, C.M. O'Hara, L. Jevitt, J.B. Patel, and B. Bozdogan (2004). Vancomycin-resistant *Staphylococcus aureus* isolate from a patient in Pennsylavania. *Antimicrobial Agents and Chemotherapy* 48:275–280.

Ternes T.A. (1998). Occurrence of drugs in German sewage treatment plants and rivers. *Water Research* 32:3245–3260.

Ternes T.A. (2001). Pharmaceuticals and Metabolites as Contaminants of the Aquatic Environment. In: C.G. Daughton and T.L. Jones-Lepp (Eds.). *Pharmaceuticals and Personal Care Products in the Environment. Scientific and Regulatory Issues.* ACS Symposium Series 791. American Chemical Society, Washington, D.C., pp. 39–54.

Ternes T.A., M. Bonerz, and T. Schmidt (2001). Determination of neutral pharmaceuticals in wastewater and rivers by liquid chromatography—electrospray tandem mass spectrometry. *Journal of Chromatograpy A* 938:175–185.

Ternes T.A. and R. Hirsch (2000). Occurrence and behavior of x-ray contrast media in sewage facilities and the aquatic environmet. *Environmental Science and Technology* 34:2741–2748.

Ternes T.A., P. Kreckel, and J. Mueller (1999b). Behaviour and occurrence of estrogens in municipal sewage treatment plants—II. Aerobic batch experiments with activated sludge. *Science of the Total Environment* 225:91–99.

Ternes T.A., M. Meisenheimer, D. McDowell, F. Sacher, H.-J. Brauch, B. Haist-Gulde, G. Preuss, U. Wilme, and N. Zulei-Seibert (2002). Removal of pharmaceuticals during drinking water treatment. *Environmental Science and Technology* 36:3855–3863.

Ternes T.A., J. Stüber, N. Herrmann, D. McDowell, A. Reid, M. Kampmann, and B. Teiser (2003). Ozonation: A tool for removal of pharmaceuticals, contrast media and musk fragrances from wastewater. *Water Research* 37:1976–1782.

Ternes T.A., M. Stumpf, J. Mueller, K. Haberer, R-D. Wilken, and M. Servos (1999a). Behaviour and occurrence of estrogens in municipal sewage treatment plants—I. Investigations in Germany, Canada and Brazil. *Science of the Total Environment* 225:81–90.

Tešar T., V. Foltán, L. Langšadl, and K. Baňasová (2004). Utilization of antibiotics within Europe and the trends of consumption in the Slovak Republic. *Acta Facultatis Pharmaceuticae Universitis Comenianae.* Tomus LI 2004. Available at: http://www. fpharm.uniba.sk/Root/Acta_Facultaties/LI/AF25.pdf; downloaded 3/1/2007.

Testa B., A. Pagliara, and P.A. Carrupt (1997). Molecular behaviour of cetirizine. *Clinical and Experimental Allergy* S2:S13–18.

Thanuttamavong M., K. Yamamoto, J.I. Oh, K.H. Choo, and S.J. Choi (2002). Rejection characteristics of organic and inorganic pollutants by ultra low-pressure nanofiltration of surface water for drinking water treatment. *Desalination* 145:257–264.

Thiele, S. (2000). Adsorption of the antibiotic pharmaceutical compound sulfapyridine by a long-term differently fertilized loess Chernozem. *Journal of Plant Nutrition and Soil Science* 163:589–594.

Thiele-Bruhn S. (2003). Pharmaceutical antibiotic compounds in soils—A review. *Journal of Plant Nutrition and Soil Science* 166:145–167.

Thiele-Bruhn S., T. Seibicke, H.-R. Schulten, and P. Leinweber (2004). Sorption of sulfona-mide pharmaceutical antibiotics on whole soils and particle-size fractions. *Journal of Environmental Quality* 33:1331–1342.

Thomas K.V., M.R. Hurst, P. Matthiessen, M. McHugh, A. Smith, and M.J. Waldock (2002). An assessment of in vitro androgenic activity and the identification of environmental andro-gens in United Kingdom estuaries. *Environmental Toxicology and Chemistry* 21:1456–1461.

Threlfall E.J., L.R. Ward, J.A. Skinner, and B. Rowe (1997). Increase in multiple antibiotic resistance in nontyphoidal Salmonella from humans in England and Wales: A comparison of data for 1994 and 1996. *Microbial Drug Resistance* 3:263–266.

Tixier C., H.P. Singer, S. Canonica, and S.R. Müller (2002). Phototransformation of triclosan in surface waters: A relevant elimination process for this widely used biocide-laboratory studies, field measurements, and modeling. *Environmental Science and Technology* 36:3482–3489.

Tixier C., H.P. Singer, S. Ollers, and S.R. Müller (2003). Occurrence and fate of carbamazepine, clorofibric acid, diclofenac, ibuprofen, ketoprofen, and naproxen in surface water. *Environmental Science and Technology* 37:1061–1068.

Tolls J. (2001). Sorption of veterinary pharmaceuticals in soils: A review. *Environmental Science and Technology* 35:3397–3406.

Tomlinson T.G., A.G. Boon, and C.N.A. Trotman (1966). Inhibition of nitrification in activated sludge processes of sewage disposal. *J. Appl. Bacteriol.* 29:266–291.

Turner J.V., D.J. Maddalena, and S. Agatonovic-Kustrin (2004). Bioavailability prediction based on molecular structure for a diverse series of drugs. *Pharmaceutical Research* 21:68–82.

Tyler C.R., R. van Aerle, T.H. Hutchinson, S. Maddix, and H. Trip (1999). An in vivo testing system for endocrine disruptors in fish early life stages using induction of vitellogenin. *Environmental Toxicology and Chemistry* 18:337–347.

UCS (2001). Hogging it!: Estimates of antimicrobial abuse in livestock. Percent of all antibiotics given to healthy livestock. Union of Concerned Scientist. website details??

UCS (2003). Hogging it!: Estimates of antimicrobial use in livestock. Available at the Union of Concerned Scientists website: http://www.ucsusa.org/food_and_environment/antibiotic-resistance/index.

Ueda, T., K. Hata, and Y. Kikuoka (1996). Treatment of domestic sewage from rural settlement by a membrane bioreactor. *Water Science and Technology* 34:189–196.

UNEP (2002). *Environmentally Sound Technology for Wastewater and Stormwater Management.* IWA Publishing, Osaka, Japan.

U.S. EPA (1995). A Guide to the Biosolids Risk Assessment for the EPA Part 503 Rule. EPA832-B-93-005. United States Environmental Protection Agency Office of Waste Management, Washington, D.C.

U.S. EPA (1999). Analysis of GAC effluent blending during the ICR treatment studies. Office of Water (4601). EPA 815-C-99-002.

U.S. EPA (2000). National Management Measures to Control Non-Point Pollution from Agriculture. Office of Water, Non-Point Source Control Branch, Draft Report.

U.S. EPA (2003). Drinking water contaminant candidate list factsheet. Available at: www.epagov/safewater/ccl/ccfls.html.

U.S. FDA (1998). Guidance for Industry: Environmental Assessment of Human Drug and Bilogics Applications. CMC 6 Revision 1, Rockville, MD: Food and Drug Adminstration. Available at: http://www.fda.gov/cder/guidance/1730fnl.pdf; accessed September 18, 2007.

U.S. FDA (2006). Actonel. Actonel with Calcium. Available at: http://www.fda.gov/medwatch/safety/2006/Jan_PI/Actonel_Calcium_PI.pdf#search=%22actonel%20calcium%20risedronate%20usp%22; accessed 10/04/2006.

Vandenberg G.F., D. Adriaen, T. Verslycke, and C.R. Janssen (2003). Effects of 17α-ethinylestradiol on sexual development of the amphipod *Hyallela azteca. Ecotoxicology and Environmental Safety* 54:216–222.

van der Bruggen B., J. Schaep, D. Wilms, and C. Vandecasteele (1999). Influence of molecular size, polarity and charge on the retention of organic molecules by nanofiltration. *Journal of Membrane Science* 156:29–41.

Van de Waterbeemd H. (2002). Physicochemical Properties. In: F.D. King (Ed.). *Medicinal Chemistry: Principles and Practices*. The Royal Society of Chemistry, Cambridge, UK, pp. 195–214.

Vane J.R. (1971). Inhibition of prostaglandin synthesis as a mechanism of action for aspirin-like drugs. *Nature* 231:232–235.

Vane J.R. and R.M. Botting (1998). Mechanisms of action of nonsteroidal anti-inflammatory drugs. *American Journal of Medicine* 104:2S–8S.

Van Leeuwen W., H. Verbrugh, J. van der Velden, N. van Leeuwen, M. Heck, and A. van Belkum (1999). Validation of binary typing for *Staphylococcus aureus* strains. *Journal of Clinical Microbiology* 37:664–674.

van Maanen J.M.S., H.J. Albering, T.M.C.M. de Kok, S.G.J. van Breda, D.M.J. Curfs, I.T.M. Vermeer, A.W. Ambergen, B.H.R. Wolffenbuttel, J.C.S. Kleinjans, and H. M. Reeser (2000). Does the risk of childhood diabetes mellitus require revision of the guideline values for nitrate in drinking water? *Environmental Health Perspectives* 108:457–461.

Veith G.D., D.L. Defor, and B.V. Bergstedt (1979). Measuring and estimating the bioconcentration factor of chemicals in fish. *Journal Fish Research Board of Canada* 26:1040–1048.

Velicer C.M., S.R. Heckbert, J.W. Lampe, J.D. Potter, C.A. Robertson, and S.H. Taplin (2004). Antibiotic use in relation to the risk of breast cancer. *Journal of American Medical Association* 291:827–835.

Vinge E., G. Nergelius, L.G. Nilsson, and L. Lidgren (1979). Pharmacokinetics of cloxacillin in patients undergoing hip or knee replacement. *European Journal of Clinical Pharmacology* 52:407–411.

Volmer D.A., B. Mansoori, and S.J. Locke (1997). Study of 4-quinolone antibiotics in biological samples by short-column liquid chromatography coupled with electrospray ionization tandem mass spectrometry. *Analytical Chemistry* 69:4143–4155.

Voss J.G. (1975). Effects of an antibacterial soap on the ecology of aerobic bacterial flora of human skin. *Applied Microbiology* 30:551–556.

Vyas K.P., R.A. Halpin, L.A. Geer, J.D. Ellis, L. Liu, H. Cheng, C. Chavez-Eng, B.K. Matuszewski, S.L. Varga, A.R. Guiblin, and J.D. Rogers (2000). Disposition and pharmacokinetics of the antimigraine drug, trizatriptan, in humans. *Drug Metabolism and Disposition* 28:89–95.

Wagner J. (2001). Membrane Filtration Handbook Practical Tips and Hints. Available at: http://www.osmonics.com/library/1229223-%20Lit-%20Membrane%20Filtration%20Handbook.pdf; accessed 12/02/2006.

Walker S.R., M.E. Evans, A.J. Richards, and J.W. Paterson (1972). The fate of (^{14}C)disodium cromoglycate in man. *Journal of Pharmacy and Pharmacology* 24:525–531.

Wallis L.R. and S. Sciacca (2003). Anticoagulant and Antiplatelet Drugs. In: K. Gutierrez and S.F. Queener (Eds.). *Pharmacology for Nursing Practice*. Mosby, St. Louis, MO, pp. 693–709.

Walsh C. (2003). Antibiotics: Actions, Origins, Resistance. ASM Press, Washington, D.C.

Warrington S.J., R.A. Ronfeld, K.D. Wilner, and J.D. Lazar (1992). Human pharmacokinetics of sertraline. *Clinical Neuropharmacology* 15(Suppl. 1):54.

Watts M.W., D. Pascoe, and K. Carroll (2002). Population responses of the freshwater amphipod Gammarus pulex (L.) to an environmental oestrogen, 17α-ethinylestradiol. *Environmental Toxicology and Chemistry* 21:445–450.

Webb S., T. Ternes, M. Gibert, and K. Olejniczak (2005). Indirect Human Exposure to Pharmaceuticals via Drinking Water. In: D.R. Dietrich, S.F. Webb, and T. Petry (Eds.). *Hot Spot Pollutants: Pharmaceuticals in the Environment*. Elsevier, New York, pp. 79–94.

Webb S.F. (2001). A Data-Based Perspective on the Environmental Risk Assessment of Human Pharmaceuticals. III—Indirect Human Exposure. In: K. Kümmerer (Ed.). *Pharmaceuticals in the Environment*. Springer, New York, pp. 221–230.

Weerasinghe C.A. and D. Towner (1997). Aerobic biodegradation of virginiamycin in soil. *Environmental Toxicology and Chemistry* 16:1873–1876.

Wegener H.C. (2003). Ending the use of microbial growth promoters is making a difference. *ASM News* 69:442–448.

Weirup M. (2001). The Swedish experience of the 1986 year ban of antimicrobial growth promoters, with special reference to animal health, disease prevention, productivity, and usage of antimicrobials. *Microbial Drug Resistance* 7:183–190.

Weisblum B. (1995). Erythromycin resistance by ribosome modification. *Antibiotic Agents and Chemotherapy* 39:577–585.

Welton L.A., L.A. Thai, M.B. Perri, S. Donabedian, J. McMahon, J.W. Chow, and M.J. Zervos (1998). Antimicrobial resistance in enterococci isolated from turkey flocks fed virginiamycin. *Antimicrobial Agents and Chemotherapy* 42:705–708.

Werner G., I. Klare, and W. Witte (1998). Association between quinupristin/dalfopristin resistance in gylcopeptide-resistant *Enterococcus faecium* and the use of additives in animal feed. *European Journal of Clinical Microbiology and Infectious Diseases* 17:401–402.

West C.W. and G.T. Ankley (1998). A laboratory assay to assess avoidance of contaminated sediments by the freshwater oligochaete Lumbricus variegatus. *Archives of Environmental Toxicology* 35:20–24.

Westergaard K., A.K. Müller, S. Christensen, J. Bloem, and S.J. Sørensen (2001). Effect of tylosin as a disturbance on the soil microbial community. *Soil Biology and Biochemistry* 33:2061–2071.

Wetzstein H-G., N. Schmeer, and W. Karl (1997). Degradation of the fluoroquinolone enrofloxacin by the brown rot fungus *Gloeophyllum striatum*: Identification of metabolites. *Applied and Environmental Microbiology* 63:4272–4281.

Whitener C.J., S.Y. Park, F.A. Browne, L.J. Parent, K. Julian, B. Bozdogan, P.C. Appelbaum, J. Chaitram, L.M. Weigel, J. Jernigan, L.K. McDougal, F.C. Tenover, and S.K. Fridkin (2004). Vancomycin-resistant *Staphylococcus aureus* in the absence of vancomycin exposure. *Clinical Infectious Diseases* 38:1049–1055.

WHO (2001). Monitoring antimicrobial usage in food animals for the protection of human health. Report of the WHO Consultation. Olso, Norway. Report# WHO/CDS/CSR/EPH/2002.11.

Widemann B.C. and P.C. Adamson (2006). Understanding and managing methotraxate nephrotoxicity. *Oncologist* 11:694–703.

Willems R.J.L., J. Top, M. van Santen, D.A. Robinson, T.M. Coque, F. Baquero, H. Grundmann, and M.J.M. Bonten (2005). Global spread of vancomycin-resistant *Enterococcus faecium* from distinct nosocomial genetic complex. *Emerging Infectious Diseases* 11:821–828.

Williams D.H. and J.P. Waltho (1988). Molecular basis of the activity of antibiotics of the vancomycin group. *Biochemical Pharmacology* 317:133–141.

Wilson E.O. (1999). *The Diversity of Life*. Norton, New York.

Wilson B.A., V.H. Smith, F. deNoyelles Jr., and C.K. Larive (2003). Effects of three pharmaceuticals and personal care products on natural freshwater algal assemblages. *Environmental Science and Technology* 37:1713–1719.

Wilson J.G. and H.C. Wilson (1943). Reproductive capacity in adult rats treated prepubertally with androgenic hormone. *Endocrinology* 33:350–353.

Wise R. (2002). Antimicrobial reistance: Priorities for action. *Journal of Antimicrobial Chemotherapy* 49:585–560.

Wise R. (2003). Maximizing efficacy and reducing the emergence of resistance. *Journal of Antimicrobial Chemotherapy* 51(Suppl. S1):37–42.

Witte W. (1998). Medical consequences of antibiotics in agriculture. *Science* 279:996–997.

Witte W. (2001). Nosocomial infections. *International Journal of Medical Microbiology* 291:1–2.

Xia K., A. Bhandari, K. Das, and G. Pillar (2005). Occurrence and fate of pharmaceuticals and personal care products (PPCPs) in biosolids. *Journal of Environmental Quality* 34:91–104.

Xing B. and J.J. Pignatello (1997). Dual-mode sorption of low-polarity compounds in glassy poly(vinyl chloride) and soil organic matter. *Environmental Science and Technology* 31:792–799.

Xu P., J.E. Drewes, C. Bellona, G. Amy, T-U. Kim, M. Adam, and T. Heberer (2005). Rejection of emerging organic micropollutants in nanofiltration—reverse osmosis membrane applications. *Water Environment Research* 77:40–48.

Yamamoto H., H.M. Liljestrand, Y. Shimizu, and M. Morita (2003). Effects of physical-chemical characteristics on the sorption of selected endocrine disruptors by dissolved organic matter surrogates. *Environmental Science and Technology* 37:2646–2657.

Yamashita F., Y. Koyama, H. Sezaki, and M. Hashida (1993). Estimation of a concentration profile of acyclovir in the skin after topical administration. *International Journal of Pharmaceutics* 89:199–206.

Yang S. and K. Carlson (2004). Routine monitoring of antibiotics in water and wastewater with a radioimmunoassay technique. *Water Research* 38:3155–3166.

Yardeny Y., H. Rodriguez, S.K.-F. Wong, D.R. Brandt, D.C. May, J. Burnier, R.N. Harkins, E.Y. Chen, J. Ramachandran, A. Ullrich, and E.M. Ross (1986). The avian beta-adrenergic receptor: Primary structure and membrane topology. *Proceedings of the National Academy of Science U.S.A.* 83:6795–6799.

Yeleswaram K., D.W. Rurak, E. Kwan, C. Hall, A. Doroudian, F.S. Abott, and J.E. Axelson (1993). Disposition, metabolism, and pharmacodynamics of labetalol in adult sheep. *Drug Metabolism and Disposition* 21:284–292.

Ylitalo P., K. Holli, J. Monkkonen, H.A. Elo, A. Juhakoski, S. Liukko-Sipi, and R. Ylitalo (1999). Comparison of pharmacokinetics of clodronate after single and repeated doses. *International Journal of Clinical Pharmcology and Therapeutics* 37:294–300.

Young R.C., R.C. Mitchell, T.H. Brown, C.R. Ganellin, R. Griffiths, M. Jones, K.K. Rana, D. Saunders, I. R. Smith, N.E. Sore, and T.J. Wilks (1988). Development of a new physicochemical model for brain penetration and its application to the design of centrally acting H_2 receptor histamine antagonists. *Journal of Medicinal Chemistry* 31:656–671.

Zadoks R., W. van Leeuwen, H. Berkema, O. Sampimon, H. Verbrugh, Y.H. Schukken, and A. van Belkum (2000). Application of pulsed-field gel electrophoresis and binary typing as tools in veterinary clinical microbiology and molecular epidermiologic analysis of bovine and human *Staphylococcus aureus* isolates. *Journal of Clinical Microbiology* 38:1931–1939.

Zakowski M.I., S. Ramanathan, and H. Turndorf (1993). A 2-dose epidural morphine regimen in cesarean-section patients—pharmacokinetic profile. *Acta Anaesthesiologica Scandinavica* 37:584–589.

Zapata R., D. Martín, M-D. Piulachs, and X. Bellés (2002). Effects of hypocholesterolaemic agents on the expression and activity of 3-hydroxy-3-methylglutaryl-CoA reductase in the fat body of the German Cockroach. *Archives of Insect Biochemistry and Physiology* 49:177–186.

Zapata R., M-D. Piulachs, and X. Bellés (2003). Inhibitors of 3-hydroxy-3-methylglutaryl-CoA reductase lower fecundity in the German cockroach: Correlation between the effects on fecundity *in vivo* with the inhibition of enzymatic activity in embryo cells. *Pest Management Science* 59:1111–1117.

Zepp R.G. and D.M. Cline (1977). Rates of direct photolysis in aquatic environment. *Environmental Science and Technology* 11:359–366.

Zerulla M., R. Länge, T. Steger-Hartmann, G. Panter, T. Hutchinson, and D.R. Dietrich (2005). Morphological Sex Reversal upon Short-term Exposure to Endocrine Modulators in Juvenile Fathead Minnos (*Pimephales promelas*). In: D.R. Dietrich, S.F. Webb, and T. Petry (Eds.). *Hot Spot Pollutants: Pharmaceuticals in the Environment*. Elsevier, New York, pp. 97–113.

Zielezny Y., J. Groeneweg, H. Vereecken, and W. Tappe (2006). Impact of sulfadiazine and chlorotetracycline on soil bacterial community structure and respiratory activity. *Soil Biology and Biochemistry* 38:2372–2380.

Zoller U. (1993). Groundwater contamination by detergents and polycyclic aromatic hydrocarbons—a global problem of organic contaminants: Is the solution locally specific? *Water Science and Technology* 27:187–195.

Zou J., N.F. Neumann, J.W. Holland, M. Belosevic, C. Cunningham, C.J. Secombes, and A.F. Rowley (1999). Fish macrophages express a cyclo-oxygenase-2-homologue after activation. *Biochemistry Journal* 340(Pt 1):153–159.

Zuccato E., E.R. Bagnati, F. Fioretti, M. Natangelo, D. Calamari, and R. Fanelli (2001). Environmental Loads and Detection of Pharmaceuticals in Italy. In: K. Kümmerer (Ed.). *Pharmaceuticals in the Environment—Sources, Fate, Effects and Risks*. Springer, Heidelberg, Germany, pp. 19–27.

Zuccato E., D. Calamari, M. Natangelo, and R. Fanelli (2000). Presence of therapeutic drugs in the environment. *Lancet* 355:1789–1790.

Zurita J.L., Á. Jos, A. del Peso, M. Salguero, M. López-Artíguez, and G. Repetto (2005). Ecotoxicological evaluation of the antimalarial drug chloroquine. *Aquatic Toxicology* 75:97–107.

Zwiener C. (2007). Occurrence and analysis of pharmaceuticals and their transformation products in drinking water treatment. *Analytical and Bioanalytical Chemistry* 387:1159–1162.

Zwiener C. and F.H. Frimmel (2000). Oxidative treatment of pharmaceuticals in water. *Water Research* 34:1881–1885.

Zwiener C. and F.H. Frimmel (2003). Short-term tests with a pilot sewage plant and biofilm reactors for the biological degradation of the pharmaceutical compounds clofibric acid, ibuprofen, and diclofenac. *Science of the Total Environment* 309:201–211.

Zwiener C. and F.H. Frimmel (2004). Residues of Clofibric Acid, Ibuprofen and Diclofenac in the Aquatic Environment and Their Elimination in Sewage Treatment and Drinking Water Production. In: K. Kümmerer (Ed.). *Pharmaceuticals in the Environment: Sources, Fate, Effects and Risks*. Springer, New York, pp. 121–132.

Zwiener C., T. Glauner, and F.H. Frimmel (2000). Biodegradation of pharmaceutical residues investigated by SPE-GC/ITD-MS and on-line derivatization. *Journal of High Resolution Chromatography* 23:474–478.

Zwiener C., S. Seeger, T. Glauner, and F.H. Frimmel (2002). Metabolites of the biodegradation of pharmaceutical residues of ibuprofen in biofilm reactors and batch experiments. *Analytical and Bioanalytical Chemistry* 372:569–575.

INDEX

*Pharma-Ecology: The Occurrence and Fate of Pharmaceuticals and Personal Care Products
in the Environment.* By P.K. Jjemba
Copyright © 2008 John Wiley & Sons, Inc.

295